Environmental Education and Sustainable Development

Kamal Hirech
Editor

Environmental Education and Sustainable Development

Challenges and Prospects for the Future

palgrave
macmillan

Editor
Kamal Hirech
Higher School of Education and Training
Mohamed First University
Oujda, Morocco

ISBN 978-3-032-05759-4 ISBN 978-3-032-05760-0 (eBook)
https://doi.org/10.1007/978-3-032-05760-0

This Palgrave Macmillan imprint is published by the registered company Springer Nature Switzerland AG.
The registered company address is: Gewerbestrasse 11, 6330 Cham, Switzerland

Preface

Environmental education now holds a central place in ecological transition and sustainable development strategies. In the face of growing challenges related to climate change, biodiversity loss, ecosystem degradation, and energy crises, it has become imperative to adopt a global and integrated approach to raise awareness among present and future generations. Far from being merely an academic discipline, environmental education is a strategic tool for transforming behaviors, fostering innovation, and encouraging more responsible public policies.

This book aligns with this vision by bringing together research studies and case studies that explore the connections between education, environment, sustainability, and ecological transition. It highlights the challenges faced by educational systems, researchers, and policymakers in integrating sustainable development principles into training programs. It also proposes concrete and innovative solutions to strengthen the impact of environmental education at various levels: school, university, community, and institutional.

The book is structured around several major themes: new pedagogical approaches to environmental education, the use of technology for learning, policies and initiatives promoting the integration of sustainable development into school and university curricula, as well as the social and cultural dimensions of ecological awareness. The contributions gathered here reflect a diversity of perspectives and experiences, illustrating how different researchers, stakeholders, and regions worldwide are addressing these critical issues.

By offering an in-depth reflection and presenting concrete initiatives, this book aims to serve as a reference for educators, researchers, policymakers, and actors engaged in the ecological transition. It emphasizes the need for interdisciplinary dialogue and international cooperation to accelerate the adoption of sustainable and effective educational practices.

We hope that this publication will contribute to enriching the debate and inspiring new actions for education that serves both the environment and the well-being of future generations. The sustainable transformation of our societies depends above all on the training of enlightened citizens, aware of environmental issues and capable of acting responsibly.

Prof. Kamal Hirech
Higher School of Education and Training
Mohammed First University, Oujda, Morocco

Oujda, Morocco Kamal Hirech

Acknowledgment

We would like to express our deepest gratitude to all the contributors, reviewers, and collaborators who made this book possible. Their dedication, expertise, and commitment to advancing environmental education and sustainable development have been invaluable throughout this project.

Special thanks go to the institutions, research centers, and universities that supported the authors in their scientific and pedagogical work. We are also grateful to the editorial team and the publisher for their guidance and professionalism during the review and publication process.

This book is the result of a collective effort, inspired by the urgent need to promote a greener, more sustainable, and educationally inclusive future. We hope it serves as a valuable resource for educators, researchers, policymakers, and all those engaged in building a sustainable world through education.

Contents

Notes on Contributors

Ouali Abderrazak born in 1983, is a lecturer at the Faculty of Legal, Economic, and Social Sciences (FSJES), Cadi Ayyad University, Marrakech, Morocco. His research focuses primarily on energy and environmental economics. He has co-authored articles such as "Impact of Income and Energy Consumption on CO_2 Emissions in Morocco: Testing the Environmental Kuznets Curve." His research focuses particularly on the impact of health capital on economic growth in Morocco, which highlights the importance of health investments for economic development. He has also analyzed the effects of cost-sharing on health status based on social protection status, emphasizing the positive influence of health insurance on individuals' perception of their health. Furthermore, he has explored the dynamics of succession in family businesses where he examines inheritance-related tensions and proposes solutions centered on equity and governance. In addition, he has contributed to research on renewable energy. Collectively, these works reflect his commitment to interdisciplinary research, combining economics, innovation, and sustainable development.

El Aidi Abir was born in 1999 in Rabat, Morocco. She is a Ph.D. candidate in economics in the Department of Economics at Cadi Ayyad University, Marrakech, Morocco, and member of the Energy, Environment and Resources Economics Research Laboratory (GREER) at the Faculty of Legal, Economic and Social Sciences of Marrakech. Her research interests include energy economics, social economy, innovation, and sustainable development. She is the first author of the article "The Effects of

Technological Innovation on Sustainable Development in Morocco: Does the Transition to Social Innovation Matter?," published in 2024 in the *International Journal of Research and Innovation in Social Science*. This article examines the impact of technological innovation on sustainable development in Morocco, with a particular focus on the role of transitioning to social innovation. Additionally, she contributed to a theoretical reflection on the implications of innovation in education for sustainable development, in collaboration with N. Alaamri and E. Tarbalouti, in 2024.

Ahmed Aït Aghzzaf is Professor of Chemistry at Abdelmalek Essaâdi University, Tétouan, Morocco. He received her Ph.D. in Corrosion of Materials from Caddi Ayyad University of Marrakech and Lorraine University of Nancy in 2014. He is a leader of Nanomaterials, Technology and Innovation Team. Aït Aghzzaf's research focuses on the environment, metallic materials, and protection against corrosion. He has written numerous articles in international scientific journals. In addition to his research activities, Aït Aghzzaf teaches chemistry courses to Bachelor's and Master's students.

Abdelilah Allaoui is a Moroccan Ph.D. candidate in Communication Sciences at Mohammed V University, Rabat, Faculty of Letters and Human Sciences. His research focuses on internal communication within public institutions, particularly in the education sector. He is interested in how digital tools and artificial intelligence can enhance pedagogical effectiveness and organizational performance. He has contributed to several interdisciplinary projects linking education, information technologies, and institutional communication strategies. He is also involved in academic networks that promote educational innovation, inclusive learning environments, and effective internal communication.

Hmid Amine is a versatile and dedicated professional deeply involved in higher education and academic research in the field of management sciences. He currently holds the position of head of Pedagogical Administration, while also working as a part-time university lecturer. In parallel, he is pursuing a Doctoral Degree in Management Sciences, focusing on critical areas such as the strategic management of territories and climate change. His research explores how ecological and environmental challenges can be integrated into the strategic frameworks of organizations and territorial governance. Amine has actively participated in numerous academic conferences, where he shares his insights and findings. His work reflects a strong commitment to promoting sustainable practices and addressing contemporary challenges in both public and private sector management.

Yvonnie Antolin is a political science scholar researcher in the Department of Political Science at the College of Social Sciences of Humanities, Mindanao State University, General Santos, General Santos City, Philippines. Her research interests center on youth engagement in global climate justice, social justice, poverty alleviation, Philippine politics, ASEAN studies, and civil society. She also actively advocates for environmental awareness and is actively engaged in human rights education in Southern Philippines through her membership on the Human Rights Society of the same university.

Eveline Omagano Anyolo is Senior Lecturer in Social Science Education in the Department of Intermediate and Vocational Education at the University of Namibia, Namibia. She holds a Master's Degree in Environmental Education and a Ph.D. in Education with the focus in ESD. Anyolo's research focuses on the integration of ESD into the school and teacher education curricula and how communities facilitate sustainable ways of living in their daily lives. Anyolo has been involved in national and international projects on ESD. Among others is the latest Future Teacher Education for Sustainable Development (FUTE) project, which aims at strengthening teacher educators' and student teachers' competencies and best practices in sustainability education in Finland, Namibia, and Tanzania.

Charles Rj Arroyo is a political science researcher in the Department of Political Science at the College of Social Sciences of Humanities, Mindanao State University, General Santos, General Santos City, Philippines. His research interests center on youth engagement in global climate justice, ASEAN studies, Philippine politics, communication and media studies, local politics, and environmental studies. He also actively advocates for environmental awareness and is actively engaged in human rights education in Southern Philippines through his membership on the Human Rights Society of the same university.

Raed Atef is an independent researcher from Ramallah, the West Bank, Palestine. He specializes in consulting and research for government agencies and NGOs. He provides services such as organization development, leadership, and strategy. Previously, he worked with two universities:

An-Najah University, teaching undergraduate and gifted studies, and Al-Quds University, teaching human resources and developmental studies before transitioning to an independent career as a consultant and researcher. Atef has worked with several NGOs, such as the Jerusalem Media and Communication Center, the Technical Women Committee, and the Palestine National NGO, and was a director for the Central Elections Commission at the West Bank.

Raed Awashreh is Professor of Management and Public Administration with over 25 years of experience spanning academia and industry, including roles in NGOs, government, and the private sector. He has taught at the American University of Sharjah, the UAE, and A'Sharqiyah University, Oman. His responsibilities include teaching undergraduate and postgraduate courses, supervising theses, serving as a postgraduate thesis examiner, and actively contributing as a researcher and reviewer for several academic journals. Awashreh has written numerous articles, multiple book chapters, and books on topics such as human capital, education, governance, management, public policy, strategy, leadership, entrepreneurship, innovation, and the multidisciplinary applications of AI. He holds a Doctorate in Public Administration and Management from Flinders University, Australia, and a master's from the USA.

Amal Azeroual received a Ph.D. Degree in "Computer Science" from the National Graduate School of Arts and Crafts of Rabat (ENSAM) in the Department of Research Center STIS, M2CS, ENSIAS/Mohammed V University in Rabat, Morocco. She obtained her DESA in Computer Science, Telecommunication, and Multimedia from the Faculty of Sciences of Rabat, Mohammed V University in Rabat, Morocco. She is currently Professor of Computer Science at the Center of Guidance and Educational Planning (COPE) in Rabat, Morocco. Her research interests include artificial intelligence, big data, biomedical engineering, education, and computer science. She served as a reviewer for many international conferences and journals.

Imane Bakkali has 28 years of professional experience. She is Research Professor of Educational Technologies and the ICT and Digital Skills Modules coordinator at École Normale Supérieure (ENS) of Tétouan, UAE. She holds a Doctorate in Educational Sciences from UCA and is an African Leadership in ICT (2015). She is a member and coordinator of the "Representations, Linguistic, Technological, and Social Practices" research team. Bakkali is also the former head of the regional center of SI and was the regional manager of ICT and E-Learning at AREF TTH (1996–2018). She has been an Erasmus Mundus Alumni (EMA) since 2009.

She holds a European Erasmus Mundus "Mundusfor" Master's degree, carried out in five universities: University of Reims Champagne-Ardenne (URCA), France; University of Granada (UGR), Spain; University of Rovira i Virgili (URV), Spain; University of Porto (UP), Portugal; and Akershus University College (HIAK), Norway. She has received several grants for the realization of teaching and research mobilities within the framework of the "Investigation ÁFRICA_MED" project of the Spanish Agency for International Development Cooperation (AECID) and the framework of the "Erasmus + KA107" Program (Universidad Peruana de Ciencias Aplicadas [UPC], Universidad de León, Universidad de Cádiz [UCA]). At the scientific level, Bakkali is actively involved in research and supervises doctoral theses related to ICT, e-learning, AR, VR, AI, and digital skills. She is also a reviewer and member of scientific committees for several journals, congresses, and scientific events, and she is the author of several publications and communications.

Hicham Berbar is a Ph.D. researcher and Professor of Education Sciences, Pedagogical Engineering, and Didactics in the Normal Superior School (ENS) at Abdelmalek Essaâdi University, Tétouan, Morocco. He is the coordinator of modules (Educational Sciences, Didactic and Pedagogical Approaches and Methods, and Psychology and Sociology of Sport) taught at ENS Tétouan. Hicham Berbar focuses on training engineering, measurement, evaluation, and public management. He has several scientific papers credited to him. He has worked as a secondary education inspector at the Oriental Regional Education and Training Academy. His large body of scientific work, including newspaper articles and book chapters, attests to his significant contributions in the stated fields.

Narjiss Berbar is now pursuing a Doctoral Degree in Management Sciences at the Faculty of Legal, Economic, and Social Sciences (FSJES) at Abdelmalek Essaâdi University, Tangier, Morocco. His area of expertise is quality assessment and performance. Within the Moroccan Ministry of Health, he holds a senior executive position. It provides training for future executives in the public health sector and performs management initiatives for hospitals. Engaged in multidisciplinary projects, he is member of scientific committees for international conferences on renewable health.

Mohamed Berkani is Professor Emeritus in the Department of Chemistry and Environment at the Faculty of Science and Technology, Sultan Moulay Slimane University, Morocco. A specialist in chemistry, applied spectrochemistry, and the environment, he is a member of the Spectroscopy and Applied Chemometrics research team (LITA). His expertise extends to pollution analysis, energy recovery from municipal solid waste (including waste-derived fuel production), and landfill site selection using tools such as GIS and multi-criteria decision-making. He has also contributed to understanding the impact of the COVID-19 pandemic on waste production. As a university lecturer, he teaches applied chemistry, analytical techniques, and quality to students in the Sciences and Technology degree program.

Loubna Bouselham is an associate professor in the National School of Architecture (ENA), Oujda, Morocco, where she delivers lectures on various topics related to efficiency energy in building. She holds a Ph.D. in Solar Energy in 2019 from National School of Applied Sciences (ENSA), Oujda. Her researches focus on optimizing photovoltaic production, as well as the challenges of energy transition and innovative solutions for improving energy efficiency and management in buildings, and she has written several references and articles in these fields. She is also the author of the book *A Practical Guide for Simulation and FPGA Implementation of Digital Design*, which explores simulation methods and digital implementation applied to logic circuits. Engaged in multidisciplinary projects, he is member of scientific committees for international conferences on renewable energy and electrical engineering.

Giambattista Bufalino is Researcher in General and Social Pedagogy at the University of Catania, Italy, where he teaches intercultural education and general education. He holds a Ph.D. in Educational Research and Development from the University of Lincoln, UK. His academic career includes teaching and research experiences in several European countries, including Denmark, Malta, the UK, Slovakia, and Estonia. His main research interests focus on educational leadership, teacher training, and environmental education. He leads the Green Education Lab at the University of Catania and is currently exploring how insular and archipelagic environments shape educational practices and policies. Bufalino also serves as a Dissertation Supervisor and External Examiner at the Institute for Education in Malta. He was awarded the *Italian Prize for Pedagogy* in 2020 and the *Basile Prize* for his project on eco-narratives in education. He collaborates on national and international research initiatives and is affiliated with SIPED, SIPEGES, and BELMAS.

Mohamed Chahbi is an English teacher in Abdellah Ibrahim High School in Tétouan, Morocco. He holds a Master's Degree in Theatre and the Art of Performance, with a thesis centered on promoting the integration of individuals with special needs into society through cinema. Deeply committed to education and innovation, Chahbi is passionate about instructional engineering and fostering learners' communicative competencies. His current research investigates the impact of immersive technologies, such as virtual and augmented reality, on educational practices, specifically focusing on English language teaching and learning. By exploring how these technologies can enhance student engagement, interaction, and language acquisition, he aims to contribute to the evolution of modern pedagogical strategies. Chahbi actively participates in the academic community, bridging his artistic background and educational expertise to advocate for inclusive, tech-enhanced learning environments.

Bouamama Chera is Professor of Higher Education at the Regional Center for Education and Training Professions of the Tangier, Tétouan, Al Hoceima Region (CRMEFtth), Morocco, since 2011. Cherai holds a Bachelor's Degree in Hydrogeology from Fez Saiss Faculty of Science and Technology, Sidi Mohamed Ben Abdellah University, Fez, Morocco, in 1999. Cherai also holds a Doctorate's Degree in Applied Geology from Fez Saiss Faculty of Science and Technology, Sidi Mohamed Ben Abdellah University, Fez, Morocco in 2009. Cherai's research interests include environment, geology, geo-heritage, didactics of sciences and education.

Abbadi Driss is the Dean of the Faculty of Legal, Economic, and Social Sciences, Aïn Sebaâ, Casablanca, Morocco, as well as a distinguished professor specializing in territorial management. With a strong academic background and years of experience, he is widely recognized for his expertise in optimizing territorial governance while effectively integrating economic, social, and environmental dimensions. His work focuses on creating sustainable development strategies tailored to the specific needs of different regions. He has authored and co-authored numerous scholarly articles and books that explore innovative approaches to managing territorial resources. In addition to his research and teaching responsibilities, he plays a vital role in the academic leadership and administrative coordination of his faculty, contributing significantly to both institutional development and public policy discussions.

Mounia Rachdi Errachdi is a doctoral candidate in Innovation and Educational Intervention at the University of Vic—Central University of Catalonia, Spain. She is a lecturer at the University of Manresa in the Bachelor's Degree in Education, and also teaches the course "Education and Gender in a Changing Society" in the Master's Degree in Women's Studies, Gender, and Citizenship at the University of Barcelona. Her research focuses on science education in migration contexts, intersectionality, and inclusive pedagogies in non-formal learning environments. Mounia is a member of the Research Group on Knowledge Construction (GRECC), coordinated by the Vic Campus.

Tarbalouti Essaid born in Marrakech in 1963, is a Moroccan economist renowned for his academic contributions. Married and father of two, he holds a Ph.D. in Economics from the University of Paris II, with a thesis entitled "Economic Analysis of Business Failure Law." Before joining Cadi Ayyad University in Marrakech as Professor of Economics, Tarbalouti taught at the University of Lille II as a temporary research associate, then at the University of Paris II as a temporary assistant, and finally at Hassan I University in Settat as an assistant professor. His research interests include microeconomics, marriage economics, education economics, public choice theory, and international economics. Tarbalouti has written several articles in national and international journals, addressing topics such as the impact of economic growth and CO_2 emissions on healthcare spending in Morocco, corruption and trade, and socioeconomic factors influencing marital instability in Morocco. In addition to his research activities, he teaches courses such as "International Economic Relations" at Cadi Ayyad University.

Rock Finale holds a Degree in Landscape Architecture which was titled "Golf Course as Reclamation of Ljubljana Marsh Landfill." He first explored this topic during a study exchange in Sweden. He later expanded his research on waste management and visual art through his doctoral dissertation, "Enhancing Environmental Awareness Through Art," at the Faculty of Education, University of Ljubljana, Slovenia. The main research interests are environmental education, environmental awareness, connection with visual arts, teaching approaches, and garden design. He is currently employed as a teacher of professional theoretical and practical courses in visual merchandizing in the School of Commerce and Visual Merchandising in Ljubljana. Additionally, he serves as a lecturer in the Department of Horticulture at the Higher Vocational College, Biotechnical Centre Naklo, Slovenia, and at the Polytechnic of Rijeka in Croatia, where he teaches at the Department of Sustainable Agritourism.

Guillermina Ávila García holds a Ph.D. in Science with a specialty in educational physics from the Center for Research in Applied Science and Advanced Technology (CICATA), National Polytechnic Institute (IPN), Mexico. Her doctoral thesis received honorable mention and the award for best graduate thesis at the IPN in 2023. She is an adjunct professor at CECyT 11 and a visiting professor at CICATA, Legaria, and at CIECAS of the National Polytechnic Institute. She is a member of the Mexican Mathematics Society, the Educational Innovation and Research Network, the Repensar Seminar Network, the Mexican Computer Society, and the Inter-American Roundtable for Dialogue on Science Education in Argentina. She is a member of the National System of Researchers Level 1 in Mexico. She recently received the "Wilfrido Massieu Award for Merit as a Teacher of the Year 2025." And the 2025 "Amalia Solorzano de Cárdenas" medal, for her contributions to scientific education.

Lance Guiang is a political science scholar and researcher at Department of Political Science, College of Social Sciences of Humanities of Mindanao State University – General Santos, General Santos City, Philippines. His research interests center on youth engagement in global climate justice, political dynamics, ASEAN studies, social policies, and Philippine politics. He also actively advocates for environmental awareness and is actively engaged in human rights education in Southern Philippines through his membership on the Human Rights Society of the same university.

Mohamed Haba received a Ph.D. Degree in "Educational Sciences" from Sidi Mohamed Ben Abdellah University, Fes, Morocco. He obtained his DESA in "Social Development Management" from the Faculty of Legal, Economic and Social Sciences at Mohammed V University, Rabat, Morocco. He has professional experience as a school and vocational guidance counselor, working with students, as well as supervisory experience as a guidance inspector, after obtaining certificates from the Center of Guidance and Educational Planning (COPE) in Rabat, Morocco. He is currently Professor of "Academic and Professional Guidance" at the Center of Guidance and Educational Planning, Rabat, Morocco. His research interests include academic and professional guidance, career, transition, pathway, lifelong guidance, training engineering, and management educational system.

Mohamed El hadri holds a Baccalaureate in Experimental Sciences, a Bachelor's Degree in Physics and Chemistry, and a Professional Bachelor's Degree in Teaching Professions from Abdelmalek Essaâdi University, Morocco. El hadri received a Master's Degree in Applied Chemistry and Sustainable Development from Ibn Tofail University. El hadri is a Ph.D. student in the field of education and sustainable development at the Nanomaterials, Technology, and Innovation Group (NTI) Laboratory, Higher Normal School, Abdelmalek Essaâdi University, Tétouan, Morocco.

Abdelkader Hammaoui is a teacher and researcher in didactics, educational sciences, and communication at Mohammed First University, Oujda, Morocco. He is actively involved in several psychology training projects, particularly in Morocco and at Al-Ahliyya Amman University (AAU) in Jordan, in collaboration with academic institutions and specialized centers. He holds the position of academic director and lecturer at the École des Hautes Études en Psychologie Mathématique (EHEPM) in Paris, France, where he is responsible for coordinating and developing training activities in Morocco. He is also recognized as a psychology expert by the Centre for Study, Training, and Professional Immersion (CEFIP) in Parakou, Benin. His research and publications focus particularly on disciplinary convergences, which he explores through interdisciplinary and transdisciplinary approaches within the humanities and social sciences.

Almi Hassiba is an associate professor and Vice Head of the Department of Economic Sciences at Badji Mokhtar University, Annaba, Algeria. She specializes in insurance economics, digital transformation, Islamic finance, and sustainable development. She holds a Ph.D. in Insurance Economics from the University of Blida 2 (2021) and has extensive teaching and research experience. Almi has taught various economics and business-related courses and has written research on fintech, insurance, corporate social responsibility, and digital marketing. She actively participates in national and international conferences, and her ongoing research focuses on economic policy, insurance, and the impact of financial technology on economic growth.

Kamal Hirech is a professor in the Higher School of Education and Training at Mohammed First University, Oujda, Morocco, where he has been serving since 2021. Prior to this role, he was a professor at University of Moulay Ismail from 2018 to 2021. He is an active member of the Electrical Engineering and Maintenance Laboratory, where he engages in cutting-edge research and academic activities. Hirech has coordinated numerous scientific events focusing on environmental education, renewable energies, and the energy transition. His research primarily centers on solar renewable energies, energy storage, and the broader scope of environmental education and sustainability. He has an extensive portfolio of scientific publications, including patents, journal articles, conference papers, and book chapters, reflecting his significant contributions to these fields.

Rachid Hjaji has been a Project Manager at the Hanns Seidel Foundation in Morocco since December 2021. Before this position, he served as a Senior Project Assistant at UNOPS. He is currently a Ph.D. candidate at the Faculty of Legal, Economic, and Social Sciences at Mohammed First University in Oujda. A highly committed individual, he is an active member of the ERJAP Laboratory (Laboratory of Legal, Administrative, and Political Studies and Research) at the same faculty. He has overseen several development projects in partnership with the university, international organizations, and Moroccan civil society. Among his recent achievements are the organization and coordination of conferences on key topics such as environmental education, volunteering and sustainable development, and innovative solidarity in Morocco. He has also written scientific articles in specialized journals, addressing critical issues such as the legal framework for associations in Morocco and the new law on contractual volunteering. His work reflects his involvement in innovative projects and his growing expertise in development and public policy in Morocco.

Sanae Khali Issa is Associate Professor of Computer Science at the Faculty of Science and Technology of Tangier (FST), Abdelmalek Essaâdi University, Morocco. She has worked on the modeling of a decision support system for risk management. She is a member of the Laboratory of Informatics, Systems and Telecommunication (LIST) at the Faculty of Science and Technology of Tangier (FSTT). She is intrigued by the influence of artificial intelligence on the environment and education. She is the author of scientific publications in the fields of data analysis and computer science. She has authored numerous scientific articles on data analysis and programming systems, in addition to thesis projects on artificial intelligence. She helped organize national and international conferences and symposia.

Priyanka Kaushik is a seasoned academician with over 21 years of experience in computer science and engineering. Currently professor and head at Chandigarh University's Computer Science & Engineering (CSE) (AIML) Department, she holds a Ph.D. along with multiple technical qualifications including B.E., M.Tech., and M.Phil. Her areas of expertise include AI, Machine Learning, Cloud Computing, IoT, and mobile networking. She has authored numerous papers in IEEE and Scopus-indexed journals, holds multiple patents, and has contributed significantly to books and conferences. Recognized with several awards, she serves in key roles such as keynote speaker, conference chair, and technical committee member. She is actively involved in Ph.D. supervision, curriculum development, and mentoring, and is a certified professional in Generative AI and Deep Learning. As an Asia Intel Facilitator and an IBM-certified expert, she is committed to interdisciplinary innovation. Kaushik is an active member of IEEE, ISTE, and other professional bodies.

Kenneth Lisbo is a political science faculty member and researcher in the Department of Political Science at the College of Social Sciences of Humanities, Mindanao State University, General Santos, General Santos City, Philippines. His research focuses on political behavior, governance, environmental justice, and the intersections of media, civil society, and public policy. He also actively advocates for environmental awareness and is actively engaged in human rights education in Southern Philippines through her advisership on the Human Rights Society of the same university.

Abdelhakim Mansoureddine is a professor at the Faculty of Letters and Human Sciences (FLSH), Mohammed V University, Rabat, Morocco. He specializes in information and communication sciences, with a focus on internal and institutional communication. His research explores media discourse and its impact on organizations, as well as the cultural and symbolic dimensions of communication within institutional settings. He is particularly interested in how communication practices shape organizational identity and influence decision-making processes. He has contributed to several academic projects and publications addressing issues related to communication strategies, governance, and social dynamics in public institutions.

Christine Faith Murillo is a political science scholar and researcher in the Department of Political Science at the College of Social Sciences of Humanities, Mindanao State University, General Santos, General Santos City, Philippines. Her research interests center on youth engagement in global climate justice, social justice, Philippine politics, ASEAN studies, and public policy. She is also actively advocates for environmental awareness and is actively engaged in human rights education in Southern Philippines through her membership on the Human Rights Society of the same university.

Omar Ouhsine is a Moroccan teacher-researcher deeply committed to the environment and sustainable development and an active member of the Applied Spectro-Chemistry and the Environment at the Faculty of Science and Technology, Sultan Moulay Slimane University, Morocco. Combining scientific rigor with a passion for teaching, he tackles contemporary environmental and ecological challenges from a technical and educational perspective. At the same time, as a teacher in qualified secondary schools, he seeks to innovate by integrating the challenges of sustainable development into school life and training the younger generation in an enlightened ecological conscience. His interdisciplinary approach combines science and practical solutions to harmonize socioeconomic development and environmental preservation. Renowned for his scientific popularization, Omar Ouhsine makes a significant contribution to environmental awareness in Morocco, making sustainability an educational and societal pillar, and embodying an essential link between science and a more responsible future.

Abdellah Ouigmane is a researcher affiliated with the Faculty of Science and Technology at Sultan Moulay Slimane University, Béni Mellal, Morocco, where he is a member of the Agro-Industrial and Environmental Processes Team and has collaborated with the Applied Spectro-Chemistry and Environment Team. His research focuses on environmental sciences and waste management, exploring in particular energy recovery from municipal solid waste (such as waste-derived fuel production), GIS-based landfill site selection and multi-criteria decision-making, the impact of the

COVID-19 pandemic on waste production, and the characterization of waste dry fractions. He also uses GIS and remote sensing in his environmental studies, and actively collaborates with other researchers in this field.

Kathleen Mariel Parreñas is a political science scholar researcher in the Department of Political Science at the College of Social Sciences of Humanities, Mindanao State University, General Santos, General Santos City, Philippines. Her research interests center on youth engagement in global climate justice, ASEAN studies, communication and media studies, and local politics. She also actively advocates for environmental awareness and is actively engaged in human rights education in Southern Philippines through her membership on the Human Rights Society of the same university.

Phil Rodolf Quisel is a political science scholar researcher in Department of Political Science at the College of Social Sciences of Humanities, Mindanao State University, General Santos, General Santos City, Philippines. His research interests center on youth engagement in global climate justice, political behavior, ASEAN studies, and social policies. He also actively advocates for environmental awareness and is actively engaged in human rights education in Southern Philippines through his membership on the Human Rights Society of the same university.

Bremananth Ramachandran is a professor at A'Sharqiyah University, Oman, where he serves in the Department of Information Systems and Business Analytics within the College of Business Administration. He holds a Ph.D. in Computer Science and is an active researcher with a strong commitment to advancing knowledge in his field. Bremananth's long-term research direction aims to bridge the gap between theoretical research and practical implementation, particularly in understanding the deceptive mechanisms that underlie observable phenomena. He is especially interested in addressing emerging challenges in interdisciplinary areas, where the convergence of domains can lead to innovative solutions. In pursuit of this vision, Bremananth is focused on the development of fundamental theories, design methodologies, and application prototypes that are not only robust in theory but also viable for real-world deployment. His work continues to contribute to the evolution of research practices that are both academically rigorous and practically impactful.

Rachna Rathore is a dedicated academic and researcher in mathematics, specializing in queueing theory. Currently a Research Scholar at Jiwaji University, Gwalior, India, she holds an M.Sc. and a B.Sc. in Mathematics and allied subjects, both with distinction. With experience as an assistant professor, she is also an active IEEE member and has written over 24 Scopus-indexed papers and holds 6 patents. Her research focuses on applications of queueing models in real-life systems, including bed occupancy in healthcare and traffic congestion management. She has presented her work at numerous national and international conferences, addressing advanced mathematical topics such as fractional q-derivatives and Mittag-Leffler functions. Rachna is deeply engaged in academic discourse through seminars and webinars, consistently exploring the real-world relevance of mathematics. Her commitment to innovative research and academic excellence positions her as a promising contributor to the field of applied mathematics and stochastic modeling.

Saurabh Pratap Singh Rathore is an Executive Alumnus of IIM, Indore, with over 25 years of experience spanning industry and academia. His versatile career includes achievements in coaching, HR, pharmaceutical sales, training, and business development. As a dedicated researcher and IEEE member, he has authored 4 Scopus-indexed journal papers, 22 IEEE conference papers, and 7 book chapters with top publishers like Springer, Wiley, and IEEE Press. He holds one Indian and one German patent. As Project Director for an ICSSR-funded initiative aligning with NEP 2020, Rathore promotes quality research in Indian higher education. His contributions bridge the academic-industry gap, fostering impactful collaborations. Known for his leadership in scholarly publishing and academic events, he actively supports digital transformation, sustainable development, and interdisciplinary studies. With a passion for mentoring and institutional growth, he is a pivotal figure in advancing research ecosystems through national projects and global academic engagements.

Mohammed Rhiat has been pursuing his Ph.D. at Mohammed First University, Morocco, since 2021. He is currently a member of the Electrical Engineering and Maintenance Laboratory, where he develops innovative systems for renewable energy to promote energy transition and environmental protection. His research focuses on advancing sustainable energy technologies, and he has authored multiple research articles in the field.

Anass Saddati holds a Bachelor's Degree in French Language and Literature from the École Normale Supérieure (ENS), University Education Training (FUE) of Tétouan, Morocco. He also received a specialized Master's Degree in Educational Technology and Pedagogical Innovation from the Faculty of Educational Sciences (FSE) in Rabat. He is in his second year of doctoral research at the École Normale Supérieure of Tétouan. His main academic interests focus on integrating educational technologies in teaching and learning and pedagogical innovations aimed at improving educational practices.

Mohammed Lamarti Sefian is a Mohammadia School of Engineering graduate and Accredited Lecturer in Mathematics and Computer Science. With 27 years of experience in education, ranging from school teaching to managing the Planning Division of the Tangier-Tétouan-Al Hoceima region, he has joined the École Normale Supérieure (ENS) of Martil to embark on a new journey in teaching mathematics, computer science, and mathematics didactics. He currently supervises doctoral theses in various fields, including adaptive learning systems and virtual reality in education, for which he has written articles in reputable academic journals.

Richa Sharma is currently serving as Deputy Director (Research) at Indian Council of Social Science Research (ICSSR), New Delhi, India, overseeing research institutes and regional centers. Formerly Assistant Professor of History at Gujarat National Law University, she has developed an impressive scholarly profile with expertise in legal history, gender equality, primary education, and women empowerment. She has been invited as a keynote speaker at prestigious international platforms including the University of the Free State, South Africa; University of Helsinki, Finland; and has held teaching roles at the University of Paris and Osaka University. Her leadership at ICSSR includes initiating the 2023–2024 Short-Term Empirical Research scheme, emphasizing her commitment to advancing social science research. Sharma has authored several publications across national and international journals, conferences, and edited volumes. Her interdisciplinary impact and global academic presence highlight her significant role in enhancing social science research, policy relevance, and cross-cultural academic exchange.

Rosemary Shikangalah is a distinguished academic and researcher with extensive expertise in environmental science, soil ecology, geomorphology, and dendrochronology. With a career spanning over two decades, she has made significant contributions to the fields of environmental conservation, sustainable land use, and climate change adaptation. Her work is deeply rooted in addressing the challenges of arid and semi-arid regions, particularly in Namibia, where she has conducted groundbreaking research on bush encroachment, soil erosion, and the impacts of climate change on ecosystems.

Chardrey Tan is a political science scholar researcher in Department of Political Science at the College of Social Sciences of Humanities, Mindanao State University, General Santos, General Santos City, Philippines. His research interests center on youth engagement in global climate justice and media studies, Philippine politics, regional studies, and social justice. He also actively advocates for environmental awareness and is actively engaged in human rights education in Southern Philippines through his membership on the Human Rights Society of the same university.

Shoopala Uugulu is a senior lecturer at the University of Namibia, Namibia, mainly involved in groundwater and environmental sustainability studies in semi-arid regions. She has been involved in various projects focusing on groundwater recharge, the impact of vegetation on water systems, and sustainable land use practices in Southern Africa under climate change conditions. She has written work on topics such as groundwater quality and recharge mechanisms in Namibia, highlighting the complexities of groundwater dynamics in these arid environments. Uugulu's research often integrates hydrochemical and isotopic analysis to understand water resource management in challenging climates better. She has also written on environmental issues such as environmental contamination/pollution due to mining activities.

Mearlyn Vargas is a political science scholar researcher in the Department of Political Science at the College of Social Sciences of Humanities, Mindanao State University, General Santos, General Santos City, Philippines. Her research interests center on youth engagement in global climate justice, social justice, communication studies, Philippine politics, ASEAN studies, and environmental studies. She also actively advocates for environmental awareness and is actively engaged in human rights education in Southern Philippines through her membership on the Human Rights Society of the same university.

Hajar Zekeik is a doctoral student in education with a research focus on the integration of augmented reality (AR) and virtual reality (VR) technologies in teaching and learning environments. She holds a Master's Degree in Instructional Engineering and possesses solid expertise in designing and developing educational applications. Her work explores how immersive technologies can enhance learner engagement, motivation, and academic performance across diverse educational contexts. Hajar is particularly interested in the pedagogical potential of AR/VR to support both cognitive and non-cognitive skill development. She actively participates in conferences, hackathons, and collaborative research projects investigating innovative uses of immersive media in education. Her current doctoral research aims to develop frameworks and best practices that guide educators and developers in implementing AR/VR effectively in classrooms. Hajar combines theoretical insight with technical proficiency to bridge the gap between educational theory and technological innovation.

Abbreviations

AGFI	Adjusted Goodness of Fit Index
AI	Artificial Intelligence
AR	Augmented Reality
AU	African Union
CAQDAS	Computer-Assisted Qualitative Data Analysis Software
CAS	Chinese Academy of Sciences
CC	Climate Change
CCF	Cheetah Conservation Fund
CE	Environmental Club
CECyT 11	Centro de Estudios Científicos y Tecnológicos No. 11
COFEPRIS	Federal Commission for the Protection of Sanitary Risks
COPE	Center for Guidance and Educational Planning
COVID-19	Coronavirus Disease 2019
CSE	Centre for Science and Environment
CSP	Cleaner School Program
DRFN	Desert Research Foundation of Namibia
EC	Environmental Communication
ECEE	Early Childhood Environmental Education
ECLAC	Economic Commission for Latin America and the Caribbean
ECTA	Interdisciplinary didactic approach to teaching environmental content through visual arts
EE	Environmental Education
EI	Educational Institutions
EKT	Environmental Knowledge Test
ESD	Education for Sustainable Development
EU	European Union
EU-ETS	European Union Emissions Trading System

FLE	French Foreign Language
FLSH	Faculté des Lettres et des Sciences Humaines
GCC	Gulf Cooperation Countries
GCF	Giraffe Conservation Foundation
GFI	Goodness of Fit Index
HEI	Higher Education Institutions
HRI	Human-Robot Interaction
HSET	Higher School of Education and Training
ICT	Information and Communication Technology
IoT	Internet of Things
IPN	Instituto Politécnico Nacional
KAZA	Kavango-Zambezi Transfrontier Conservation Area
KEEP	Khomas Environmental Education Programme
L&E	Life and Earth Sciences
LISREL	Linear Structural Relationship
MET	Ministry of Environment and Tourism
MoEAC	Ministry of Education Arts and Culture
MSU-GensSan	Mindanao State University—General Santos
NACSO	Namibian Association of Community-Based Natural Resource Management Support Organisations
NaDEET	Namibia Desert Environmental Education Trust
NEEN	Namibia Environmental Education Network
NGC	National Green Corps
NGO	Non-Governmental Organization
NIED	National Institute of Educational Development
OBIS	Ocean Biogeographic Information System
OECD	Organisation for Economic Co-operation and Development
OL	Ocean Literacy
ORCID	Open Researcher and Contributor ID
PBL	Project Based Learning
PBP	Problem-Based Pedagogy
PMI	Philip Morris International
PPE	Personal protective Equipment
PSA	Public Service Announcement
SC	Science Communication
SCG	School and Career Guidance
SDGs	Sustainable Development Goals
SED	Sustainable Education Development
SEEN	Support Environmental Education in Namibia
STEAM	Science, Technology, Engineering, Arts, and Mathematics
STEM	Science, Technology, Engineering, and Mathematics
TEK	Traditional Ecological Knowledge

UN	United Nations
UNEP	United Nations Environment Programme
UNESCO	United Nations Educational, Scientific and Cultural Organization
VOC	Volatile Organic Compounds
VR	Virtual Reality
VUCA	Volatility, Uncertainty, Complexity, and Ambiguity
WHO	World Health Organization
WWF	World Wide Fund for Nature

List of Figures

List of Tables

PART I

Foundations and Educational Approaches in Environmental Education

CHAPTER 1

The Initial Teacher Training and Ecological Values: From a Cognitive Approach to a Psychoeducational Approach

Abdelkader Hammaoui and Kamal Hirech

1 Introduction

In the face of ecological urgency, Education for Sustainable Development (ESD) occupies a central place in contemporary educational policies. This is clearly evident when we observe that the majority of the Sustainable Development Goals (SDGs), set by the United Nations (UN) for 2030, primarily concern the environmental domain, which is essential for achieving true sustainable development (Hák et al., 2016). Future teachers play a key role in transmitting ecological values to new generations, meaning their initial training must integrate content and teaching practices aligned

A. Hammaoui (✉)
Higher School of Technology, Mohammed First University, Oujda, Morocco
e-mail: abdelkader.hammaoui@ump.ac.ma

K. Hirech
Higher School of Education and Training, Mohammed First University, Oujda, Morocco
e-mail: k.hirech@ump.ac.ma

K. Hirech (ed.), *Environmental Education and Sustainable Development*, https://doi.org/10.1007/978-3-032-05760-0_1

with sustainable development goals. At the Higher School of Education and Training (HSET), this requirement raises several questions: What ESD-related content is included in initial teacher training programs? Is this content sufficient to sustainably instill ecological values in future teachers? What are the challenges and opportunities to strengthen this integration and promote truly transformative, environmentally respectful education?

On the other hand, despite the growing number of ecological campaigns and educational initiatives, destructive behaviors persist: excessive consumption, resource waste, inertia in the face of environmental crises (Sachs et al., 2019). This gap between ecological awareness and real engagement raises questions about the effectiveness of cognitive approaches in driving sustainable behavioral transformation (Mayer et al., 2023). Moreover, it calls into question the effectiveness of this approach in fostering behavioral autonomy.

The hypothesis here is that awareness based primarily on a cognitive approach, while essential, is not enough to deeply instill autonomous ecological behaviors. For an individual to adopt environmentally respectful practices in a stable and lasting way, these practices must be rooted in an authentic and internal value system, solidly integrated into their psychological structure. In this regard, the psychoeducational approach inspired by Paul Diel's psychology of motivation provides a relevant framework. According to Diel and his student Stéphane Solotareff, values are not limited to external injunctions but are constructed through an introspective process in which the individual confronts their desires, contradictions, and illusions (Solotareff, 2024).

To assess the effectiveness of the current cognitive approach and explore the possibilities of transitioning to a more transformative education from the depths of the psyche, this study adopts a mixed methodology. It involves an in-depth analysis of the pedagogical content from various years and programs of the undergraduate cycle at HSET. Subsequently, the goal is to propose both a theoretical framework and a practical method to promote ecological valorization through an analytical reading of the research by Paul Diel, Jeanine Solotareff, and Stéphane Solotareff on the psychology of motivation.

2 Analysis of Training Descriptions: State of ESD at HSET

Before proceeding with the actual analysis of the different modules, it is essential to present the corpus under study as well as the methodology used in this analysis.

2.1 *Corpus and Methodology of Analysis*

The analysis of training descriptions is based on the examination of curricula from various specializations within the undergraduate education program at the Higher School of Education and Training. The relevant programs are listed in Table 1.1.

The objective is to assess the presence of themes related to Education for Sustainable Development (ESD) by identifying the knowledge and skills integrated into various disciplines. Two analytical approaches were adopted. First, a thematic analysis, aimed at identifying elements related to ESD through an in-depth review of module descriptions. Second, a quantitative analysis, which involves counting occurrences of key terms such as "*environment*", "*ecology*", and "*sustainable development*" to evaluate the prominence of these concepts within the training programs. This dual approach helps to identify general trends regarding the integration of ESD in curricula and highlights disparities between disciplines.

Table 1.1 Modules by specialization subject to research

Specialization	*Number of modules*
Primary Education	38
Primary Education—Amazigh Language Option	46
Secondary Education—English Language	38
Secondary Education—French Language	38
Secondary Education—Mathematics	38
Secondary Education Physics Chemistry	38
Secondary Education—Life and Earth Sciences (LES)	38

2.2 *Research Results and Quantitative and Qualitative Data Analysis*

The results of this research are presented in Table 1.2 and Table 1.3 that illustrate the distribution of concepts related to *environment*, *ecology*, and *sustainable development*, as well as the specific modules addressing these topics in each specialization, with representative examples. Each of these tables will be analyzed to shed light on the current state of ESD at HSET through the content of the different programs.

2.2.1 *Distribution of Environment Related Concepts*

The following table presents the occurrence of the concepts *"Environment"*, *"Ecology"*, and *"Sustainable Development"* across different teaching specializations.

In the Primary Education—Amazigh Language Option specialization, environmental concepts are present but remain limited. The word *"Environment"* appears three times, while *"Ecology"* and *"Sustainable Development"* are mentioned twice each. The content covers essential themes such as natural balance, pollution, the greenhouse effect, overgrazing, environmental preservation, and sustainable development.

In Secondary Education specializations, including English and French Languages, Mathematics, and Physics-Chemistry, references to ecological

Table 1.2 Distribution of concepts by specialization

Specialization	*No. of pages*	*No. of modules*	*Environment*	*Ecology*	*Sustainable development*
Primary Education—Amazigh Language Option	223	46	3	2	2
Secondary Education—English Language	150	38	0	0	0
Secondary Education—French Language	194	38	1	0	0
Secondary Education—Mathematics	139	38	1	0	0
Secondary Education—Physics-Chemistry	144	38	1	0	0
Secondary Education—Life and Earth Sciences (LES)	153	38	8	13	0
Primary Education	171	38	4	2	2

Table 1.3 Module content featuring ecological concepts

Specialization	*Relevant module*	*Mentioned concept*	*Content excerpt*
Primary Education—Amazigh Language Option	S1 M4: Sciences 1	Environment	Understanding concepts of natural balance and threats (pollution, greenhouse effect, overgrazing, soil degradation, etc.)
	S1 M4: Sciences 1	Sustainable Development	Environmental protection—sustainable development
Secondary Education—French Language	S3 M20: Educational Sciences	Environment	Environmental influences
Secondary Education—Mathematics	S3 M20: Educational Sciences	Environment	Environmental influences
Secondary Education—Physics-Chemistry	S3 M20: Educational Sciences	Environment	Environmental influences
Secondary Education—Life and Earth Sciences (LES)	M18 S3: Sedimentology and Stratigraphy	Environment	Sedimentary facies zones and models, continental and marine environments
	S5 M27: Applied Geology	Environment	Environmental impact
	S5 M28: General Ecology 1	Ecology	Contributing to environmental preservation, species adaptation
	M28 & M35: General Ecology 1 & 2	Ecology	Dedicated modules on general ecology
Primary Education	S1 M4: Life and Earth Sciences	Environment	Environmental protection—sustainable development
	S3 M15: Developmental Psychology	Environment	Environmental influences

concepts are almost nonexistent. Only a few disciplines, such as French Language, Mathematics, and Physics-Chemistry, mention the word "*Environment*" once, without further developing ecological notions in depth.

Conversely, the Secondary Education—Life and Earth Sciences (SVT) specialization stands out for its strong emphasis on ecological education. The word *"Environment"* appears eight times, while *"Ecology"* is mentioned thirteen times. This specialization includes specific modules such as *"General Ecology 1 and 2"*, *"Applied Geology"*, as well as themes related to environmental impact and species adaptation, demonstrating a comprehensive approach to ecological issues.

Finally, the Primary Education specialization, in a broader sense, contains some references to environmental issues, with four occurrences of the word *"Environment"*, two of *"Ecology"*, and two of *"Sustainable Development"*. The covered content integrates elements of developmental psychology and life and earth sciences but remains relatively limited in terms of an in-depth ecological approach.

2.2.2 Modules Addressing Ecological Concepts

The results of the content analysis of modules from each specialization are classified in Table 1.3.

The analysis of Table 1.3 highlights an unequal distribution of ecological concepts across the different modules of the studied specializations.

In the Primary Education—Amazigh Language Option specialization, ecological notions appear mainly in the *Sciences 1* module during the first semester. The concept of "*Environment*" is addressed from the perspective of natural balance and threats such as pollution, the greenhouse effect, and soil degradation. "*Sustainable Development*" is also mentioned in connection with environmental protection, reflecting awareness of ecological issues but through a descriptive and informative approach.

In Secondary Education, the specializations in French Language, Mathematics, and Physics-Chemistry mention the concept of "Environment" only in the Educational Sciences module (S3 M20). However, these references remain vague and are limited to studying the influences of the environment without further exploring ecological themes.

Conversely, the Life and Earth Sciences (LES) specialization stands out for its strong and structured presence of ecological concepts across several modules. The "*Environment*" is addressed from various perspectives, notably in Sedimentology and Stratigraphy (zones and sedimentary facies models) as well as in Applied Geology, where environmental impact is explicitly covered. Moreover, "*Ecology*" is deeply integrated into dedicated modules (General Ecology 1 and 2), emphasizing a specialized and detailed education on environmental preservation and species adaptation.

In Primary Education, ecological notions appear in two modules: Life and Earth Sciences (S1 M4) and Developmental Psychology (S3 M15). The first addresses environmental protection and sustainable development, while the second is limited to the influences of the environment on individual development, without a deeper ecological approach.

Overall, the analysis reveals that only the LES specialization offers a structured and in-depth education on ecological concepts. Other specializations, particularly Languages, Mathematics, and Physics-Chemistry, integrate these notions marginally and often without real pedagogical exploitation. The Primary Education specialization presents a moderate inclusion of ecological concepts, but these remain general and do not seem to constitute a central axis of training. These results indicate that the theme of Education for Sustainable Development (ESD) is largely concentrated in scientific disciplines, particularly Biology and Environmental Sciences, whereas it is almost absent in specializations related to Languages, Mathematics, and Physics-Chemistry.

2.2.3 Nature of Integrated Content

A qualitative analysis of the content reveals two main approaches to integrating *ESD*. First, an informative and descriptive approach predominates in scientific modules. This approach is based on the transmission of factual knowledge about the environment, ecosystems, and ecological issues. It favors a theoretical and disciplinary vision without necessarily engaging in reflection on individual and collective commitment.

Second, a cross-disciplinary and implicit approach, present in *Educational Sciences*, where the environment is mentioned indirectly, particularly in relation to the influences of surroundings on child development. However, this approach remains limited and does not include a deep reflection on the construction of ecological values.

In terms of disciplines, the LES specialization stands out as the most favorable to the ecological cause, with dedicated modules on Ecology and Environmental Sciences, such as General Ecology and Applied Geology. The Primary Education specialization, while less in depth, provides an introduction to environmental concepts through Life and Earth Sciences (SVT) and Educational Sciences modules. In contrast, specializations in Languages, Physics-Chemistry, and Mathematics show a notable absence of content related to ESD, limiting their potential to raise awareness among future teachers about ecological issues.

An examination of the training descriptions reveals an uneven integration of *ESD* in the *HSET* curricula. While some scientific disciplines give significant importance to environmental and ecological concepts, other specializations remain largely disengaged. Moreover, the dominant approach primarily relies on the cognitive transmission of knowledge, without fostering genuine reflection on the ecological commitment of future teachers. This analysis thus highlights the limitations of a predominantly informative approach, which will be further explored in the second part of this study.

3 Limitations of the Cognitive Approach in ESD

The analysis of the training descriptions has highlighted a predominantly cognitive approach to Education for Sustainable Development (ESD) at HSET. While some disciplines, particularly Life and Earth Sciences, incorporate environmental knowledge, the transmission remains focused on information and awareness, without truly working on the construction of ecological values. This approach has several limitations (Brown, 2023), reducing its impact on the commitment and behaviors of future teachers.

3.1 Awareness Limited to Theoretical Knowledge

Education for Sustainable Development, as integrated into curricula, is primarily based on the transmission of factual content related to the environment, ecosystems, and climate change issues (Rieckmann, 2018). This approach emphasizes the acquisition of scientific knowledge but leaves little room for the development of personal reflection and the internalization of ecological values (Ssossé et al., 2021).

The training mainly focuses on learning environmental concepts (biodiversity, pollution, renewable energies, etc.) without necessarily engaging students in a reflective process about their own relationship with the environment (Rieckmann, 2017). There is a clear predominance of information over the transformation of attitudes.

Indeed, the current approach does not incorporate tools that encourage future teachers to reflect on their role in transmitting ecological values (Thompson, 2010). Value education requires work on introspection and personal motivation, which is lacking in the analyzed content. There is no real reflection on ecological commitment. As a result, awareness remains at the level of information, without a deeper exploration of the affective and motivational dimensions that foster genuine adherence to the principles of sustainable development (Brown, 2023).

3.2 *A Weak Impact on Inner Satisfaction*

The following diagram (Fig. 1.1) summarizes the dissatisfaction with the ecological cause despite the multiplication of awareness campaigns:

This diagram highlights a problematic cycle where industrialization and over-exploitation of resources lead to growing pollution, causing environmental and societal imbalances (Elisha & Felix, 2021). This degradation results in collective discomfort, prompting researchers to delve deeper into ecological studies and raise awareness about the dangers. However, despite alarming findings, the primary response remains awareness-raising, which appears insufficient to halt the destructive dynamic. The diagram thus illustrates a loop where awareness fails to sustainably alter behaviors, and where pollution and excessive resource exploitation continue.

This situation highlights a major flaw in the cognitive approach, which primarily relies on the transmission of environmental knowledge without ensuring its conversion into concrete practices. Numerous studies in educational psychology have shown that acquiring knowledge is not enough

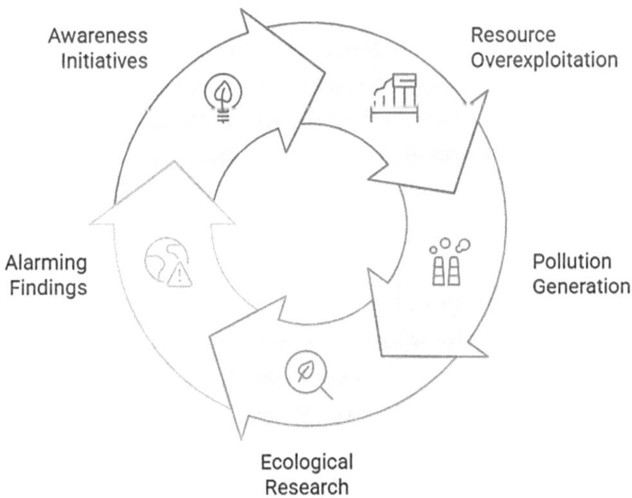

Fig. 1.1 Diagram of ecological dissatisfaction

to change attitudes and behaviors (Handtke et al., 2022). An educator informed about the issues of climate change will not necessarily develop pedagogical practices integrating Education for Sustainable Development (ESD) unless they recognize the importance of their personal commitment.

Furthermore, extrinsic motivation, which is based on institutional incentives or academic obligations, remains insufficient. When an individual acts under external pressure, their engagement remains fragile and short-lived. On the other hand, real change in practices requires intrinsic motivation, where the individual feels a personal and deep attachment to ecological values (Griffiths, 2021). Without such involvement, ESD risks remaining a peripheral theme, perceived merely as an academic requirement rather than a fundamental lever for transforming practices.

Thus, the diagram indirectly highlights the limitations of current approaches that focus solely on awareness without addressing the deep drivers of change (Pache et al., 2025). It calls for a rethinking of environmental education by integrating a psychoeducational approach that promotes voluntary and sustainable engagement, going beyond the mere transmission of knowledge to address the values and deep motivations of individuals.

3.3 *The Absence of Training in Introspection and Ecological Appreciation*

The current cognitive approach does not include support for future teachers to develop an integrated ecological consciousness. However, education in ecological values cannot be limited to intellectual awareness: it must involve affective, emotional, and experiential dimensions (Bufalino & D'Aprile, 2024).

An analysis of training descriptions reveals the absence of activities or pedagogical tools encouraging students to examine their personal relationship with the environment. ESD is mainly approached from the perspective of knowledge, without working on representations and individual values due to a lack of methodological introspection tools. Without these introspective and experiential dimensions, teacher training cannot fully foster the emergence of sustainable ecological engagement.

In short, the cognitive approach in ESD, while essential for transmitting basic knowledge, has significant limitations when it comes to fostering personal engagement and behavior changes. The absence of training in reflection on ecological values and immersive pedagogical tools prevents future teachers from truly owning the challenges of sustainable development. These findings highlight the need for a paradigm shift toward a psychoeducational approach, focused on intrinsic motivation and the appreciation of ecological behaviors. This approach will be developed in the third part of this study.

4 Toward a Psychoeducational Approach Based on the Psychology of Motivation

The construction of ecological values requires a deeper foundation, involving an internal transformation of individuals. From this perspective, a psychoeducational approach inspired by the psychology of motivation offers a relevant framework to sustainably embed these values in the psychological functioning of future teachers (Suzuki & Tanoue, 2020).

4.1 Appreciation and Ecological Values

The psychoeducational approach is based on the idea that the sustainable adoption of ecological values can only be achieved through introspective work that allows for the evaluation and prioritization of deep motivations. Paul Diel emphasizes that: "*Valorization can be either true or false*" (Diel, 1962, p. 9). People often resist recognizing their mistaken valorization, justifying it erroneously. By hiding it from conscious awareness, they push it into the subconscious, where it transforms into false motivation.

This quote from Paul Diel highlights the psychological process by which an individual can adopt values or motivations without fully being aware of them or questioning their legitimacy. This means that a person can assign value to an idea, behavior, or goal, whether correctly or incorrectly. However, they often struggle to recognize their mistakes in judgment. When they make an error in this valuation, they unconsciously seek to justify their choice in a biased way. By avoiding a conscious and objective analysis of this mistake, they relegated this justification to the subconscious level, which transforms this erroneous valuation into an illusory motivation that guides their actions without their full awareness.

In other words, if ecological awareness is solely based on external norms and institutional discourses, it risks failing to anchor itself in authentic motivation. It is, therefore, essential to guide future teachers through an introspective process that allows them to move beyond false valuations and to structure their own ecological values.

Paul Diel explains here a fundamental psychological mechanism: when we value something erroneously, we tend to avoid admitting our mistake. To protect our self-esteem, we create justifications that reassure us but are not necessarily objective. For example, placing excessive importance on social status or ostentatious consumption at the expense of deeper ecological values is a false motive that generates false inner valuations that are unjustifiably defended. These individual valuations can end up becoming social values defended by a group of people.

On the other hand, by refusing to examine these justifications clearly, we allow them to take root in our subconscious, where they influence our choices and behaviors without our knowing. Thus, we may believe we are motivated by noble values, for example, acting for the environment, while in reality; our motivation is based on an erroneous valuation such as seeking social recognition by displaying a façade of ecological commitment. This process must be understood within the framework of ecological value education: as long as individuals are unaware of their false motivations and do not question them deeply, their environmental engagement risks remaining superficial and fragile.

4.2 The Value-Enhancing Function of Ecology in Education

Genuine education in ecological values cannot, therefore, be limited to external injunctions. It must enable individuals to integrate these values into their personal framework through lived and reflective experience. Stéphane Solotareff (2024, p38) also emphasizes this process of value construction: "*the valorizing function assigns a satisfaction value to the outcomes of our experiences*". This value is then organized within our personal hierarchy of values, which acts as a reference framework guiding individual decision-making.

It should be noted that each individual evaluates their experiences by attributing a certain value to them based on the satisfaction they provide. These values are not isolated; they are organized within a personal hierarchy, thus forming an internal reference framework. This framework then serves as a guide for decision-making and behavior orientation.

Solotareff S. therefore highlights a fundamental process in the construction of values: we do not adopt our values in an abstract or theoretical manner, but through our lived experiences and how we feel about them.

When an experience provides us with deep satisfaction and is consistent with our search for the meaning of life, we tend to attribute greater value to it.

Over time, these successive evaluations organize into a personal value system that influences our choices and actions (Solotareff, 2017). For example, a person who experiences genuine satisfaction in adopting an ecological lifestyle because they find meaning and coherence with their deep convictions will be more likely to persist in this commitment. Conversely, if adopting ecological behaviors is perceived as an external constraint imposed by society, it will not be permanently integrated into the person's framework.

This perspective is essential in ecological value education: for an individual to sustainably adopt a value, it must be experienced and validated internally through an experience that provides real satisfaction. Simple awareness-raising or external injunctions are not enough to deeply anchor a value in the consciousness and behavior of an individual. True transformation, therefore, requires introspective work that allows one to distinguish authentic valuations from illusory ones.

4.3 *Objective Introspection and the Construction of Guiding Values for Sustainable Ecological Education*

Diel specifies that introspection becomes objective when an individual is willing to uncover the underlying causes of emotional and subconscious confusion, and when they have the courage to bring their false motivations into conscious awareness (Diel, 1962, p. 10). Thus, introspection plays a key role in the awareness of our deep motivations. As Paul Diel explains, an individual can only achieve true understanding of themselves by exploring the hidden causes of their emotions and identifying their erroneous motivations. In other words, it is a process of self-analysis that allows one to move beyond superficial justifications and recognize the psychological resistances that prevent us from adopting virtuous behaviors (Solotareff, 1991).

Applied to education for sustainable development, this principle means that simple awareness of ecological issues is not enough. It is essential to lead individuals to reflect on their own blockages and contradictions in relation to ecological values. For example, a person may intellectually recognize the importance of reducing their plastic consumption but continue to use plastic bags out of habit or convenience. Sincere introspection would allow them to understand that they minimize the impact of their actions (Solotareff, 2016) or hide behind the idea that “others pollute much more”.

The psychoeducational approach relies on this introspection to help each individual deconstruct their resistances and align their ecological values with their daily actions. For instance, a teacher may be convinced of the importance of nature preservation, but if they are not aware of their own inconsistencies, such as excessive paper use or energy waste, they will not be able to embody these values in an authentic and inspiring way for their students.

Thus, a true ecological transition cannot happen without structured inner work. By becoming aware of our deep motivations and harmonizing our values with our actions, we go beyond simple awareness and engage in a personal and collective transformation that is sustainable.

As Solotareff S. (2024, p. 84) states, a guiding value is one that can be embraced by every member of a society or group living together, without endangering the survival of that group. He defines a guiding value as a principle universally applicable within a group or society without threatening its balance or survival. In other words, a guiding value must be adoptable by everyone without generating destructive conflicts or compromising social cohesion.

In the context of the environment, a guiding value could be respect for nature. If this value is adopted by all, it does not create any imbalance within society; on the contrary, it fosters better coexistence between individuals and their environment. For example, a community that values water preservation will encourage each member to avoid waste, use sustainable resources, and respect aquatic ecosystems. This value, when applied to all, does not endanger society, but instead strengthens its resilience in the face of environmental challenges.

On the other hand, a value that would not truly be a “guide” could be an extreme or exclusive principle that divides rather than unites. For instance, advocating for a total rejection of technology in the name of protecting nature could become a barrier to development and create conflicts within society. Therefore, an ecological guiding value must be compatible with collective survival and enable a harmonious transition to more sustainable behaviors.

In this perspective, values such as ecological responsibility, intergenerational solidarity, or even happy sobriety can be considered guiding values, as they are applicable to all and contribute to the longevity of the group without generating insurmountable tensions. They allow for a gradual transformation of behaviors without compromising social organization, but rather guiding it toward an evolution that is more respectful of the environment.

As a result, lived experiences accumulate and gradually build a value system, based on the association between an action and the satisfaction it provides. Through memory, this system allows us to reproduce or avoid certain behaviors. This process of feedback, also known as the feedback loop, plays an essential role in our direction and life choices (Solotareff, 2024, p. 41).

5 Proposal for a Module on the Integration of Ecological Values at HSET Through the Psychoeducational Approach

To implement the psychoeducational approach outlined in the previous chapter, inspired by the motivation psychology developed by Diel and Solotareff, we propose a module dedicated to ecological causes. This module will enable future teachers to guarantee the ecological transition at a profound level (Schlosberg et al., 2019).

5.1 Module Overview

- Module Title: Education on Ecological Values and Environmental Responsibility;
- Module Type: Cross-disciplinary;
- Institution: Higher School of Education and Training (ESEF);
- Duration: 30 hours (semester);
- Target Audience: Students in initial teacher training, across all fields;
- Semester: Preferably the first semester, as it is important to introduce students to ecological commitment early in their training;
- General Objective of the Module: To train teachers capable of transmitting and embodying ecological values in their professional practice and daily lives;
- Learning Objectives: By the end of the module, students will be able to:

- Understand and explain the fundamental ecological values;
- Develop an eco-responsible attitude in both their personal and professional lives;
- Integrate ecological values into their teaching practices;
- Design projects and educational activities related to sustainable development;
- Develop critical and introspective thinking about their own behaviors and values.

5.2 *List of Targeted Values*

We propose a list of ecological values that cover the essential dimensions of a sustainable environmental ethic (Jayal, 2020). This list provides a comprehensive framework for thinking about and promoting an integrated ecological ethic, ranging from individual actions to collective and philosophical issues.

5.2.1 Fundamental Values

- Respect for Nature: Consider nature as a shared heritage to be preserved;
- Ecological Responsibility: Acknowledge our impact on the environment and act accordingly;
- Ecological Balance: Strive to maintain harmony between living beings and their ecosystems;
- Sustainability: Consider the long-term effects of our actions to preserve resources for future generations;
- Precaution: Adopt a cautious attitude toward environmental risks and avoid actions with irreversible consequences.

5.2.2 Relational and Ethical Values

- Intergenerational Solidarity: Act to ensure a viable future for coming generations;
- Environmental Justice: Guarantee equitable access to natural resources for all;
- Ethics of Life: Respect all forms of life and avoid destructive or cruel practices toward animals;
- Cooperation: Encourage collective and collaborative actions for environmental protection;
- Empathy Toward Nature: Develop sensitivity and emotional connection with the environment.

5.2.3 *Values Related to Lifestyles and Behaviors*

- Voluntary Simplicity: Reduce consumption to minimize our ecological footprint;
- Moderation: Use natural resources thoughtfully and balanced;
- Autonomy: Develop self-sufficient lifestyles that respect the environment (e.g., food self-sufficiency, renewable energy);
- Repair and Reuse: Promote repair rather than replacement to limit waste;
- Circular Economy: Reduce, reuse, recycle to minimize waste and the extraction of new resources.

5.2.4 *Cognitive Values*

- Critical Thinking: Develop autonomous reflection in response to environmental discourses and proposed solutions;
- Scientific Curiosity: Be interested in ecological knowledge and scientific advancements in sustainability;
- Eco-citizenship: Actively engage in ecological initiatives and make informed decisions;
- Knowledge Transmission: Share good environmental practices with others.

5.2.5 *Spiritual and Philosophical Values*

- Humility Toward Nature: Recognize that humans are not above nature, but are a part of it;
- Gratitude Toward Nature: Appreciate the benefits of the Earth and take care of it;
- Harmony with the Environment: Live in alignment with the natural cycles and rhythms of the planet;
- Ecological Wisdom: Make decisions based on a holistic and long-term view of the environment.

5.3 *Module Content*

Unit 1: Foundations and Ethics of Ecology (6 hours):

- History and evolution of ecological thought;
- Principles of sustainability and ecological balance;
- Fundamental ecological values (respect, responsibility, precaution, environmental justice, etc.);

- Philosophy and spirituality of nature (humility, gratitude, ecological wisdom).
  Evaluation Activity: Critical analysis of environmental discourses and debates on human responsibilities toward nature.

Unit 2: Eco-citizenship and Collective Engagement (6 hours):

- The concept of eco-citizenship and the role of education in ecological transition;
- Intergenerational solidarity and environmental justice;
- Cooperation and collective actions for the environment;
- The link between ecological values and community development.
  Evaluation Activity: Case study on successful ecological projects at local and international levels.

Unit 3: Lifestyle and Individual Responsibility (6 hours):

- Voluntary simplicity and happy sobriety;
- Circular economy and waste reduction;
- Eco-consumption and responsible eating;
- The impact of daily habits on the environment Practical Workshop: Experimenting with a zero-waste lifestyle for a week and reflective feedback.

Unit 4: Introspection and Transformation of Behaviors (6 hours):

- The role of objective introspection in adopting ecological values (Diel and Solotareff);
- Psychological resistance to change and strategies to overcome them;
- Development of ecological consciousness and harmonizing values with actions;
- How to avoid repressing ecological responsibility;
- How to integrate these values into pedagogy Psychoeducational Workshop: Identifying individual resistances to ecology and working on intrinsic motivation.

Unit 5: Pedagogy and Transmission of Ecological Values (6 hours)

- How to integrate ecological values into the teaching of different subjects?
- Active pedagogy and education for sustainable development;
- Creation of ecological educational resources;
- Development of educational projects with ecological impact.

Final Project: Design of an eco-responsible educational project and presentation to a jury.

6 Conclusion

Ecological education cannot be limited to mere awareness-raising activities or purely cognitive transmission of environmental knowledge. It must be rooted in introspective work, allowing individuals to identify and overcome their false valuations in order to integrate authentic ecological values.

The psychoeducational approach offers guidance to future educators in this process. By structuring their relationship with ecological values, they will be better equipped to direct their own actions in support of the ecological cause and to deliver a sustainable development education that transcends social norms, becoming a truly integrated commitment full of meaning.

The analysis of initial teacher training content at HSET highlights a worrying deficiency in terms of education for sustainable development and ecology. Not only are these topics entirely absent from several academic tracks, but when addressed, the approach remains predominantly cognitive and informational. However, this limited approach does not foster a deep internal commitment from future teachers, thus making it difficult to transmit true ecological values to their students.

In response to this gap, our study proposes a psychoeducational approach inspired by Paul Diel's psychology of motivation and further developed by Stéphane Solotareff. This approach emphasizes the essential role of objective introspection and the progressive structuring of values through the valorization of lived experiences. It moves beyond simple awareness to embed ecological values in a sustainable inner dynamic, based on deep motivation and reasoned satisfaction.

To achieve this, we propose an innovative training module designed to go beyond mere awareness and initiate an inner transformation in future educators. This module combines complementary approaches: experiential learning through situational exercises and field experiences; active pedagogy, promoting interaction through debates, case studies, and collaborative projects; a psychoeducational approach based on objective introspection and inspired by the psychology of motivation; and an interdisciplinary approach, integrating ecological values into various disciplines.

In short, this study provides a concrete response to the limitations of current training by proposing an immersive and introspective approach. It engages future teachers in a process of sustainable ecological valorization, structuring their values and strengthening their deep motivation. Thus prepared, they will be capable of delivering authentic ecological education, sincere and capable of creating change for future generations.

References

Brown, A. L. (2023). ESD "in the Making"? Potentials and limitations of educational making for education for sustainable development. *MedienPädagogik: Zeitschrift für Theorie und Praxis der Medienbildung, 56*, 385–406. https://doi.org/10.21240/mpaed/56/2024.03.01.X

Bufalino, G., & D'Aprile, G. (2024). Fostering ecological awareness and inclusion through narrative pedagogy: Insights from the 'eco-narrations' projects. *University of South Florida (USF) M3 Publishing, 3*, 10. https://doi.org/10.5038/9781955833127

Diel, P. (1962). *Psychologie de la motivation: théorie et application thérapeutique*. Presses universitaires de France.

Elisha, O. D., & Felix, M. J. (2021). Destruction of coastal ecosystems and the vicious cycle of poverty in Niger Delta Region. *Journal of Global Agriculture and Ecology, 11*(2), 7–24.

Griffiths, T. G. (2021). Education to transform the world: Limits and possibilities in and against the SDGs and ESD. *International Studies in Sociology of Education, 30*(1–2), 73–92. https://doi.org/10.1080/09620214.2020.1854829

Hák, T., Janoušková, S., & Moldan, B. (2016). Sustainable development goals: A need for relevant indicators. *Ecological indicators, 60*, 565–573. https://doi.org/10.1016/j.ecolind.2015.08.003

Handtke, K., Richter-Beuschel, L., & Bögeholz, S. (2022). Self-efficacy beliefs of teaching ESD: A theory-driven instrument and the effectiveness of ESD in German teacher education. *Sustainability, 14*(11), 6477. https://doi.org/10.3390/su14116477

Jayal, N. G. (2020). Balancing political and ecological values. In *Political Theory and the Environment* (pp. 65–88). Routledge.

Mayer, M., Burger, M., & Vollan, B. (2023). A behavioral (economics) framework for sustainable transformation. In *Behavioural economics and the environment* (pp. 8–30). Routledge.

Pache, A., Gey, N., Lausselet, N., Mühlematter, Y., Blandenier, G., Massiot, P., & Gay, P. (2025). Research approaches in ESD/ESE: Reflections of Swiss researchers. *Environmental and Sustainability Education in Francophone Europe*, 144–159.

Rieckmann, M. (2017). *Education for sustainable development goals: Learning objectives*. UNESCO.

Rieckmann, M. (2018). Learning to transform the world: Key competencies in Education for Sustainable Development. *Issues and trends in education for sustainable development, 39*(1), 39–59.

Sachs, J. D., Schmidt-Traub, G., Mazzucato, M., Messner, D., Nakicenovic, N., & Rockström, J. (2019). Six transformations to achieve the sustainable development goals. *Nature Sustainability, 2*(9), 805–814. https://doi.org/10.1038/s41893-019-0352-9

Schlosberg, D., Bäckstrand, K., & Pickering, J. (2019). Reconciling ecological and democratic values: Recent perspectives on ecological democracy. *Environmental Values, 28*(1), 1–8. https://doi.org/10.3197/096327119X15445433913

Solotareff, J. (1991). *L'aventureintérieure: la méthode introspective de Paul Diel*. Payot.

Solotareff, J. (2016). *Psychanalyse introspective, déterminismes et libération*. Micro Application Editions.

Solotareff, J. (2017). *Intuition et méthode: Psychanalyse introspective*. Micro Application Editions.

Solotareff, S. (2024). *Les valeurs et le fonctionnementpsychique*. ESKA MA.

Ssossé, Q., Wagner, J., & Hopper, C. (2021). Assessing the impact of ESD: Methods, challenges, results. *Sustainability, 13*(5), 2854. https://doi.org/10.3390/su13052854

Suzuki, S. I., & Tanoue, A. (2020). Psychoeducation. In *Encyclopedia of behavioral medicine* (pp. 1760–1761). Springer International Publishing.

Thompson, M. (2010). Where reflection and motivation meet. *Professional Development in Education, 36*(3), 393–397.

CHAPTER 2

Introducing Environmental Education Skills: Analysis of Stakeholders' Perceptions

Hicham Berbar, *Narjiss Berbar*, *Loubna Bouselham*
Sanae Khali Issa, *and Kamal Hirech*

H. Berbar (✉)
Normal Superior School of Tétouan, Abdelmalek Essaâdi University, Tetouan, Morocco
e-mail: h.berbar@uae.ac.ma

N. Berbar
Faculty of Legal, Economic and Social Sciences, Tangier, Abdelmalek Essaâdi University, Tangier, Morocco
e-mail: narjiss.berbar@etu.uae.ac.ma

L. Bouselham
National School of Architecture, Oujda, Morocco
e-mail: l.bouselham@enaoujda.ac.ma

S. K. Issa
Faculty of Science and Technology of Tangier, Abdelmalek Essaâdi University, Tetouan, Morocco
e-mail: skhaliissa@uae.ac.ma

K. Hirech
Higher School of Education and Training, Mohammed First University, Oujda, Morocco
e-mail: k.hirech@ump.ac.ma

K. Hirech (ed.), *Environmental Education and Sustainable Development*, https://doi.org/10.1007/978-3-032-05760-0_2

1 Introduction

Given that awareness and action regarding environmental issues mostly rely on education, these competencies should be included in the existing curriculum. An interdisciplinary strategy that amalgamates elements from several fields can enhance the effectiveness of fostering an integrated understanding of environmental concerns among young students. To facilitate this integration, however, challenges such as curriculum change and teacher training must be resolved. In the field of education, education for sustainable development (ESD) is now firmly integrated into the curricula of several countries (Gey et al., 2023) (e.g., Switzerland, France, Quebec, etc.). It is therefore a form of interdisciplinary education wherein several disciplines collaborate to achieve a common objective (Merritt et al., 2022). ESD pertains to the personal development of pupils. Consequently, in ESD, educators must assist their students in developing cognitive frameworks that enable a nuanced understanding of events and attitudes conducive to intentional behaviors (Pellaud et al., 2021). Considering the environmental challenges confronting Morocco, the incorporation of environmental protection competencies into its primary educational curriculum is essential.

In light of several educational reforms aimed at enhancing scientific learning, environmental education (EE) has been recognized as a crucial element of the comprehensive educational framework in Morocco (El Moussaouy et al., 2014). Highlighting the necessity of equipping students with the skills to understand and address environmental issues, these modifications establish the foundation for integrating environmental protection into the curriculum. By integrating environmental concepts into subjects such as science, geography, and social studies through an interdisciplinary approach, educators may assist students in connecting their knowledge to real-world applications.

The effective incorporation of environmental protection competencies necessitates an evaluation of the existing curriculum. The current Moroccan primary school curriculum is often overloaded with material, thus hindering educators from allocating sufficient time to environmental education (El-Batri et al., 2019). This load sometimes means insufficient covering of essential courses, preventing pupils from comprehending environmental challenges. The curriculum's structure must be evaluated to eliminate non-essential courses to make room for environmental education. Students' participation with the subject topic increases their understanding and dedication to sustainable living. Beyond formal schooling, extracurricular activities allow pupils to address environmental challenges. Much research shows that extracurricular environmental activities improve environmental education (El Batri et al., 2022). Educational institutions may promote environmental stewardship through environmental groups, tree-planting projects, and community cleanups. These projects reinforce classroom knowledge and encourage student, school, and community participation, amplifying environmental education outside the classroom.

An essential element in successfully integrating environmental conservation into the primary school curriculum appears to be teacher training. Many educators nowadays lack the specific expertise required to teach environmental issues and use interdisciplinary methods (Amahmid et al., 2019). To assist educators in effectively integrating environmental education into their curricula, professional development opportunities focused on environmental education methodologies must be incorporated into teacher training programs. The quality of education on significant issues may be greatly enhanced by using knowledgeable and assured educators in environmental concepts. Moreover, international policies such as those of UNESCO emphasize the necessity for educational institutions globally to incorporate sustainability as a fundamental principle (Zenasni et al., 2024). Aligning Moroccan elementary school curricula with global educational programs helps lay the groundwork for environmental protection. Including environmental education in the national agenda helps pupils understand local and global environmental issues and develop a feeling of responsibility for future generations. Interdisciplinary approaches allow Morocco to incorporate environmental conservation into its elementary school curriculum, which is feasible. By tackling workloads and teacher training, Morocco can raise a generation of environmentally conscious pupils who will benefit their communities and the earth. This research addresses the problems and potential of EE integration in Moroccan primary schools. Multidisciplinary and qualitative methods and suitable assessment tools are used to investigate stakeholders' viewpoints.

The current study examines the opportunities and challenges of integrating EE skills into the Moroccan primary school curriculum. We examine stakeholders' perspectives using an interdisciplinary approach and qualitative technique, offering appropriate assessment tools. This research was conducted during the 2023–2024 academic year in eastern Morocco.

2 Theoretical Framework

2.1 *Integrating Environmental Education for Sustainable Development*

Recent studies demonstrate that multidisciplinary approaches facilitate the successful integration of EE and ESD into school curricula (Boyle & Anderson, 2020). Creative pedagogical approaches should emphasize project-based learning, the use of instructional tools, and educational outdoor events. These innovative teaching strategies enhance student engagement and facilitate a deeper understanding of environmental issues. Project-based learning enables students to investigate real-world issues, thereby fostering critical thinking and teamwork skills essential for tackling complex environmental challenges.

Cross-curricular skills, commonly known as "soft skills" or "twenty-first century skills," encompass a range of qualities that facilitate learning, teamwork, and problem-solving across various contexts (Faubert, 2009; Maghnouj et al., 2018). The abilities encompass critical thinking, creativity, communication, and teamwork (Ardoin et al., 2020). In education, these abilities are essential tools for enabling students to develop the competencies necessary to tackle contemporary issues, including environmental challenges. Recent studies emphasize the importance of integrating transversal skills into the curriculum to prepare students for the challenges of contemporary society (Knight et al., 2019). This process relies heavily on environmental education (EE), which equips students with the skills required for individual and collective action, as well as for addressing current environmental challenges (Gey et al., 2023). Previous studies confirm that the integration of technology into EE and ESD enhances learning outcomes. Digital technologies facilitate interactive and immersive experiences that enhance students' understanding of environmental issues. Virtual reality simulations enable students to explore ecosystems and assess the impact of human activities on these systems, thereby fostering empathy and a sense of environmental responsibility.

2.2 *Developing Cross-disciplinary Skills in Environmental Education*

Within the realm of environmental education, the cultivation of cross-disciplinary skills holds significant value. This type of knowledge encompasses an understanding of environmental issues, the ability to address problems associated with them, and the practice of sustainable behaviors (Ernst & Burcak, 2019; Lippmann, 2021). These skills serve as essential instruments for empowering students to comprehend environmental challenges and engage in responsible actions. In this context, the significance of metacognition is paramount. According to Ober et al. (2020), metacognitive skills allow learners to reflect on their learning processes and manage their actions across various contexts. This self-awareness plays a vital role in promoting autonomy among students as they engage with intricate environmental challenges. Furthermore, the advancement of these interdisciplinary skills is progressively recognized as a way to prepare students with the essential tools for engaging in democratic processes concerning environmental governance. As noted by Carney (2022), education that fosters critical thinking and active citizenship is crucial for empowering individuals to participate in sustainable practices and shape policy decisions impacting their communities and the environment. Studies indicate that integrating cross-curricular skills into environmental education can enhance students' environmental literacy and equip them with the necessary tools to grapple with the challenges of sustainability (Ardoin & Bowers, 2020). The incorporation of project-based learning within environmental education has been linked to enhanced collaboration and communication abilities among students (Jackson et al., 2021; Merritt et al., 2022). Students collaborate to address local environmental challenges, cultivating a sense of initiative and accountability. This participatory approach empowers students to take initiative as responsible caretakers of their environment.

2.3 Pedagogical Approaches to Skills Development

In the realm of environmental education, various pedagogical approaches have demonstrated efficacy in fostering interdisciplinary skills. Inquiry-based learning (IBL) promotes the formulation of questions by students, facilitating the investigation of topics and the derivation of solutions through experiential learning. Inquiry-based learning (IBL) fosters critical thinking, creativity, and collaborative skills in students through participation in discussions and the expression of diverse perspectives (Keith et al., 2021). An educational methodology that emphasizes experiential learning is referred to as "experiential learning," as described by Frensley et al. (2022) and Kolb and Kolb (2017). Studies demonstrate that students engaged in outdoor education programs cultivate a stronger connection to nature and an increased awareness of environmental issues (Chawla, 2020). These interactions promote the formation of an environmental ethic and a dedication to sustainability by strengthening emotional ties to the environment (Bezeljak et al., 2023; Knight et al., 2019). Service-learning integrates academic courses with community engagement, demonstrating improvements in cross-disciplinary skills and addressing local environmental challenges (Křepelková et al., 2020). Service projects allow students to utilize their knowledge in addressing real-world issues, promoting social responsibility and collaboration. This educational approach incorporates experiential learning to connect theoretical knowledge with practical application. Various theoretical frameworks advocate for the integration of interdisciplinary skills within environmental education. The constructivist perspective, based on the theories of Piaget and Vygotsky, asserts that knowledge is created through social interaction and active engagement with the environment (Piaget, 1970). This perspective illustrates the value of collaborative learning experiences in fostering interdisciplinary skills, as students participate in meaning-making and comprehension alongside their peers. Furthermore, transformative learning theory, introduced by Fleming (2021), represents another pertinent framework. This viewpoint highlights the importance of critical thinking for both societal advancement and individual growth. Transformative learning in environmental education allows students to critically assess their beliefs and values about the environment, thereby improving their comprehension of sustainability issues and providing them with essential tools for advocacy and action (Chodkiewicz & Boyle, 2017) (Waters & Loton, 2019).

The ecological worldview framework underscores the interconnectedness of human and environmental systems, demonstrating the importance of a comprehensive understanding of environmental issues. This perspective prompts students to critically examine the interactions among social, economic, and environmental factors, thereby fostering skills like systems thinking and ethical reasoning, which are crucial for addressing the intricate challenges of sustainability.

3 Materials and Methods

The thematic analysis is based on the data collected. It should be stressed, however, that researchers cannot shirk their epistemological and theoretical obligations. Moreover, the fact that the data have been coded does not mean that they are devoid of epistemic content (Braun & Clarke, 2006). This study is part of an action-research project being carried out in Morocco during the 2023–2024 school year. The objective of this study is to examine the perceptions of stakeholders in the educational sector (teachers, school principals, and inspectors) concerning the integration of environmental education and sustainable development (EE) skills into the elementary school curriculum.

To assure understanding, all questions were crafted in French, a language recognized as accessible to the intended participants. To enhance the study's rigor, a selection of probing questions was presented in Arabic to aid participants in comprehending the fundamental intent of each inquiry. As the participants responded to the questions, the interviewer provided explanations and clarifications for any challenging phrases or ideas, guaranteeing the participants' comprehension throughout the research. Furthermore, when participants offered responses beyond the study's parameters, the interviewer intervened to realign the interview with the study's aims. This method was employed to ensure that research participants understood the intentions of the questions and could provide replies aligned with their thoughts and opinions. However, five individuals who persisted in struggling with understanding and answering the questions were eliminated from the research sample before the data analysis was conducted. The interview discussed in this chapter had two halves. The initial segment was a series of inquiries, but the subsequent portion, referred to as the "pre-interview," encompassed details on the interviewers' professional experience and demographic information, including age and gender. The interview was conducted in French.

3.1 *Participants*

The study involved 30 educational players, including 25 teachers and 5 educational inspectors. Participants were selected on the basis of their experience in primary education and their involvement in environment-related projects. The empirical data presented in this study were methodically collected during 30 in-depth interviews with key players working in

Table 2.1 Demographic characteristics of participants

Demographic characteristics	*Measures*	*Frequency*
Participants	Teachers	25
	Pedagogical inspectors	5
Gender	Men	18
	Woman	12
Delegations	Berkane	10
	Guercif	4
	Oujda	12
	Taourirt	4
Years of experience	5–9 years	3
	10–15 years	11
	16–20 years	10
	21–25 years	4
	26–30 years	2

schools in the Oriental region under the authority of the Moroccan Ministry of National Education, Preschool Education, and Sport. Two interviews were excluded from the analysis, as they did not include all the planned questions (Table 2.1). The sample comprised individuals with more than five years' professional experience. Table 2.1 illustrates the distribution of the sample according to various criteria, such as status, function, gender, and seniority.

The participation of these professionals was completely voluntary. Following a thorough explanation of the study's purpose and objectives, the participants exercised their autonomy in determining their participation in the research. Those who concurred with the study's aims and expressed a desire to participate were assured that their responses during the interview would be completely anonymous. The study was conducted only with individuals who met these criteria, thereby ensuring the confidentiality and integrity of the data obtained.

3.2 Data Collection

The current study was conducted using a qualitative, predominantly inductive approach aligned with an interpretivist epistemology (constructivism). In pursuit of this objective, semi-structured interviews were carried out with pertinent stakeholders to investigate their views on quality in secondary education. Subsequently, interviews were carried out with the pertinent audience, and their responses were meticulously transcribed into written form. In the subsequent initial phase, the responses underwent classification, sorting, and coding by two doctoral experts (H.B. and N.B.) in collaboration. A cohort of 30 uniform national education professionals engaged in schools was chosen (Alam, 2020). The phenomenon reached saturation with the 18th respondent, and the sample size adhered to the recommendations provided by Guraliuk et al. (2023). It has been determined that saturation is achieved with 12 or more participants when dealing with homogeneous groups of informants. Nevertheless, it has been observed that, for qualitative studies, a sample of ten people may be sufficient, particularly in the presence of homogeneous informants (Crabtree & Miller, 1999). In addition, some researchers have suggested that a sample of 5–8 people might be appropriate for non-homogeneous informants, while a sample of 12–20 people would be recommended for informants of this type.

As part of this study, participants were invited to take part in a series of in-depth individual interviews. These interviews, which took place in a climate of trust and transparency, focused on perceptions of socio-emotional skills, subjective feelings of competence (or self-efficacy), and awareness of individual responsibility for work. Participants were asked to provide their informed opinions on quality-related factors, constraints, and solutions in their respective schools. A series of four fundamental inquiries were directed at the participants, with a focus on the objectives of the interview. The interview was structured by the employment of subsequent inquiries (two follow-up questions) in instances where the interviewee reiterated information, became repetitive, or deviated from the subject matter. It is imperative to emphasize that these inquiries are not intended as a rigid protocol, but rather as a guideline. This roadmap's goal is to align the question order with the study's aims. The duration of each interview varies from 30 minutes to a minimum of 25–35 minutes, with the interview commencing with a concise explanation of its purpose and concluding with the interviewee expressing gratitude for our contribution. In the current phase of the study, the interviews were subjected to the constant comparison method of analysis. As has been previously noted, the comprehension of situations and the verbal expression thereof are inextricably linked. Therefore, in the presentation and analysis of qualitative data, it falls upon the researcher to identify patterns within the words and to present them for the examination of others, all the while maintaining fidelity to the original experiences of the participants.

Prior to the start of each interview, participants provided their explicit consent. Approval for the study was granted by the local ethics committee, which included the thesis director, the co-directors, and academic experts from the home laboratory. The study was conducted in accordance with the 1964 Declaration of Helsinki. However, it is worth pointing out certain limitations of the interview that was out. A primary concern in any interview setting is the possibility that the responses provided by the interviewee may be influenced by their behavior and attitude. Secondly, it is difficult to guarantee that participants remain focused on the subject under discussion and that they do so throughout the meeting.

3.3 Data Analysis

Two rounds of data analysis were conducted. The initial phase involved conducting a content analysis to identify recurring themes and the perceptions of stakeholders. The second phase involved a semantic analysis focused on exploring lexical domains and cognitive representations related to EE. A systematic and unbiased approach was employed to perform content analysis aimed at identifying and quantifying observed instances (Chawla, 2020).

The second stage involved the application of NVivo 14, a tool designed for qualitative data analysis that assists in the processing of qualitative data. The results were obtained in two stages. At the first stage, subjectivism was proposed by the researcher to ensure the impact of the researcher's specificity and expertise. At the second stage, the thematic analysis process (Braun & Clarke, 2006) was used for the data analysis. This process includes the following steps: becoming familiar with the data, creating initial codes, searching for themes, examining them, defining and naming them, and producing the report.

3.3.1 Data Transcription

The present study was carried out in the Oriental region of Morocco with a sample of 30 participants, including male and female teachers and inspectors of the primary cycle of national education. They were contacted and interviewed either individually and directly (face-to-face) or via a focus group of up to two people. We opted for the second technique because of the availability of specific actors. However, it was only used on two occasions. The researchers collected the data under consideration via three means: first, semi-structured interviews; secondly, classroom observations;

and thirdly, documentary analysis. Interviews were recorded, transcribed, and analyzed using NVivo14 software for in-depth qualitative analysis. Classroom observations provided data on teachers' pedagogical practices, while documentary analyses provided information on existing curricula and available teaching resources.

In the present study, the empirically obtained data were transcribed following a thorough analysis and correction of the text by the consulting researchers. The data, which were subsequently imported into the software, were then interpreted through various coding processes. Initially, axial coding was performed on the transcripts. Following this, homogeneous themes were isolated. These themes were then linked to research questions and two subquestions, in addition to relationships with codes. It is noteworthy that CAQDAS software has emerged as a significant advancement in the domain of computer-assisted qualitative data analysis. This software's primary strength lies in its ability to supersede the conventional pen-and-paper method by furnishing straightforward yet efficacious mechanisms for the management and analysis of data. In this study, the researcher initiated the process with open coding, generating initial codes from extracts of the original responses present in the interview transcripts. The set of codes identified was then organized into coherent groups, referred to as "themes."

When transcribing the interviews, a paper notebook was used to ensure that the participants' words were faithfully reproduced. Verbatim accounts and texts were obtained, and in some cases, rephrasing was carried out using linking words to avoid a spoken language style and preserve the original content. At the end of each interview, the participants' statements were recorded on a paper draft, then subjected to linguistic and grammatical corrections. They are then entered into an Excel 2021 spreadsheet for further analysis.

3.3.2 Coding and Categorization

A rigorous methodology was used to analyze the qualitative data. This procedure involved identifying themes in participants' responses, then coding and categorizing them. The analysis process was carried out with meticulous attention, examining the content of the responses line by line. This analysis enabled the assignment of codes to words or phrases, which were identified as discrete units of data associated with a central idea. Quality perceptions were then grouped into the most relevant categories, directly related to the questions asked in the structured interview. A manual content analysis was carried out for each of the five interview questions. To this end, a word frequency search was carried out on the 100 most frequently repeated four-letter words in each interview response, to familiarize ourselves with the content. Subsequently, a word tree analysis was conducted on the most frequent word to enhance comprehension of its dimensions and context. The word frequency query and word tree analysis constituted the foundation of the manual content analysis, which entailed coding each response following a meticulous line-by-line examination. The codes were subsequently categorized into broader themes or categories, adhering to established content analysis conventions (Table 2.2).

In our research, we opted for a categorical thematic analysis approach, based on interviews with stakeholders. This methodology forms part of the broader framework of methodological procedures for content analysis, which represents a set of techniques for analyzing communications. This methodological approach aims to apply objective procedures to describe texts, obtain criteria, classify them, and interpret them with inference, as indicated by the author (Alam, 2020).

4 Findings and Discussion

4.1 Cognitive Competences

We have opted for a thematic analysis approach, the aim of which is "to identify the semantic units that make up the discursive universe of the utterance" (Negura, 2006, p. 4). Thus, the various statements collected will be dissected through various analytical prisms to categorize discourse elements according to predefined themes. In particular, the analysis will focus on the teachers' feelings and on more specific aspects linked to their pedagogical approaches (habits, expectations, etc.) during the interviews. It will also look at students' ability to assess themselves accurately and demonstrate progress in mastering certain skills as part of the co-evaluation process. Other objectives will be precisely defined once the quantitative results have been received (Albarello, 2007). The results show that educational players perceive EE as an essential tool for raising students' awareness of environmental issues. However, they point to several challenges, including a lack of teacher training, a lack of suitable teaching resources, and the difficulty of integrating EE into existing disciplines. For example,

Table 2.2 Coding process

Question	*Extract*	*Identified elements*	*Coded factor*
Axis 1: Cognitive skills (e.g., "How do you deal with uncertainty or paradoxes in the classroom?")	"As part of our work, we propose to organize debates on topical issues of major importance, such as climate change. These exchanges will enable us to work on argumentation and provide concrete evidence for our analyses"	Critical discussion and analysis of paradoxes: understanding current environmental and societal issues	Uncertainty management
Axis 2: Metacognition and responsibility (e.g., "What strategies do you use to develop student autonomy?")	"As part of the curriculum and pedagogical program, students are invited to plan their weekly tasks in a logbook. They assess their progress through self-assessment exercises, carried out every Friday"	Self-evaluation, planning	Empowering students
Axis 3: Socio-emotional and collaboration (e.g., "How do you integrate empathy into group projects?")	"When setting up work groups, each student is expected to assume a specific role. By way of illustration, this role may consist of being responsible for carrying out specific tasks, or acting as reporter. We also recommend encouraging mutual assistance between group members"	Defined roles, cooperation	Structured collaboration
Axis 4. Self-knowledge and values (e.g., "How do you approach ethics in ESD?")	"Our approach is to examine concrete situations, such as water wastage in schools, in order to engage in ethical reflection"	Local case studies, ethical reflection	Values education

one teacher commented, "We lack the teaching materials to teach the environment effectively. We often have to improvise with what we have."

As part of this study, an in-depth analysis was carried out to identify the codes, or subthemes, that structure the data collected. These codes were associated with the main research questions, helping to guide the interpretation and discussion of the results. In addition, responses, quotations, or open nodes were meticulously analyzed and justified, bringing additional depth to the examination of the data. It is important to acknowledge that respondents provided various homogeneous pieces of information for each code, thereby ensuring the reliability and consistency of the data set. Following this, the subthemes were methodically grouped into four overarching themes by the two researchers. This process was conducted in accordance with the research questions and subquestions, while taking into account the thematic fields of the lexical meanings of the recorded codes.

In the context of the Moroccan educational reform, this study is part of a process to integrate cross-disciplinary skills and Education for Sustainable Development (ESD) into primary school curricula. Conducted in three pilot schools, it seeks to evaluate the perceptions of educational players (principals, teachers, inspectors) regarding the implementation of these skills. The results, organized into four axes (cognitive, metacognitive, socio-emotional, and self-awareness) in Table 2.3, highlight innovative practices while highlighting challenges specific to the Moroccan context.

4.2 *Cognitive Competences*

As part of their pedagogical practice, teachers have put forward interdisciplinary projects, such as "Renewable Energies in Our Region," during which students analyze socio-economic and environmental impacts. This systems approach, as Jickling and Sterling (2017) point out in their work on transformative learning, proves essential for tackling complex problems. Nevertheless, managing paradoxes, as exposed in the "Protecting endangered species vs. Developing infrastructure" debate, highlights tensions between environmental ethics and human needs. These findings are consistent with the observations of Maddamsetti (2023), who indicates that ethical dilemmas are effective catalysts for the development of critical thinking, provided that appropriate pedagogical support is provided. The results of the study reveal that cross-curricular competencies are relevant to understanding ESD issues in Morocco, particularly in a context of ecological transition and educational reform. Nevertheless, their integration remains restricted by structural constraints, such as a lack of teacher training and rigid curricula. Boyle and Anderson (2020) emphasize that there must be systemic reform to support education for sustainability. Although mind maps have proved effective in synthesizing complex problems (e.g., the polluted water cycle), their use remains sporadic. This observation echoes the criticisms made by Chey (2024) concerning the difficulty of integrating systemic tools into often fragmented curricula.

Table 2.3 Summary of cross-disciplinary skills and teaching practices in Education for Sustainable Development (ESD) in the Moroccan context

Types of competences	*Cross-disciplinary skills and abilities*	*ESD competences (curriculum)*
Cognitive skills	*Systems thinking:* Teachers have stressed the importance of projects like "Renewable energies in our region" (crossing physics and geography), where students analyze socio-economic and environmental impacts *Managing paradoxes:* A recurring debate on "Protecting endangered species vs. developing infrastructure" has revealed tensions between environmental ethics and human needs *Synthesis:* Use mind maps to summarize complex problems (e.g., polluted water cycle)	*Relationship to knowledge* Building interdisciplinary knowledge from different perspectives
Motivational and metacognitive skills	*Autonomy:* Digital reflexive notebooks where students document their strategies for reducing their carbon footprint (e.g., "I chose to come by bike 3 times a week") *Collective responsibility:* "Our eco-responsible class" project with monthly waste monitoring, correlated with a 40% reduction in plastics used" *Taking the initiative*	*Responsibility* Developing a sense of belonging to the world
Socio-emotional and relationship competences	*Inter-school collaboration:* "School life" activities between schools, with assigned roles (scientific, political, civic) *Empathy:* "Walk in the footsteps of a climate refugee" workshop, where students write fictional logbooks *Self-knowledge* *Applied ethics:* Structured discussions on dilemmas such as, leading to reflections on personal values *Self-confidence:* Participation in a poster competition on biodiversity, where students expressed increased pride in their returns	*Collaboration* Jointly addressing sustainability issues *Participation:* Contribute to collective processes *Values:* Reflect on your own values and those of others

4.3 *Metacognitive Competences*

Digital reflexive notebooks, in which students document their strategies for reducing their carbon footprint, illustrate a metacognitive approach aligned with the work of Balsiger et al. (2017) and Flavell (1976). These tools encourage autonomy and responsibility, as demonstrated by the "Our Eco-Responsible Classroom" project, which reduced plastic use by 40%. These findings concur with those of Jickling and Sterling (2017), who highlight the importance of experiential learning in consolidating ESD skills. However, the expression of individual initiative is restricted by institutional constraints, lack of time and resources. This finding corroborates the observations of Kopnina (2018), who advocates for greater support from school management to free up time for ESD.

The study highlights the effectiveness of interdisciplinary projects and metacognitive tools, such as reflective notebooks, in developing students' autonomy and responsibility. Nevertheless, the implementation of these practices is often compromised by logistical constraints and a shortage of resources. These findings concur with those of AMRANI et al. (2021), who argue for stronger institutional support for ESD in Morocco.

4.4 *Socio-emotional Competences*

In the simulation of "School Life Activities" between schools, students developed relational and empathic skills by taking on various roles (scientists, politicians, citizens). These results are consistent with those of Petrucci and Roos (2020), who demonstrate that role-playing encourages empathy and cooperation. In addition, the "Walk in the Footsteps of a Climate Refugee" workshop was designed to elicit positive emotions, thereby enhancing student engagement. However, cooperation between schools remains limited, mainly due to logistical and legislative constraints (insurance and coverage) and the burden of school curricula. These limitations are also highlighted by Dewey (1986), who advocates a more open and collaborative pedagogy.

Structured debates on his dilemmas enabled students to reflect on their personal values and ethics. These findings are in line with those of Audigier (2012) and Barratt Hacking et al. (2017), who highlight the importance of citizenship education in developing ethical awareness.

To overcome these challenges, it seems necessary to consolidate teacher training and promote collaborative and reflective pedagogy, as suggested by Jickling and Sterling (2017). Initiatives such as climate summit simulations and empathy workshops demonstrate that experiential learning is capable of transforming students' perceptions and behaviors.

5 Conclusion

Within the Moroccan curriculum, curricula based on competency- and value-based approaches, as well as an innovative pedagogical model focused on active learning, have been implemented. Previous studies (Balaji & Anbalagan, 2017; Sahidullah, 2022) have revealed similar findings. Integrating concepts related to environmental education in various contexts enables students to develop a deeper understanding and illustrate the links between these concepts (Havu-Nuutinen et al., 2018; Sra, 2020). An in-depth analysis of these programs revealed that the theme of the environment encompasses a wide range of topics that are suitable for transdisciplinary and interdisciplinary approaches to teaching.

This study's results will first allow for the accurate assessment of teachers' needs and preferences, as well as the efficacy of these teaching methods. Secondly, it will be possible to rework this tool by placing it under the aegis of a more specific framework of reading tools, and in particular those relating to the assessment of cross-curricular skills, following the example of the Triple-T model (Brockwell, 2019). This research work emphasizes the value of interdisciplinarity and innovative pedagogical approaches in fostering effective environmental education. These approaches not only raise students' awareness of environmental issues but also encourage them to adopt eco-responsible behaviors in their daily lives. Project-based learning, for example, enables students to engage in concrete actions to protect the environment, boosting their motivation and commitment (Abdulwahed & Hasna, 2018).

Introducing environmental education (EE) skills into the Moroccan elementary school curriculum represents a major challenge but also an opportunity to train aware and responsible citizens. This study proposes concrete avenues for overcoming the challenges identified and promoting effective implementation of EE in Moroccan elementary schools. The results of this research contribute to enriching the debate on education for sustainable development and offer practical tools for those involved in education.

Future research should focus on identifying best practices for integrating cross-curricular skills into environmental education programs, particularly in diverse educational contexts. In addition, longitudinal studies tracking the impact of cross-curricular skills on students' environmental attitudes and behaviors over time are needed to better understand the long-term benefits of environmental education.

References

Abdulwahed, M., & Hasna, M. O. (2018). The impact of engineering innovation and eco-design competition on environmental awareness, and academic performance: A global qualitative analysis in Europe, Asia, and the Middle East. *International Journal of Environmental and Science Education.*

Alam, M. K. (2020). A systematic qualitative case study: Questions, data collection, NVivo analysis and saturation. *Qualitative Research in Organizations and Management: An International Journal, 16*(1), 1–31. https://doi.org/10.1108/QROM-09-2019-1825

Albarello, L. (2007). *Apprendre à chercher: L'acteur social et la recherche scientifique (2ème édition).* https://dial.uclouvain.be/pr/boreal/object/boreal:79967

Amahmid, O., El Guamri, Y., Yazidi, M., Razoki, B., Kaid Rassou, K., Rakibi, Y., Knini, G., & El Ouardi, T. (2019). Water education in school curricula: Impact on children knowledge, attitudes and behaviours towards water use. *International Research in Geographical and Environmental Education, 28*(3), 178–193. https://doi.org/10.1080/10382046.2018.1513446

Amrani, N., Sadik, M., & Hemmi, M. (2021). Gouvernance des collectivités territoriales au Maroc: Une approche prospective à la lumière des expériences internationales. *Alternatives Managériales Economiques, 3*(1), 144–165.

Ardoin, N. M., & Bowers, A. W. (2020). Early childhood environmental education: A systematic review of the research literature. *Educational Research Review, 31*, 100353. https://doi.org/10.1016/j.edurev.2020.100353

Ardoin, N. M., Bowers, A. W., & Gaillard, E. (2020). Environmental education outcomes for conservation: A systematic review. *Biological Conservation, 241*, 108224. https://doi.org/10.1016/j.biocon.2019.108224

Audigier, F. (2012). Les Éducation à…: Quels significations et enjeux théoriques et pratiques? Esquisse d'une analyse. *Recherches en didactiques, 13, 25–38.*

Balaji, P., & Anbalagan, A. (2017). A study on environmental awareness among rural and urban Secondary school students in Thiruvallur district. *International Educational Scientific Research Journal, 3*(6), 26–31.

Balsiger, J., Förster, R., Mader, C., Nagel, U., Sironi, H., Wilhelm, S., & Zimmermann, A. B. (2017). Transformative learning and education for sustainable development. *GAIA – Ecological Perspectives for Science and Society, 26*(4), 357–359. https://doi.org/10.14512/gaia.26.4.15

Barratt Hacking, E., Blackmore, C., Bullock, K., Bunnell, T., Donnelly, M., & Martin, S. (2017). *The international mindedness journey: School practices for developing and assessing international mindedness across the IB continuum*. International Baccalaureate Organisation.

Bezeljak, P., Torkar, G., & Möller, A. (2023). Understanding Austrian middle school students' connectedness with nature. *The Journal of Environmental Education, 54*(3), 181–198. https://doi.org/10.1080/00958964.2023.2188577

Boyle, C., & Anderson, J. (2020). The justification for inclusive education in Australia. *PROSPECTS, 49*(3-4), 203–217. https://doi.org/10.1007/s11125-020-09494-x

Braun, V., & Clarke, V. (2006). Using thematic analysis in psychology. *Qualitative Research in Psychology, 3*(2), 77–101. https://doi.org/10.1191/1478088706qp063oa

Brockwell, A. J. (2019). *Measuring what matters?: Exploring the use of values-based indicators in assessing Education for Sustainability*. Wageningen University. https://doi.org/10.18174/476056

Carney, S. (2022). Reimagining our futures together: A new social contract for education: par l'UNESCO, Paris, UNESCO, 2021, 186 pages, ISBN 978-92-3-100478-0. *Comparative Education, 58*(4), 568–569. https://doi.org/10.1080/03050068.2022.2102326

Chawla, L. (2020). Childhood nature connection and constructive hope: A review of research on connecting with nature and coping with environmental loss. *People and Nature, 2*(3), 619–642. https://doi.org/10.1002/pan3.10128

Chey, S. (2024). Singapore's 21st-century competencies through the complexity lens. In *Curriculum innovation in East Asian schools*. Routledge.

Chodkiewicz, A. R., & Boyle, C. (2017). Positive psychology school-based interventions: A reflection on current success and future directions. *Review of Education, 5*(1), 60–86. https://doi.org/10.1002/rev3.3080

Crabtree, B., & Miller, W. (1999). *Doing qualitative research*. Sage Publications.

Dewey, J. (1986). *Experience and Education, 50*(3), 241–252.

El Batri, B., Maskour, L., Ksiksou, J., Jeronen, E., Ismaili, J., Alami, A., & Lachkar, M. (2022). Teaching environmental themes within the "Scientific Awakening" course in Moroccan Primary School: Approaches, methods and difficulties. *Education Sciences, 12*(11), Article 11. https://doi.org/10.3390/educsci12110837

El Moussaouy, A., Abderbi, J., & Daoudi, M. (2014). Environmental education in the teaching and the learning of scientific disciplines in Moroccan High Schools. *International Education Studies, 7*(4), 33–46.

El-Batri, B., Alami, A., Zaki, M., & Nafidi, Y. (2019). Extracurricular environmental activities in Moroccan Middle Schools: Opportunities and challenges to promoting effective environmental education. *European Journal of Educational Research, 8*(4), 1013–1028. https://doi.org/10.12973/eu-jer.8.4.1013

Ernst, J., & Burcak, F. (2019). Young children's contributions to sustainability: The influence of nature play on curiosity, executive function skills, creative thinking, and resilience. *Sustainability, 11*(15), 4212. https://doi.org/10.3390/su11154212

Faubert, V. (2009). School evaluation: Current practices in OECD countries and a literature review.

Flavell, J. H. (1976). Metacognitive aspects of problem solving. In *The nature of intelligence* (pp. 231–236). Routledge.

Fleming, T. (2021). Toward a critical theory of transformative learning: Learning from Oskar Negt. *International Journal of Adult Education and Technology (IJAET), 12*(1), 1–16. https://doi.org/10.4018/IJAET.2021010101

Frensley, B. T., Stern, M. J., Powell, R. B., & Blackwell, M. J. (2022). Exploring the links between teaching approaches and student outcomes at a residential setting. *Environmental Education Research, 28*(6), 826–844. https://doi.org/10.1080/13504622.2022.2044454

Gey, N., Pellaud, F., Blandenier, G., Lepareur, C., Massiot, P., Shankland, R., & Gay, P. (2023). Assessment of cross-cutting competences in education for sustainable development. *Environmental Education Research, 29*(5), 766–782. https://doi.org/10.1080/13504622.2022.2136362

Guraliuk, A., Varava, I., Holovko, S., Shapenko, L., & Oleshchenko, V. (2023). Expert assessment of the quality of remote educational resources. *International Journal of Engineering Pedagogy (iJEP), 13*(1), 34–44. https://doi.org/10.3991/ijep.v13i1.36121

Havu-Nuutinen, S., Kärkkäinen, S., & Keinonen, T. (2018). Changes in primary school pupils' conceptions of water in the context of Science, Technology, and Society (STS) instruction. *International Research in Geographical and Environmental Education, 27*(2), 118–134.

Jackson, S., Peterson, Lawson, Olson, & Joseph. (2021). Lasting conservation and science-related outcomes associated with science education, environmental education, and outdoor science education. *Children, Youth and Environments, 31*(1), 116. https://doi.org/10.7721/chilyoutenvi.31.1.0116

Jickling, B., & Sterling, S. (2017). *Post-sustainability and environmental education: Framing issues.* Springer.

Keith, R. J., Given, L. M., Martin, J. M., & Hochuli, D. F. (2021). Urban children's connections to nature and environmental behaviors differ with age and gender. *PLOS ONE, 16*(7), e0255421. https://doi.org/10.1371/journal.pone.0255421

Knight, A. T., Cook, C. N., Redford, K. H., Biggs, D., Romero, C., Ortega-Argueta, A., Norman, C. D., Parsons, B., Reynolds, M., Eoyang, G., & Keene, M. (2019). Improving conservation practice with principles and tools from systems thinking and evaluation. *Sustainability Science, 14*(6), 1531–1548. https://doi.org/10.1007/s11625-019-00676-x

Kolb, A. Y., & Kolb, D. A. (2017). Experiential learning theory as a guide for experiential educators in higher education. *Experiential Learning and Teaching in Higher Education, 1*(1), Article 1. https://doi.org/10.46787/elthe.v1i1.3362

Kopnina, H. (2018). Education for sustainable development (ESD): The turn away from 'environment' in environmental education? In *Environmental and sustainability education policy* (pp. 135–153). Routledge.

Křepelková, Š. D., Krajhanzl, J., & Kroufek, R. (2020). The influence of interaction with nature in childhood on future pro-environmental behavior. *Journal of Baltic Science Education, 19*(4), 536–550. https://doi.org/10.33225/jbse/20.19.536

Lippmann, M. (2021). Inquiry-based learning in psychology. In *International handbook of psychology learning and teaching* (pp. 1–30). Springer. https://doi.org/10.1007/978-3-030-26248-8_59-1

Maddamsetti, J. (2023). Elementary preservice teachers' navigation of dilemmatic spaces in linking critical pedagogical thinking with practice during practicum. *Teachers and Teaching, 29*(4), 439–457. https://doi.org/10.1080/13540602.2023.2191185

Maghnouj, S., Belanger, J., Clarke, M., Fordham, E., Kitchen, H., & McGregor, I. (2018). *Examens de l'OCDE du cadre d'évaluation de l'éducation: Maroc: OECD Reviews of Evaluation Framework Education: Morocco.*

Merritt, E. G., Stern, M. J., Powell, R. B., & Frensley, B. T. (2022). A systematic literature review to identify evidence-based principles to improve online environmental education. *Environmental Education Research, 28*(5), 674–694. https://doi.org/10.1080/13504622.2022.2032610

Negura, L. (2006). L'analyse de contenu dans l'étude des représentations sociales. *SociologieS*. https://doi.org/10.4000/sociologies.993

Ober, T. M., Brooks, P. J., Homer, B. D., & Rindskopf, D. (2020). Executive functions and decoding in children and adolescents: A meta-analytic investigation. *Educational Psychology Review, 32*(3), 735–763. https://doi.org/10.1007/s10648-020-09526-0

Pellaud, F., Shankland, R., Blandenier, G., Dubois, L., Gey, N., Massiot, P., & Gay, P. (2021). The competencies that school-leavers should possess in order to meet the challenges of the 21st century. *Frontiers in Education, 6*. https://doi.org/10.3389/feduc.2021.660169

Petrucci, F., & Roos, E. (2020). *Atteinte des compétences fondamentales dans le canton de Genève : Que nous enseignent les enquêtes COFO 2016 et 2017?* SRED.

Piaget, J. (1970). *Science of education and the psychology of the child* (D. Coltman, Trans., p. 186). Orion.

Sahidullah, A. (2022). Environmental awareness among High School Students: A study in Majuli District of Assam, India. *Ecology, Environment and Conservation*, 347–351.

Sra, H. K. (2020). Environmental awareness of government and non-government college students of Chandigarh. *International Journal of Research in Education Methodology, 11*, 31–35. https://doi.org/10.24297/ijrem.v11i.8659

Waters, L., & Loton, D. (2019). SEARCH: A meta-framework and review of the field of positive education. *International Journal of Applied Positive Psychology, 4*(1-2), 1–46. https://doi.org/10.1007/s41042-019-00017-4

Zenasni, S., Kuppens, T. E., Vaesen, J., Surmont, J., & Stiers, I. (2024). Conceptualizing education for sustainable development in urban secondary schools. *Education and Urban Society*. https://doi.org/10.1177/00131245241238001

CHAPTER 3

ECTA Approach to Teaching Environmental Content

Rock Finale

1 Introduction

The integration of natural sciences, particularly environmental awareness, with visual arts in primary schools has gained attention as a multifaceted approach to enhancing students' understanding and appreciation of both disciplines.

The innovative approach of teaching environmental content through visual arts (ECTA) posits that engaging students in artistic activities related to environmental themes can significantly enhance their understanding of ecological issues. This is supported by research indicating that integrating environmental education into art curricula serves as a catalyst for strengthening a sense of responsibility toward nature among young students (Zhongbin, 2024). By employing creative methods, such as painting, sculpting, and poster-making, teachers can facilitate a deeper emotional connection to environmental issues, thereby promoting ethical creativity (Hagerman & Rundgren, 2022). This aligns with the findings of Inwood

R. Finale (✉)
University of Applied Sciences of Rijeka Department of Sustainable Agritourism, Rijeka, Croatia

K. Hirech (ed.), *Environmental Education and Sustainable Development*, https://doi.org/10.1007/978-3-032-05760-0_3

and Kennedy (2020), who argue that art education can serve as a form of environmental activism, encouraging students to explore and communicate their understanding of ecological challenges through creative projects.

The literature suggests that combining arts and sciences in educational settings can lead to enhanced student engagement and creativity. For instance, Sutjonong et al. (2022) highlight the importance of creative teaching in increasing student creativity, which is crucial for effective learning in both the arts and sciences. The argumentation method further supports this notion by demonstrating that creative pedagogical strategies can significantly improve students' problem-solving abilities in various subjects, including environmental science (Doğan & Sır, 2022). The method fosters students' ability to articulate their thoughts and participate in discussions within a collaborative learning environment, which is essential for understanding complex environmental issues.

The role of teachers in facilitating this integration cannot be overstated. Teachers who adopt a creative teaching identity can effectively bridge the gap between disciplines, allowing for a more holistic educational experience (Gilbert & Dobson, 2024). Chen and Yuan (2021) emphasize that teachers' creative teaching practices can significantly enhance students' learning experiences, making them more meaningful and engaging. This is particularly relevant in the context of environmental education, where the need for innovative teaching methods is paramount to address pressing ecological challenges.

Incorporating eco-art projects into the curriculum is one effective way to achieve this integration. Through eco-art projects, students can investigate environmental problems, thereby enhancing their understanding of ecological issues while developing their creative skills (Tsevreni, 2022). Ecological art can serve as an interdisciplinary connection that not only raises environmental awareness but also encourages students to adopt sustainable behaviors (Şeren & Gül, 2022). This aligns with the ECTA approach framework, which advocates for the use of art as a medium to explore and communicate environmental themes.

Eco-art and different teaching approaches like place-based teaching demonstrate how local environmental issues can be addressed through creative projects, encouraging students to develop empathy for their surroundings (Sunassee & Bokhoree, 2021). This approach not only enhances students' artistic skills but also instills a sense of responsibility toward their environment, aligning with the goals of the ECTA approach. Creative teaching methods can significantly improve students' engagement and learning outcomes, particularly for those with special educational needs (Govindasamy et al., 2023). This underscores the potential of the ECTA approach to cater to diverse learning needs while promoting environmental awareness through artistic expression.

Moreover, the use of innovative teaching approaches supported also by diverse technologies, such as AI, in enhancing creative teaching practices is increasingly recognized. The use of digital tools and platforms can facilitate collaborative art projects that address environmental problems, allowing students to engage with their peers and the broader community (Demčenko et al., 2022). This not only enhances their artistic skills but also enriches a sense of collective responsibility toward environmental stewardship.

By employing creative teaching methods and interdisciplinary projects, teachers can inspire a deeper understanding of ecological issues while promoting creativity and critical thinking skills. Such integrative approaches not only enhance student engagement but also contribute to the development of responsible and environmentally conscious citizens.

The ECTA approach specifically emphasizes the intersection of art and environmental education as a means to cultivate a deeper understanding of ecological challenges (Finale, 2021). The arts can effectively connect students with their natural surroundings to stimulate their sense of responsibility toward environmental preservation (Ison & Bramwell-Lalor, 2023). Art projects can, therefore, actively and effectively raise awareness about environmental issues, not only among students but also within the broader community (Marković & Petrović, 2021). This aligns with the notion that art serves as a powerful medium for communication and advocacy in environmental education.

2 Visual Art and Environmental Awareness

Visual arts have long been a means of expressing emotional connections to the environment, allowing artists to translate their responses to nature through colors, shapes, and textures into various artistic forms. Historically, natural materials like branches and soil have been used, reflecting the deep relationship between humans and their surroundings (Varnell et al., 2024). In education, art serves as an effective tool for synthesizing knowledge across domains, with nature as a primary inspiration (Deer et al., 2024).

In Finland, environmental education in the 1990s emphasized the role of visual arts in a connection to nature. Art education encourages students to explore their environment, transforming sensory experiences into creative expressions (Breukel & Cieraad, 2024). Rauhala (2003) highlighted how art allows individuals to engage with sensory, emotional, and cognitive aspects, enhancing reflections on environmental impact, which traditional science education often lacks (Zhongbin, 2024). This experiential learning encourages environmental awareness and encourages students to confront their perceptions and emotions about ecological issues (Iwasaki, 2022). The integration of visual arts in curricula supports emotional and cognitive development, providing a holistic understanding of the environment (Vasilaki, 2024).

While scientists approach the environment systematically, artists engage subjectively, offering emotional insights that resonate deeply with audiences (Hsiao, 2023). Research shows that visual arts activities can reduce stress in children and enhance their interest in learning (Moula et al., 2022). Eco-art projects raise environmental awareness and promote sustainable behaviors (Riet et al., 2020). Technology further enhances art education, enabling collaborative projects addressing environmental themes and collective responsibility for environmental stewardship (Zaeri & Roozafzai, 2024).

Integrating visual arts into environmental education enhances students' understanding of ecological issues and stimulates emotional connections to nature, cultivating a generation of environmentally conscious individuals.

3 Pedagogical Strategies in the ECTA Approach

3.1 Gestalt Pedagogy, Problem-Based Pedagogy, Cross-curricular Connections, and Interdisciplinary Integration

This paragraph presents the four most used teaching approaches in Slovenian primary schools. These four approaches were also used as part of the ECTA approach. Cross-curricular connections are widely recommended in every subject knowledge catalog.

(1) Gestalt pedagogy emphasizes holistic student development by integrating experience, emotions, and the environment. It prioritizes experiential learning, where active engagement supports deeper insights (Coudray, 2020) and a personal connection to ecological issues (Zocher & Hougham, 2020). Critical reflection enables stu-

dents to analyze experiences and consider multiple perspectives on environmental challenges (Byrnes et al., 2022), enhancing problem-solving skills. Collaborative learning further strengthens teamwork and communication, and shared responsibility for sustainability. In environmental education, Gestalt pedagogy promotes emotional engagement, experiential learning, reflection, and collaboration, deepening students' commitment to sustainability.

(2) Problem-based pedagogy (PBP) uses real problems as the foundation for learning, critical thinking, collaboration, and active engagement in environmental education. It encourages students to apply knowledge to real-world ecological issues, deepening their understanding of sustainability (Namukasa et al., 2023). Collaboration is central, as group work enhances teamwork and learning through diverse perspectives (Millanzi & Kibusi, 2021). PBP also integrates technology, providing resources for research and problem-solving (Niemelä, 2021). However, it can be resource-intensive for teachers (Elo & Nygren-Landgärds, 2020), and some students may struggle in self-directed environments (Pham & Ünaldı, 2021).

(3) Cross-curricular connections in environmental education enhance engagement and deepen understanding by linking subjects to develop a comprehensive ecological perspective. Integrating science, mathematics, arts, and social studies helps students develop critical thinking and problem-solving skills (Tofel-Grehl et al., 2021). Collaboration across subjects promotes diverse perspectives and practical application of knowledge, improving both social skills and environmental literacy (Mäkelä et al., 2022). However, effective integration requires extensive planning and teacher collaboration, which may not always be feasible (Mobjörk et al., 2020). Aligning curricula and assessments can also be challenging due to standardized testing constraints (Zhao et al., 2023).

(4) Interdisciplinary integration promotes a holistic learning approach by combining knowledge from multiple disciplines, enhancing engagement with complex issues (Dmitrievna, 2024). In environmental education, it supports a deeper understanding of ecological systems through science, social studies, and the arts. This method develops critical thinking, creativity, and problem-solving skills while making learning more relevant. However, challenges include coordinating educators, resource allocation, and aligning with standardized testing (Kluger & Bartzke, 2020). Power dynamics

among disciplines may also create imbalances in educational priorities (Niemelä, 2021). Despite these challenges, interdisciplinary integration is essential for preparing students to tackle real-world problems.

It is important to know the basis of each approach or model of teaching and the advantages and disadvantages that they may bring into teaching. Each approach contributes to the acquisition of environmental content, but not all are suitable for all students. The teacher must, therefore, constantly adapt to the classroom situation, especially to students with special needs.

3.2 *Innovative Teaching of Environmental Content with Visual Arts (ECTA)*

The ECTA approach blends project-based learning with a problem-solving focus on environmental issues, particularly waste and its reuse in art. This pedagogical strategy encourages student self-expression through artistic mediums like drawing and design while promoting autonomy and engagement. It emphasizes a structured learning process that facilitates lasting knowledge application in everyday life. Key components of the ECTA approach include (1) Supportive Emotional and Social Climate: Teachers create a positive and creative atmosphere, providing additional support for students with special needs, essential for artistic exploration. (2) Holistic and Independent Research: Students engage in independent and collaborative research, guided by Gestalt pedagogy, to explore waste materials and their potential reuse in art. This approach encourages self-responsibility and mutual respect among students while linking art with environmental science. (3) Monitoring and Evaluation: Teachers guide the creation of artworks, using collaborative assessment to deepen students' understanding of artistic standards and encourage self-evaluation.

The ECTA curriculum consists of nine teaching units, totaling 40 school hours. The first three units (12 hours) focus on the impact of civilizational development on the environment, particularly municipal waste. The remaining six units (28 hours) combine environmental themes with visual arts, enriching students' artistic experiences and environmental awareness (Fig. 3.1). The ECTA approach improves artistic skills while deepening the understanding of environmental issues, empowering students to address waste and sustainability challenges.

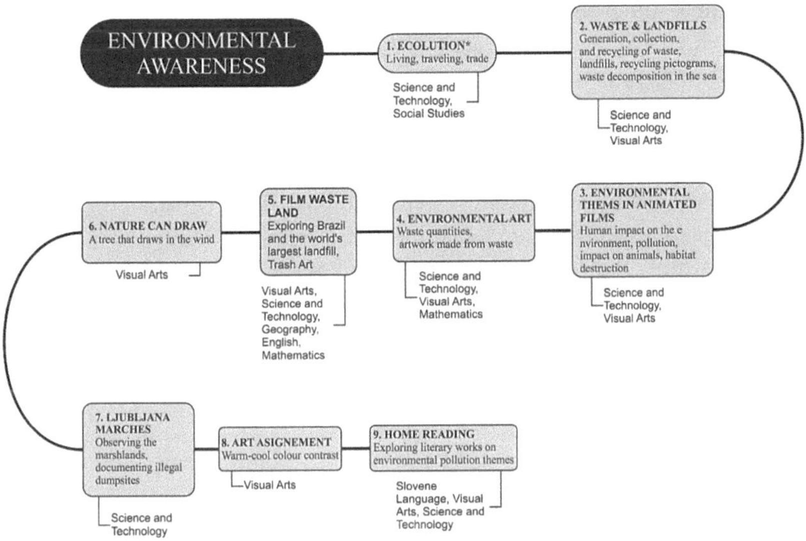

Fig. 3.1 Teaching units and the cross-curricula connections of the ECTA approach. (Source: Finale)

Before and after the implementation of the ECTA approach, students took a test of their environmental knowledge.

The term "ecolution" in the first unit has been introduced as a novel concept and serves as the title for the initial teaching unit within the curriculum framework. This term encapsulates the intersection of ecological principles and evolutionary concepts, highlighting the dynamic between humans, with their way of living, traveling, and trade, and their influence on the environment. The adoption of "ecolution" as a pedagogical term aims for a comprehensive understanding of ecological systems and the evolutionary processes that shape today's living principles, shaping landscapes and influencing biodiversity.

Integrating environmental education across various subjects is crucial for the understanding of environmental issues among students. The curriculum outlined in the teaching unit emphasizes cross-curricular links that enhance students' awareness and engagement with environmental themes. Each unit is designed to not only impart knowledge but also to cultivate pro-environmental attitudes and behaviors, which are essential for sustainable development.

The first unit, “Ecolution,” links science and technology with social studies, focusing on living, traveling, and trade. This interdisciplinary approach is supported by research indicating that environmental education significantly enhances students’ environmental knowledge and attitudes, which in turn influences their pro-environmental behaviors (Liao & Li, 2019). Such connections between subjects can stimulate a holistic understanding of how human activities impact the environment, thereby encouraging responsible citizenship (Obasi & Osah, 2022).

In this unit, the students were given instructions to find and examine the development of man in the individual areas of (1) living, (2) movement, and (3) trading, and at the same time to examine what impact man has had on the environment throughout history. At the end of the assignment, the students presented the results of their own research to their classmates. A guided discussion followed.

The second unit, “Waste and Landfills,” combines science and technology with visual arts to explore waste generation, collection, recycling, and landfill issues. This unit highlights the importance of visual representation in understanding complex environmental issues. Studies have shown that integrating the arts into environmental education can enhance students’ engagement and understanding of scientific concepts, making them more relatable and impactful. Furthermore, visual arts can serve as a powerful medium for communicating environmental messages (Safitri et al., 2020).

For this unit, the teacher prepared a PowerPoint presentation in which he presented the municipal waste disposal site to the students, its operation, cross-section, and all the consequences it has on the environment. The students later visited the largest Slovenian waste disposal site near the capital.

In "Environmental Themes in Animated Films," the curriculum examines human impacts on the environment through popular media. This approach aligns with findings that narrative-based environmental education can effectively improve environmental awareness and attitudes among children (Yang et al., 2022). By analyzing animated films, students can critically engage with themes of pollution and habitat destruction, and gain a deeper understanding of these pressing issues.

In this unit, the emphasis was on environmental themes in cartoons and animated films. The students watched the animated film *WALL-E* and after the viewing held an in-depth debate in which they emphasized the presented aspects of the pollution of the Earth, the migration of man into space, and his return to the alarm when the first new plant was detected on earth.

The unit "Environmental Art" emphasizes the intersection of visual arts and mathematics, particularly in the context of waste quantities and artwork created from recycled materials. This integration not only enhances creative expression but also reinforces mathematical concepts through practical applications. Research supports the notion that STEAM (Science, Technology, Engineering, Arts, and Mathematics) education can cultivate well-rounded individuals who are capable of innovative thinking and problem-solving in environmental contexts (Lindsay, 2021). For this unit, the teacher prepared a PowerPoint presentation with photographs of various works of art that belong to the so-called field of Trash art, Land art, and Environmental art. Among the most moving for students was the installation of 1000 ice sculptures in Germany by environmental artist Nela Azevedo. After the presentation, we talked with the students about environmental pollution and the possibilities of using waste as a means of artistic creation and environmental awareness at the same time. The students also highlighted the appropriateness of the ice sculptures, which also drew attention to the ever-increasing climate change and the melting of glaciers and ice at the poles.

The unit entitled "Nature Can Draw" emerged as particularly engaging for students. Prior to the workshop, the teacher introduced the guiding question: Can nature draw? The students spent one school hour thinking and talking about whether and how nature can draw and create. Most of the conversation revolved around the autumn colors of the leaves and the sunrises and sunsets. In principle, the students were thinking correctly, but when they arrived at the weeping willow in front of the school, they were surprised. The students were given the task of attaching different pens to the branches, following the example of the British artist Tim Knowles, with which the branches, swaying in the wind, drew on the base they placed below.

The curriculum also includes practical fieldwork, as seen in the "Ljubljana Marshes" unit, where students observe marshlands and document illegal dumpsites. Such hands-on experiences are vital for developing environmental literacy, as they allow students to connect theoretical knowledge with real-world applications (Wu et al., 2023). Engaging with local ecosystems promotes a sense of responsibility and stewardship toward the environment, which is essential for cultivating future generations of environmentally conscious citizens (Zainuri et al., 2022).

On a walk through the Marsh, the students learned about its characteristics, the use of the area throughout history, and the various ways of its pollution. During the walk, they entered the locations of discovered black waste dumps into a special application. Next to each landfill, they also carried out an inventory of the observed waste, which they identified.

4 Methodology, Sample, and Instruments

4.1 *Methodology*

The study utilized quantitative and qualitative research, specifically a causal-experimental educational research methodology (Stuart & Rubin, 2008; Sagadin, 2003). Within this framework, participants were randomly assigned to either an experimental or a control group based on the results of a pre-test of environmental knowledge. This randomization serves to minimize bias and facilitates more reliable conclusions regarding causality. Such an approach is prevalent in pedagogy research as it effectively elucidates how specific interventions impact educational outcomes. A notable advantage of the causal-experimental design is its ability to enhance the reliability of research findings, thereby yielding credible insights into the effectiveness of the intervention (Lin et al., 2021). The primary focus of this investigation was to assess the efficacy of the developed ECTA approach in understanding environmental content, enhancing environmental awareness, and increasing environmental knowledge among fourth-grade elementary school students. As delated by Mesec (1998), a quantitative and qualitative research methodology was employed, utilizing pre- and post-tests to evaluate changes in environmental knowledge and semi-structured interviews with all teachers of experimental classes.

4.2 *Sample*

The study involved 154 fourth-grade elementary school students, 72 (47%) male and 82 (53%) female. The average age of the students at the time of the study was nine years. All students were assigned codes that they used throughout the study. Parents also gave consent for their children to participate in the research.

The study also involved three teachers with 39, 35, and 26 working years.

At a 5% risk level, it was found that there was no statistically significant difference (p = 0.167) in the success rates on the environmental knowledge test between the students of planned groups. Therefore, the students were randomly divided into experimental and control groups. The experimental group consisted of 37 (24.03%) male students and 39 (25.32%) female students, while the control group comprised 35 (22.73%) male students and 43 (27.92%) female students. All students are from the Central Slovenian region, from the suburbs of Slovenia's capital city, Ljubljana.

4.3 *Instruments*

For the survey, we used an environmental knowledge test (EKT) that was administered before and after teaching using the ECTA approach.

The EKT was based on the science and technology curriculum, covering topics such as waste separation, recycling, pollution, water consumption, and sustainable development. It was designed at a higher difficulty level, expecting students to meet these standards after learning through the ECTA approach. The test consisted of five tasks, including multiple-choice, matching, and open-ended questions. The first task assessed the knowledge of compost through multiple-choice questions with several correct answers. The second task required students to classify waste by identifying different types and assigning them to the correct category. The third task involved matching images, such as a compost bin and recycling symbols, with their meanings. The fourth task was a free-response question where students calculated water consumption for a four-member family. They identified the activity that used the most water and proposed ways to reduce consumption. The fifth task required students to explain the concept of illegal landfills. The maximum possible score on the test was 21 points.

For the quantitative part of the research, we had a questionnaire for teachers of the experimental group that consisted of 15 questions that were placed into seven meaningful sets from which the most important were (1) teaching methods, (2) ECTA approach, and (3) the difference between nature conservation and environmental protection.

5 Results

5.1 The Results of EKT Tests

The EKT test indices of difficulty (*It*), discriminability (*Id*) (Table 3.1), and reliability (Cronbach-α coefficient) were calculated and are 0.71, which indicates acceptable reliability of the test. In the belief that students on the EKT would demonstrate a higher level of acquired knowledge after the implemented approach, the same knowledge test was administered as before the introduction of the ECTA approach.

All tasks are within the recommended range in terms of difficulty; the recommended threshold value of the difficulty index is between 0.10 and 0.90 (Pallant, 2020). The difficulty of the EKT is acceptable, as the difficulty index of the entire test is in the experimental group (*It* = 0.58) and in the control group (*It* = 0.54). The resolution index (*Id*) separates the success of solving more difficult tasks from less difficult ones in terms of success in the entire test. Tasks with a higher difficulty index value distinguish better between weaker and better students. Tasks with an index greater than 0.2 resolution points are acceptable (Pallant, 2020).

Most tasks were very well resolved in both groups (*Id* > 0.4). In the repeated test after teaching, the easiest task for students in both groups was, as expected, the second task, in which students named and sorted

Table 3.1 The content framework of EKT and the corresponding difficulty indices (*It*) and clarity (*Id*) according to the group

	Group			
	Experimental		*Control*	
Questions in EKT	I_t	I_d	I_t	I_d
1 What is compost?	0.57	0.48	0.52	0.68
2 Name the waste and sort it accordingly.	0.88	0.42	0.92	0.43
3 Use lines to connect the descriptions in the first column with the corresponding labels in the second column.	0.73	0.41	0.72	0.56
4 Amount of water used by a family of four in one day: 4a Which activity uses the most water?	0.24	0.85	0.16	0.57
4b Amount of water used by a family of four in one day: How can you contribute to reducing water consumption?	0.24	0.58	0.16	0.58
5 Explain what an illegal landfill is.	0.79	0.56	0.73	0.41
Total	0.58	0.55	0.54	0.54

Source: Finale (2021)

waste. The most difficult task for students in both groups was the fourth task, in which they calculated water consumption per family member and the entire family and gave open-ended answers on which of the listed activities the family uses the most water in one day.

The content validity of the knowledge test was checked by two experts in the field of science teaching and two teachers with 26 and 35 years of work experience teaching in the third and fourth grades of primary school. Their comments were taken into account in the final design of the test.

The objectivity of the knowledge test was ensured by appropriate spatial conditions, precise instructions for solving the problem, and objective evaluation of the solutions. An appropriate coder was created for the answers to the open-ended questions, which unified the definition of the correct answer.

Objectivity was ensured when solving the test before and after implementing the approach with the same time limit of one school hour and in the unambiguous evaluation of the results.

The paired *t*-test compared the answers to the EKT test before and after teaching with the ECTA approach, which shows that the level of knowledge of the experimental group students after the implemented ECTA teaching approach is higher and statistically significant for EKT ($p \leq 0.001$).

The level of knowledge of the control group students is not statistically significant for the EKT ($p = 0.07$). There is no statistically significant change in any of the variables for the control group students.

A *t*-test of the equality of the averages of the EKT before teaching revealed that there was no statistically significant difference between the intended groups (experimental and control). For EKT ($t(151) = -1.39$, $p = 0.17$) (Table 3.2).

Table 3.2 *T*-test of the equality of the averages of EKT grades of all students ($N = 154$) before teaching

				Levene's test for equality of variances		*t-test for equality of means*		
	Group	*M*	*SD*	*F*	*P*	*t*	*df*	*2p*
EKT	Experimental	1.2	0.46	0.48	0.49	−1.39	151	0.17
	Control	1.1	0.44					

Source: Finale

Comparison of mean values in achievements in the EKT and standard deviations, before teaching, between the groups (experimental and control) did not show significant differences; the differences are also statistically insignificant. According to the results, students were divided into experimental ($N = 75$) and control ($N = 78$) groups. According to the results of the knowledge tests, classes were formed that, as a whole, represented the experimental and control groups.

A paired *t*-test was performed on the scores of the EKT test before and after teaching with the ECTA approach in the experimental group (Table 3.3).

The paired *t*-test showed that the level of environmental knowledge at EKT in the group improved statistically significantly and had a medium effect size ($t(74) = 3.49$, $2p \leq 0.001$, $\eta^2 = 0.49$). The knowledge of recycling pictograms and environmental art on the EAT also improved ($t(74) = 19.13$, $2p \leq 0.001$, $\eta^2 = 3.09$). A paired *t*-test of the EKT before and after teaching was also conducted in the control group (Table 3.4).

The test showed that the level of knowledge in the group did not improve statistically significantly ($t(77) = -1.86$, $2p = 0.07$), although, within the group, the level of knowledge improved by 2.5%, and the students did not have a negative grade in the test.

The χ^2 test was used to examine changes in EKT before and after teaching in the experimental ($N = 75$) and control group ($N = 78$) students. Table 3.5 shows that the knowledge of the experimental group students' EKT improved statistically significantly after teaching ($p = 0.02$).

In the experimental group, there is a statistically significant change in knowledge, compared to the initial state, in the third, fourth, and fifth tasks ($\chi^2(4, N = 75) = 0.65$, $p \leq 0.05$). There is also a statistically significant change in the entire knowledge test. In the control group, 54 (69%) students correctly solved the third task after teaching, which is statistically significant compared to the previous state ($\chi^2(4, N = 78) = 8.56$, $p \leq 0.05$,

Table 3.3 Paired t-test of EKT within the experimental group

				Paired t-test			
		M	*SD*	*t*	*df*	*2p*	η^2
EKT	Before teaching	0.42	1.22	−3.49	74	**≤ 0.001**	0.49
	After teaching	0.58	1.53				

Source: Finale

Table 3.4 Paired t-test of the EAT tests within the control group

				Paired t-test			
		M	*SD*	*t*	*df*	*2p*	η^2
EKT	Before teaching	0.34	1.62	−1.86	77	0.07	–
	After teaching	0.35	1.64				

Source: Finale

Table 3.5 χ^2 test of change at EKT before and after teaching according to group

	Group							
	Experimental			η^2	Control			η^2
Questions in EKT	χ^2	*df*	*p*		χ^2	*df*	*p*	
1 What is compost?	0.02	4	0.99	–	0.12	3	0.09	–
2 Name the waste and sort it accordingly.	1.12	2	0.44	–	1.15	2	0.75	–
3 Use lines to connect the descriptions in the first column with the corresponding labels in the second column.	8.52	4	**0.05**	0.69	8.56	4	**0.05**	0.51
4 Amount of water used by a family of four in one day: 4a Which activity uses the most water?	4.90	3	**0.04**	0.47	4.95	2	0.58	–
4b Amount of water used by a family of four in one day: How can you contribute to reducing water consumption?	3.56	1	**0.05**	0.41	2.76	1	0.65	–
5 Explain what an illegal landfill is.	0.68	4	**0.05**	≤ 0.01	3.68	4	0.95	–
Total	0.65	4	**0.02**	0.09	2.65	3	0.52	–

Source: Finale

$\eta^2 = 0.51$). The results of the effect size measure for this task show that the difference between the knowledge test before and after the ECTA teaching approach is medium.

Students in the control group did not solve the other tasks statistically significantly better after teaching ($p \leq 0.09$).

5.2 *The Results of the Semi-structured Interview*

The answers of the three teachers are comprehensive and reveal their interest in the environment and nature and environmental issues.

The first section outlines teaching methods and lesson preparation. Science and technology lessons often include visits to collection centers, while fine arts classes focus on environmental projects using waste materials, with student artworks displayed in the department hallway. Teachers highlighted visits to a local beekeeper and forester to educate students on pollution's impact. They also enhance environmental lessons with visual materials, films, and interactive activities. However, their knowledge of the subject varies: one of the teachers feels confident in her expertise, while the other two rely on online sources and newspapers but seek further training. They all feel a lack of direction. To engage students, teachers use personal examples and historical photographs illustrating environmental changes in their communities, creating awareness of human impact on nature.

In the second section, we wanted to learn more about the implemented ECTA approach, the appropriateness of the student's age, the course, and possible changes that the teachers might introduce. It can be seen from the teachers' responses that the students were enthusiastic about the approach, and so were the teachers. All three teachers found the ECTA approach appropriate for the student's age because it was very concrete and at the level of students who are not yet capable of abstraction.

In the third section, we sought the difference between phrases nature conservation and environmental protection. Apart from one teacher, who answered flawlessly and explained the difference, we encountered a minor obstacle. The other two teachers were not able to immediately explain what the difference was, but based on the question, they concluded that there was a difference and that they were not synonyms. They managed to correctly summarize the essence of the difference.

6 Conclusion

Exploring environmental problems through different and new approaches, such as ECTA, is close to students. Students learn that we come from nature and that we are a part of it. Historically, man has risen above nature and become superior, and nature is not only at man's disposal for resource exploitation and economic calculation. Nature is not only the origin of man, but it is the space in which we live. The inspiration necessary for artistic creation has always come and still comes from observing nature.

For the research, we designed an innovative didactic approach to teaching ECTA, which was applied in the classroom and addresses the content area of waste with an emphasis on environmental pollution, separate waste collection, its recycling, and reuse for artistic purposes. The ECTA approach is designed with a combination of Gestalt and problem-based pedagogy and with cross-curricular and interdisciplinary integration. The ECTA approach is based on experiential learning, thereby developing a positive attitude toward the environment.

Before teaching, the initial knowledge of the students in the experimental and control groups was checked with the knowledge test of environmental content (EKT). The *t*-test confirmed a statistically insignificant difference between the groups ($2p = 0.17 > 0.05$). A comparison of the results of the knowledge tests EKT before and after teaching with the ECTA approach in the experimental group showed a statistically significant difference (EKT $2p \leq 0.001$). There was no statistically significant change in the control group (EKT $p = 0.07$).

Despite the numerous advantages of the ECTA approach, it is important to note that its implementation in Slovenia presents challenges. This approach is more readily adopted in the lower classes of primary education, where instruction typically occurs under the purview of a single educator for the entire class. This organizational structure facilitates the integration of the ECTA framework, allowing for a cohesive and uninterrupted pedagogical strategy (Finale, 2021). In higher class levels, particularly in contexts where subject specialization is prevalent, coordination and collaboration among multiple educators become necessary. This may complicate the practical application of the ECTA approach, as it requires collective efforts from teachers across different subjects to maintain its interdisciplinary and experiential focus (Finale, 2021).

By using various approaches or parts of them that have been recognized as good, we have designed a unique approach to teaching environmental content through art by integrating them into the ECTA teaching approach. Statistical analysis of knowledge tests before and after the implementation of the ECTA approach showed its positive effects.

6.1 *Research Limitations*

Within the framework of the research, it was possible to identify some shortcomings, mainly related to the selection of the sample, since only two primary schools participated. The experimental group included teachers who wanted to participate, but the sample was not random.

The research also has a shortcoming in its duration because it was adapted to the schedule of six grades at two primary schools. The instruments in the research were used without changes before and after teaching with the ECTA approach.

It is also important to point out the teacher's adequate knowledge, continuous improvement, and awareness that innovations enable students to develop an interest in various subjects, especially science and fine arts. At the same time, it is important to emphasize that when connecting subjects, none should be in a subordinate position, especially not fine arts.

In addition, we are explaining that in Slovenia, the average age of primary school teachers is 46 years, with 39% of all teachers being over 50 years of age (Delo, 2025). This indicates that many teachers are nearing retirement, which may affect the renewal of pedagogical approaches and teaching methods, especially in the field of environmental and ecological education. Research shows that many teachers lack the necessary knowledge and experience to teach environmental topics, which presents a significant challenge at a time when climate change and sustainable development issues are becoming increasingly relevant (Kovač & Jovanović, 2023). The lack of specific knowledge in ecology and environmental issues often means that teachers require further education and training to integrate these topics into the curriculum. While many teachers still use traditional teaching methods focused on direct knowledge transfer, modern approaches, such as problem-based learning (PBL) and interdisciplinary integration, are increasingly necessary to address complex environmental challenges. In addition, teachers need support in adapting lesson plans and methods to incorporate sustainability themes, including topics on pollutants, climate change, and biodiversity conservation (Sah et al., 2024).

For successful integration of environmental content into the learning process, it is essential to provide quality educational programs that allow teachers to continuously update their knowledge and adapt to new challenges. This includes collaboration with NGOs, training through online courses, and facilitating participation in interdisciplinary projects that promote ecological awareness (Zelenjak, 2022).

References

Breukel, K., & Cieraad, E. (2024). Using light as a medium to convey its dark side a light festival case study. https://doi.org/10.20944/preprints202407.1536.v1

Byrnes, D., Blum, L., & Walker, W. (2022). Undisciplining environmental communication pedagogy: Toward environmental and epistemic justice in the interdisciplinary sustainability classroom. *Sustainability*. https://doi.org/10.3390/su15010514

Chen, H., & Yuan, Y. (2021). The study of the relationships of teacher's creative teaching, imagination, and principal's visionary leadership. *Sage Open, 11*(3). https://doi.org/10.1177/21582440211029932

Coudray, C. (2020). Theory and praxis in experiential education: Some insights from gestalt therapy. *Journal of Experiential Education, 43*(2), 156–170. https://doi.org/10.1177/1053825920904387

Deer, G., Wu, H., Zhang, L., Tadesse, E., Khalid, S., Duan, C., et al. (2024). Effect of out-of-school visual art activities on academic performance. The mediating role of socioeconomic status. *Plos One, 19*(5), e0298901. https://doi.org/10.1371/journal.pone.0298901

Delo. (2025). Kadrovska stiska je ena večjih težav v slovenskih šolah. Delo.si. Retrieved March 3, 2025, from https://www.delo.si/novice/slovenija/kadrovska-stiska-je-ena-vecjih-tezav-v-slovenskih-solah/

Demčenko, O., Bondar, Y., Shykyrynska, O., Komarivska, N., Lyubchak, L., Стахова, I., et al. (2022). Development of youth's creative abilities schoolchildren by byod technology. *Society Integration Education Proceedings of the International Scientific Conference, 1*, 341–350. https://doi.org/10.17770/sie2022vol1.6892

Dmitrievna, G. (2024). Interdisciplinary integration of science education at higher education institutions. *Journal of Education and Culture Studies, 7*(2), 43–54. https://doi.org/10.22158/jecs.v7n2p43

Doğan, A., & Sır, H. (2022). Development of primary school fourth-grade students' fraction calculation strategies through the argumentation method. *Journal of Education and Learning (Edulearn), 16*(2), 262–272. https://doi.org/10.11591/edulearn.v16i2.20511

Elo, J., & Nygren-Landgärds, C. (2020). Teachers' perceptions of autonomy in the tensions between a subject focus and a cross-curricular school profile: A case study of a Finnish upper secondary school. *Journal of Educational Change, 22*(3), 423–445. https://doi.org/10.1007/s10833-020-09412-0

Finale, R. (2021). Enhancing environmental awareness through art. Doctoral dissertation, p. 285, Faculty of Education, University of Ljubljana, Ljubljana.

Gilbert, F., & Dobson, T. (2024). Towards boundary crossing: Primary and secondary school teachers teaching creative writing and its redrafting. *English Teaching Practice & Critique*. https://doi.org/10.1108/etpc-03-2024-0039

Govindasamy, P., Abdullah, N., & Ibrahim, R. (2023). Employing modified Design and Development Research (DDR) approach to develop creativity clay module to teach Special Educational Needs (SEN) students with learning disabilities. *International Journal of Academic Research in Business and Social Sciences, 13*(3). https://doi.org/10.6007/ijarbss/v13-i3/15773

Hagerman, F., & Rundgren, S. (2022). A literature review on teaching ethical creativity in primary education. *Multidisciplinary Journal of School Education, 11*(2 (22)), 235–250. https://doi.org/10.35765/mjse.2022.1122.11

Hsiao, M. (2023). Environmental consciousness on sustainable development in contemporary art design. *International Journal of Art Innovation and Development, 4*(1) https://doi.org/10.38007/ijaid.2023.040116

Inwood, H., & Kennedy, A. (2020). Conceptualising art education as environmental activism in preservice teacher education. *International Journal of Art & Design Education, 39*(3), 585–599. https://doi.org/10.1111/jade.12308

Ison, M., & Bramwell-Lalor, S. (2023). The arts in environmental education: Connecting learners with their talents and nature. *Environmental Education Research, 29*(7), 964–979. https://doi.org/10.1080/13504622.2023.2205062

Iwasaki, S. (2022). Effects of environmental education on young children's water-saving behaviors in Japan. *Sustainability, 14*(6), 3382. https://doi.org/10.3390/su14063382

Kluger, K., & Bartzke, S. (2020). The challenges of interdisciplinary teaching in higher education: A case study. *Educational Studies, 46*(3), 365–378. https://doi.org/10.1080/03055698.2020.1789201

Kovač, P., & Jovanović, T. (2023). Environmental education in Slovenian schools: Challenges and solutions. *Journal of Environmental Education, 34*(2), 55–68. https://doi.org/10.1080/00958964.2023.2128701

Liao, C., & Li, H. (2019). Environmental education, knowledge, and high school students' intention toward separation of solid waste on campus. *International Journal of Environmental Research and Public Health, 16*(9), 1659. https://doi.org/10.3390/ijerph16091659

Lin, T. J., Wang, L. C., & Lee, H. W. (2021). Fostering collaborative learning and evidence-based problem-solving through a digital health intervention: A randomized controlled trial. *Health Education Research, 36*(3), 121–133. https://doi.org/10.1093/her/cyab019

Lindsay, S. (2021). Integrating microscopy, art, and humanities to power steam learning in biology. *Invertebrate Biology, 140*(1). https://doi.org/10.1111/ivb.12327

Mäkelä, T., Fenyvesi, K., Kankaanranta, M., Pnevmatikos, D., & Christodoulou, P. (2022). Co-designing a pedagogical framework and principles for a hybrid stem learning environment design. *Educational Technology Research and Development, 70*(4), 1329–1357. https://doi.org/10.1007/s11423-022-10114-y

Marković, S., & Petrović, J. (2021). Raising environmental awareness through art projects. *Reciklaza I Odrzivi Razvoj, 14*(1), 11–17. https://doi.org/10.5937/ror2101011m

Mesec, B. (1998). *Uvod v kvalitativno raziskovanje v socialnem delu.* Visoka šola za socialno delo Univerze v Ljubljani.

Millanzi, J., & Kibusi, S. (2021). Exploring the effect of problem-based facilitatory teaching approach on motivation to learn: A quasi-experimental study of nursing students in Tanzania. *BMC Nursing, 20*(1), 1–9. https://doi.org/10.1186/s12912-020-00509-8

Mobjörk, M., Wachtel, A., & Smith, D. (2020). Sustainable development and cross-disciplinary research education: Challenges and opportunities for learning. *Högre Utbildning, 10*(2), 121–134. https://doi.org/10.23865/hu.v10.1942

Moula, Z., Palmer, K., & Walshe, N. (2022). A systematic review of arts-based interventions delivered to children and young people in nature or outdoor spaces: Impact on nature connectedness, health and wellbeing. *Frontiers in Psychology, 13.* https://doi.org/10.3389/fpsyg.2022.858781

Namukasa, I., Bertrand, M., Tangredi, D., Roy, J., Barek, H., Campigotto, R., & Aryee, K. (2023). Reimagining the mathematics curriculum through a cross-curricular and maker education lens. *International Journal of Teacher Education and Professional Development, 6*(1), 1–22. https://doi.org/10.4018/ijtepd.324166

Niemelä, M. (2021). Subject matter specific curriculum integration: A quantitative study of Finnish student teachers' integrative content knowledge. *Journal of Education for Teaching International Research and Pedagogy, 48*(2), 228–240. https://doi.org/10.1080/02607476.2021.1989288

Obasi, K., & Osah, S. (2022). Curriculum development planning in environmental education for developing environmental citizenship among primary school pupils in rivers state. *International Journal of Instructional Technology and Educational Studies, 3*(2), 34–41. https://doi.org/10.21608/ihites.2021.107723.1080

Pallant, J. (2020). SPSS survival manual: *A step by step guide to data analysis using IBM SPSS statistics* (7th ed.). Routledge. https://doi.org/10.4324/9781003117452

Pham, P., & Ünaldı, A. (2021). Cross-curricular collaboration in a clil bilingual context: The perceptions and practices of language teachers and content subject teachers. *International Journal of Bilingual Education and Bilingualism, 25*(8), 2918–2932. https://doi.org/10.1080/13670050.2021.1995320

Rauhala, O. (2003). *Nature, science and art.* Otava Publishing Company.
Riet, P., Jitsacorn, C., & Thursby, P. (2020). Approaches to visual arts-based data collection with hospitalized children: A discussion of lessons learnt and suggestions for practice. *Journal of Advanced Nursing, 77*(1), 492–498. https://doi.org/10.1111/jan.14615
Safitri, D., Putra, Z., & Sujarwo, S. (2020). Ecolabel in environmental education. *Proceedings of the International Conference On Social Studies, Globalisation And Technology (ICSSGT 2019).* https://doi.org/10.2991/assehr.k.200803.032
Sagadin, J. (2003). *Statistične metode za pedagoge.* Obzorja d.o.o.
Sah, F., Sasikirana, H. N., & Pujiani, T. (2024). The implementation of project-based learning in developing 21stcentury skills in EFL class. Jadila: *Journal of Development and Innovation in Language and Literature Education, 4*(4), 257–272. https://doi.org/10.52760/jadila.v4i4.797
Şeren, N., & Gül, S. (2022). An innovative and interdisciplinary perspective on environmental issues through ecological art projects: A mixed methods study. *Romanian Review of Geographical Education, 11*(1), 73–91. https://doi.org/10.23741/rrge120225
Stuart, E. A., & Rubin, D. B. (2008). Best practices in quasi-experimental designs: A guide for policy research. In *Evaluating the impact of intervention* (pp. 453–469). Press.
Sunassee, A., & Bokhoree, C. (2021). A designed eco-art and place-based curriculum encouraging students' empathy for the environment. *Ecologies, 2*(3), 248–267. https://doi.org/10.3390/ecologies2030014
Sutjonong, W., Salim, R., & Safitri, S. (2022). Teachers' self-efficacy as a mediator of their perception and behavior regarding creative teaching for elementary school students. *Mimbar Sekolah Dasar, 9*(1), 161–173. https://doi.org/10.53400/mimbar-sd.v9i1.44253
Tofel-Grehl, C., Searle, K., Hawkman, A., Hansen, T., & Lott, K. (2021). Crafting circuits: Integrating culturally responsive teaching and current events into science. *Science and Children, 58*(4), 81–86. https://doi.org/10.1080/00368148.2021.12291660
Tsevreni, I. (2022). Encounters among environmental education and eco-art in the Anthropocene. *International Journal of Art & Design Education, 41*(3), 482–495. https://doi.org/10.1111/jade.12417
Varnell, S., Brown, E., Holochwost, S., Allen, A., Malatesta, J., Garnett, M., et al. (2024). Visual arts activities relate to interest and cortisol for children in head start preschool. *Psychology of Aesthetics Creativity and the Arts.* https://doi.org/10.1037/aca0000711
Vasilaki, V. (2024). Pre-school teachers' views on the impact of visual arts education on early childhood education. *Innovare Journal of Education*, 1–9. https://doi.org/10.22159/ijoe.2024v12i4.51710

Wu, Y., Wan, J., & Yu, W. (2023). Impact of environmental education on environmental quality under the background of low-carbon economy. *Frontiers in Public Health, 11*. https://doi.org/10.3389/fpubh.2023.1128791

Yang, B., Wu, N., Tong, Z., & Sun, Y. (2022). Narrative-based environmental education improves environmental awareness and environmental attitudes in children aged 6–8. *International Journal of Environmental Research and Public Health, 19*(11), 6483. https://doi.org/10.3390/ijerph19116483

Zaeri, P., & Roozafzai, Z. (2024). Technology-enhanced art and sustainable discourse practices. *International Journal of Culture and Art Studies, 8*(2), 109–120. https://doi.org/10.32734/ijcas.v8i2.16654

Zainuri, N., Rahman, N., Halim, L., Chan, M., & Bazari, N. (2022). Measuring pro-environmental behavior triggered by environmental values. *International Journal of Environmental Research and Public Health, 19*(23), 16013. https://doi.org/10.3390/ijerph192316013

Zelenjak, S. (2022). Professional development for sustainable teaching practices. *Education and Sustainability Review, 7*(3), 24–38. https://doi.org/10.1080/23500051.2022.2097820

Zhao, L., Zhao, B., & Li, C. (2023). Alignment analysis of teaching–learning–assessment within the classroom: How teachers implement project-based learning under the curriculum standards. *Disciplinary and Interdisciplinary Science Education Research, 5*, 13. https://doi.org/10.1186/s43031-023-00078-1

Zhongbin, Z. (2024). Innovative approaches to art education by integrating environmental awareness in art curriculum in Beijing. *International Journal of Academic Research in Progressive Education and Development, 13*(1). https://doi.org/10.6007/ijarped/v13-i1/20657

Zocher, J., & Hougham, R. (2020). Implementing ecopedagogy as an experiential approach to decolonizing science education. *Journal of Experiential Education, 43*(3), 232–247. https://doi.org/10.1177/1053825920908615

CHAPTER 4

Opportunities and Challenges in Integrating Augmented and Virtual Reality in Environmental Education

Hajar Zekeik, *Mohamed Chahbi*, *Anass Saddati*, *Mohammed Lamarti Sefian*, *and Imane Bakkali*

1 Introduction

Environmental education (EE) is crucial in fostering awareness and understanding of ecological issues among students and empowering them to become informed citizens and stewards of the environment (Imran et al., 2024). In general, EE is "*the process that allows individuals to explore environmental issues, sustainability awareness, engage in problem-solving, and take action to improve the environment*" (Mahavidyalaya Champadanga, 2022).

H. Zekeik (✉) • M. L. Sefian • I. Bakkali
Applied Mathematics and Computer Science Team, Normal Higher School, Abdelmalek Essaadi University, Tetouan, Morocco
e-mail: hajar.zekeik1@etu.uae.ac.ma; lamarti.mohammed.sefian@etu.uae.ac.ma; i.bakkali@uae.ac.ma

M. Chahbi • A. Saddati
Representations, Linguistic, Technological, and Social Practices Team, Normal Higher School, Abdelmalek Essaadi University, Tetouan, Morocco
e-mail: mohamed.chahbi@etu.uae.ac.ma; anass.saddati@etu.uae.ac.ma

K. Hirech (ed.), *Environmental Education and Sustainable Development*, https://doi.org/10.1007/978-3-032-05760-0_4

There is an intense call for better pedagogy in accommodating students to these very complex global issues, which are complicated by climate change, loss of biodiversity, and rising pollution (Suarlin, 2023). Conventional teaching methods often fail to capture students' interests and foster deep comprehension of environmental concepts.

New technologies, such as AR and VR, have created vast opportunities for EE. Living through immersive experiences engrosses students in active engagement with virtual representations of environmental phenomena to better understand and relate to abstract concepts (Udaya Shankar et al., 2023). These technologies offer opportunities for experiential learning while promoting sustainability awareness by engaging students in simulations and scenarios that address human impact on the environment.

Integrating EE into educational contexts can effectively support the attainment of SDGs (Sustainable Development Goals) targets, highlighting its strategic importance in sustainable development efforts (Velempini, 2025).

Indeed, the potential of AR and VR seems promising, but there are challenges in ensuring their successful incorporation into EE. Technological barriers, including the lack of essential resources and infrastructure as well as insufficient educator training, hinder the implementation of these technologies in curricula.

This study is grounded in Experiential Learning Theory, as proposed by Kolb in 1984, and in Transformative Learning Theory, a concept introduced by Jack Mezirow in 1978. Experiential learning, as the name suggests, is a process of learning through direct experience. The learning aspect involves more than just experience; it also requires reflection. A comprehensive understanding of experiential learning involves reflecting on students' learning experiences. It encompasses various forms of student engagement and emphasises learning by doing. In this hands-on approach, students take responsibility for their learning, assuming control over both the learning experience and their reflections on it (Morris, 2020), which aligns well with the immersive nature of AR and VR technologies.

Transformative Learning highlights how critical reflection can lead to profound shifts in learners' perspectives, beliefs, and behaviours through the learning process (Kuriakou, 2023), which corresponds to the power of AR and VR as tools that help students not only understand, but also reconsider and change their attitudes and actions towards sustainability. Together, these frameworks provide a strong foundation for exploring how AR and VR can deepen students' understanding of environmental issues and foster meaningful changes in their attitudes and actions towards sustainability.

In this chapter, we examine the possibilities and challenges of integrating AR and VR into EE. To guide this discussion, the following research questions were addressed:

RQ1: What benefits do AR and VR technologies bring to enhance EE?
RQ2: What challenges do educators face regarding AR and VR integration into teaching practices?
RQ3: What practical solutions can be implemented to overcome the above challenges and pave the way for the beneficial use of AR and VR in EE?

Through a qualitative analysis of the existing literature, case studies, and semi-structured interviews conducted with middle school students to explore their perceptions of using AR and VR in EE, we aimed to identify the key benefits of these technologies, the obstacles educators face, and practical solutions to overcome these challenges. Consequently, we propose that by addressing these barriers and leveraging the unique capabilities of AR and VR, educators can significantly enhance the effectiveness of EE and prepare students for future ecological problems.

2 Methodology

2.1 Research Design

This study used a qualitative approach to integrate AR and VR into EE. Specifically, it examines how these technologies impact students' understanding of complex concepts, their engagement with environmental issues, and their overall learning experience. Student interviews were incorporated into the literature review and case study analysis to provide empirical evidence on the concrete impact of AR/VR on environmental awareness among middle school students.

2.2 Data Collection Methods

2.2.1 Literature Review

A literature review was conducted to gather knowledge on the applications of AR and VR in EE across all educational levels. To ensure a thorough understanding of current practices and challenges, academic databases such as Google Scholar, ERIC, and IEEE Xplore were searched using keywords including "AR in environmental education," "Virtual reality learning experiences," and "AR/VR in pollution awareness." This review attempted to identify studies that assess the effectiveness of AR/VR, related challenges, and specific applications in the context of EE.

2.2.2 Case Study Analysis

To provide a practical perspective, this study includes an analysis of existing case studies that demonstrate how AR and VR have been applied to education regarding pollution and other environmental issues. Each case study was evaluated to assess the level of student engagement, depth of conceptual understanding achieved, and learning outcomes documented by the educators. By analysing case studies, we identified common challenges, success factors, and best practices for using immersive technologies in EE.

2.2.3 Middle School Student Interviews

Semi-structured interviews were conducted with 20 middle school students involved in a VR/AR activity designed to demonstrate the effects of pollution. The interviews were conducted with the technical support of an English teacher who guided the students through the questions and recorded their answers. The questions focused on students' engagement, understanding of pollution concepts, and perceptions of VR/AR as an educational tool. Each interview lasted approximately 10–15 minutes and was conducted in a comfortable setting to ensure honest responses.

2.3 Sampling Strategy

Case studies were selected using purposive sampling to provide an information-rich example relevant to EE and sustainability in practice. Additionally, semi-structured interviews were conducted with middle school students to explore their perceptions and attitudes towards the use of AR and VR in EE. The selection was guided by the relevance of both the case and the interview to experiential learning, which emphasises active engagement and reflection, and transformative learning, which seeks to foster critical shifts in students' perspectives and behaviours through innovative pedagogical approaches.

2.4 *Data Analysis*

2.4.1 *Literature and Case Study Analysis*

Literature and case studies were analysed to extract overarching themes related to AR/VR applications, regardless of educational level, allowing for a near-general understanding of the role of AR/VR in EE. Themes such as engagement, pedagogical challenges, and technological barriers were identified and synthesised across studies to map these understandings against specific insights from middle school interviews.

2.4.2 *Interview Analysis (Middle School Focus)*

Middle school student interviews were audio-recorded, transcribed literally, and then thematically hand-coded. A deductive-inductive approach was applied to the coding process:

- **Deductive coding:** Precedence categories (e.g., engagement, understanding, and collaboration with AR/VR tools) were employed based on the literature findings.
- **Inductive coding:** Emerging themes were identified by sequentially reading the transcripts and noting recurring patterns.

Manual coding was conducted by carefully reading and annotating the transcripts and grouping similar responses under thematic categories. To ensure reliability, a second review was performed to refine the coding framework. The final analysis highlights middle school students' engagement, knowledge of environmental issues, and their interaction with AR/VR platforms, demonstrating the potential of such technologies to enhance learning about pollution and environmental issues.

2.5 *Data Validity*

To ensure the validity of our findings, we employed multiple methodological approaches, including a comprehensive literature review, case study analyses, and semi-structured interviews with middle school students. The literature review provided a theoretical foundation, while case studies from prior research helped contextualise our results within existing frameworks. Additionally, our semi-structured interviews offered in-depth, qualitative insights, ensuring that the data accurately reflected the authentic perspectives of participants. To enhance reliability, we used triangulation by cross-verifying themes from interviews, with patterns identified in the literature and case studies. Furthermore, we implemented member checking, allowing participants to review and confirm their responses, thereby minimising misinterpretations. These strategies collectively strengthen the credibility and trustworthiness of our findings.

2.6 *Ethical Considerations*

Informed consent was obtained from all participants and their guardians before conducting interviews with middle school students. Confidence and voluntary participation were ensured in accordance with the ethical guidelines for educational research involving minors.

3 Results

3.1 *RQ1: What Benefits Do AR and VR Technologies Bring to Enhance EE?*

Considering the literature review and potential insights from the case study, this section explores various instances of how AR and VR can enhance EE.

3.1.1 *Enhanced Motivation and Engagement*

Multiple studies have confirmed that AR and VR have the potential to enhance learning because of their interactive nature and visually stimulating experiences that attract the attention and motivation of students (Kavanagh et al., 2017; Rodrigues & Pombo, 2024). As defined by Kavanagh et al. (2017), motivation is the drive or incentive that encourages individuals to participate actively in specific activities. These technologies create immersive learning environments that enable students to engage with complex topics such as environmental issues in ways that traditional teaching methods cannot. For instance, research conducted by Porro et al. (2022) emphasised AR applications, such as the **Resized**

Plastic app, which emphasises microplastics. This tool allows students to experience and interact with environmental problems in real-time. AR motivates students to become interactive in the learning process, encouraging them to engage in hands-on practices related to environmental issues, which helps them develop more sustainable behaviours.

In addition, **EduCITY** is a mobile AR game tested with secondary students at Aveiro University that showcases the potential of AR in engaging students in sustainability education, deepening their knowledge of environmental issues, and motivating them to take action (Rodrigues & Pombo, 2024).

A three-month study with 400 college students found that VR significantly boosted engagement and awareness of environmental issues. By immersing students in real-world challenges, VR fosters a deeper emotional connection with sustainability. The study concluded that, compared to other technologies, VR is particularly effective in increasing environmental awareness and motivating students to take action. As their understanding grows, so does their involvement in sustainable behaviours and environmental activism (Cao & Jian, 2023).

Ultimately, AR and VR transform EE into an engaging, hands-on experience that reflects the core principles of the Experiential Learning Theory. By immersing students in interactive environments, these technologies promote active participation and encourage reflection on their experiences, which are the key components of experiential learning. Additionally, this process aligns with transformative learning theory, as the critical reflection fostered through AR and VR experiences can lead to shifts in students' perspectives, beliefs, and behaviours regarding sustainability. As a result, these immersive tools not only deepen the understanding of environmental issues, but also motivate students to take meaningful action towards addressing ecological challenges.

3.1.2 A Deeper Understanding of Complex Concepts

VR and AR create an immersive experiential learning environment for teaching complex environmental concepts. At the Harvard Graduate School of Education, the EcoLearn team uses immersive VR and AR technology to facilitate learning about the complex causal dynamics within ecosystems (Cho & Park, 2023). This research details the development of a VR simulation that enables students to engage directly with environmental issues, making the complexities of global warming and its impacts more accessible. By experiencing environmental dynamics first, students develop a deeper understanding of these issues, which can inspire them to adopt more sustainable behaviours.

In the study by Ou et al. (2021), the researchers highlighted that recent advances in VR technology, such as 360° panoramic systems, have been effectively utilised to create immersive virtual environments for EE. For instance, a VR-based virtual wetland ecological system developed for autonomous exploration enables learners to engage in inquiry-based learning using affordable devices, such as cardboard VR headsets. This system provides detailed information on the ecology of wetlands and the native species found in the Hsinchu coastal area of northern Taiwan, including birds, crabs, and plants. Through panoramic photos, videos, and descriptive texts, students can gain a deeper understanding of the habits, food sources, and biological functions of these organisms in their ecological contexts.

Moreover, the research highlights that using AR-based instruction in EE enhances students' understanding of concepts, boosts their achievement, and strengthens their knowledge retention (Ladykova et al., 2024). For example, the findings of this study demonstrate that an AR-based learning tool for carbon compounds, when combined with environmental literacy, enhances students' comprehension of concepts related to carbon compounds. It further emphasises how AR can benefit students in understanding complicated environmental issues with interactivity, visualisations, and experiences that traditional methods will not achieve (Ardyansyah & Rahayu, 2023).

Such immersive experiences facilitate a deeper understanding of complex ecosystem concepts and interactions, supporting experiential learning by allowing students to actively explore and reflect on environmental processes.

3.1.3 Promoting Sustainability and Environmental Awareness

AR and VR technologies are essential for raising awareness about sustainability and the environment. A great example is **Aire**, an iOS-based augmented reality app that brings the invisible issue of air pollution to life. By visualising this complex scientific concept, **Aire** helps users see and understand the environmental impact of air pollution, making the problem more tangible and engaging (Ladykova et al., 2024). Furthermore, AR helps students grasp complex concepts and encourages more responsible behaviour towards ecosystem care. This reflects the transformative learning theory's emphasis on shifting perspectives to foster environmental responsibility while also drawing on experiential learning theory's focus on active and immersive engagement with learning content. This integration highlights a growing trend in utilising technology to educate and inspire environmentally conscious behaviour.

3.1.4 *Applications of AR and VR in EE: Case Studies*

In addressing RQ1, which investigates the advantages of AR and VR in EE, an example case study supports the value of AR in EE, indicating that AR improves students' knowledge, attitudes, and behaviour towards the environment.

For example, this case study by Ozturk and Akcay (2023) explored the use of AR in EE among primary school students. The study, which involved 17 third-grade students over six weeks, examined the AR application's effect on students' knowledge, attitudes, behaviours, and environmental awareness. The results from semi-structured interviews showed that students had positive feelings about AR, claiming that it assisted their learning and helped them conceptualise environmental matters better. Interestingly, they wanted to use AR applications for other subjects, believing this would allow them to engage more effectively. This case study highlights several key benefits of integrating AR with EE. It not only improved their understanding of environmental issues, but also enhanced their attitudes towards both living and non-living beings, as well as environmental problems. This evidence supports the idea that AR effectively promotes environmental awareness, behaviours, and attitudes towards sustainability.

Similarly, another study by Cao and Jian (2023) examined the use of VR to enhance environmental awareness among college students. Involving 400 students, this study demonstrated that VR-based environmental education significantly enhanced students' understanding of environmental issues and promoted a conservationist value system. The VR activities helped provoke students to engage in environmental advocacy and adopt pro-environmental behaviours. Statistical analysis revealed that VR played a key role in motivating students to take action on ecological sustainability even after the VR-based intervention was completed. This demonstrates the potential of VR as a powerful tool for promoting EE and influencing students' environmental attitudes and behaviours.

3.2 RQ2: What Challenges Do Educators Face Regarding AR and VR Integration into Teaching Practices?

This section presents the challenges associated with implementing AR and VR technologies in education, specifically the technological, pedagogical, and infrastructural obstacles identified in the literature, as well as potential insights from case studies.

3.2.1 Financial and Technological Challenges

The major problem that has been identified concerning integrating AR and VR into teaching practices is that the inclusion of hardware and software for AR and VR education applications is very costly (Al-Ansi et al., 2023). The cost of the hardware and software required for AR and VR technologies can be a significant hindrance for various educational institutions, as they may lack the necessary budget to acquire these resources. In addition, the software used to develop an immersive experience may require frequent updates, thus adding to the economic burden (Al-Ansi et al., 2023). Significant financial barriers are imposed by the high costs of investing in AR/VR devices, maintaining hardware, and upgrading software and licenses (Beetul et al., 2023). These challenges have limited the widespread adoption of AR/VR technologies in professional and educational contexts.

3.2.2 Pedagogical Challenges

Although VR have been integrated into traditional educational paradigms with some degree of success, researchers agree that maximising their learning potential requires a robust pedagogical framework explicitly tailored for VR (Lege & Bonner, 2020). Moreover, it is crucial to design and implement educational programs specifically tailored to the capabilities and requirements of these AR and VR technologies (Elmqaddem, 2019). A significant challenge in integrating VR into education is determining how to use technology effectively to enhance learning rather than merely replicating traditional classroom methods. To address this, initiatives have been made to create VR-specific pedagogies, such as Southgate's Actioned Pedagogy for Immersive Learning, which offers educators a comprehensive framework for considering key elements, such as the roles of teachers, students, and technology in immersive learning environments (Lege & Bonner, 2020).

3.2.3 Professional Development Needs

A significant challenge faced by educators in AR and VR is the inherent complexity of these technologies. As these technologies are relatively new, teachers often struggle to use them effectively without prior experience or knowledge (Al-Ansi et al., 2023). This lack of familiarity can hinder the adoption and successful integration of AR and VR in the educational setting. Furthermore, many institutions fail to provide ongoing support and training, which are essential for the effective implementation of these technologies in the curriculum (Muniandy, 2024). Without adequate training programs and access to educational resources, users' skills and confidence in utilising immersive technologies, such as AR and VR, can remain underdeveloped, leading to significant barriers to widespread adoption (Kurniya Rachmawati & Tornando, 2024). Additionally, educators face difficulties in mastering AR and VR teaching methods and effectively managing the technology without proper training. Therefore, educational institutions must take responsibility for providing professional development programs that enable teachers to fully leverage AR/VR innovations to ensure that both educators and students benefit from these transformative technologies (Muniandy, 2024).

3.3 RQ3: What Practical Solutions Can Be Implemented to Overcome the Above Challenges and Pave the Way for the Beneficial Use of AR and VR in EE?

Practical solutions must be adopted to overcome the identified challenges in integrating AR and VR into EE. These solutions involve enhancing teacher training, promoting collaboration with technology providers, and developing immersive projects that engage students in environmental learning. By implementing these solutions, educators can overcome the barriers and harness the potential of AR and VR in EE.

3.3.1 Comprehensive Teacher Training

Professional development programs are necessary to equip educators with the skills required to use AR and VR tools effectively in the classroom (Lege & Bonner, 2020). One key solution is to provide teachers with robust professional development skills. Educators need to be trained in the operation of AR/VR technologies and integrate them into the curriculum in a pedagogically meaningful manner. Comprehensive training programs should also help familiarise users with and decrease apprehensions about AR/VR technologies, ensuring that educators and students feel comfortable interacting with these tools.

Effective training programs need to cover multiple aspects:

- Technical Skills and Familiarisation
  - Teachers should learn to operate AR and VR hardware and software, navigate immersive environments, and solve standard technical problems.
  - Workshops will offer hands-on experience with popular tools, such as Google Earth VR, for exploring ecosystems and AR applications for analysing soil quality.
- Curriculum Integration
  - Educators need support in integrating AR and VR into lesson plans, aligning with EE objectives.
  - Training modules feature activities such as virtual clean-ups and AR field trips to explore the impact of pollution on various areas.
- Pedagogical Approaches
  - Teachers should explore how AR and VR enhance experiential and inquiry-based learning.
  - Teachers could practice designing activities in which students role-play as environmental scientists in virtual climate simulations.
- Ongoing Support and Resources
  - Creating a support network with access to online repositories of AR/VR lesson plans and tutorials will ensure ongoing professional development.
  - Partnerships with AR/VR providers could help teachers stay updated with the latest tools and applications.

3.3.2 Overcoming Financial and Technological Barriers in AR/VR Integration

Addressing the financial and technological barriers to AR/VR integration is essential for ensuring the broader adoption of AR and VR in educational contexts, particularly in EE. Technology companies should prioritise the development of cost-effective, user-friendly AR/VR hardware (Kurniya Rachmawati & Tornando, 2024). Affordable tools encourage broader adoption, especially in educational settings where budgets are often constrained. Furthermore, Companies must enhance hardware and software reliability by addressing performance issues, improving compatibility, and creating intuitive user interfaces (Kurniya Rachmawati & Tornando, 2024). This will lower technical barriers for educators and students.

Collaboration or partnerships between educational institutions and technology providers would bring subsidised hardware and software, making high-end devices more affordable. Such partnerships would enable schools to develop AR/VR solutions tailored to meet specific educational needs. Additionally, such initiatives will encourage the development and sharing of reusable AR and VR content, making it more cost-effective to create exclusive materials. The education sector has a repository of immersive content that can serve different subjects.

In the coming years, it will be essential for governments and stakeholders to prioritise the development of digital infrastructure to support immersive technologies, including measures to ensure connectivity and sufficient computational power in schools, particularly in underserved areas. Another solution to reduce financial barriers is the compatibility of AR with readily available mobile smartphones, which are rapidly becoming commonplace in educational settings (Beetul et al., 2023). AR applications can be seamlessly accessed on these devices, making immersive learning experiences more accessible and engaging. Moreover, AR can be paired with low-cost VR solutions, such as Google Cardboard, to provide students with an immersive VR-like experience at a lower cost than traditional VR headsets (Beetul et al., 2023). This strategy enables institutions to provide comprehensive, interactive educational content while minimising the capital required for hardware investment, thereby promoting the wider-scale adoption of AR and VR technologies in classrooms and educational programs.

3.4 Student Perspectives on AR and VR in EE

As part of the research, the significant results of semi-structured interviews with 20 middle school students revealed key insights into how AR and VR impact engagement, comprehension, understanding, and perceptions of EE.

According to Table 4.1, the students' responses to the AR and VR pollution learning activities highlight strong engagement, characterised by excitement, curiosity, immersion, and enjoyment. Many students reported enjoying their first experience with VR, finding it fun and innovative. The use of VR headsets provided a setting in which students could see and feel environmental issues, developing a sense of immediacy and empathy towards topics such as marine pollution. The immersive nature of VR technology enables students to feel "present" in the pollution scenarios, thereby enhancing their engagement and immersion. Curiosity drove the students to explore the virtual environment, while the creativity of the activity was appreciated as an engaging and novel way to learn. In general, these responses showed that the students were committed, thoroughly enjoyed the experience, and found it interesting, highlighting the novelty and interactivity of the learning environment. This emphasises even more how AR and VR help students become more motivated and engaged in EE.

As shown in Table 4.2, the VR/AR activity on pollution was effective in helping students understand complex environmental concepts, particularly regarding the harmful effects of plastic pollution on marine life. Many

Table 4.1 Student reactions to AR and VR in pollution learning activities

Theme	*Description*	*Examples of student responses*
Excitement	Positive emotional reaction to the novelty of AR/VR	"It was new and exciting to use VR," "It was amazing and exciting!"
Curiosity	Desire to explore the AR/VR environment	"I felt curious," "I wanted to look everywhere to see what was happening."
Immersion	Feeling "present" within the pollution scenarios	"It felt realistic, like I was really in the ocean," "It felt strange but interesting."
Creativity	Appreciation for the novel learning approach	"I thought it was a creative way to learn," "It was a new and exciting way to learn."
Enjoyment	Fun and positive engagement with the activity	"I thought it was cool and fun," "I liked it a lot," "I thought it was very cool!"
Surprise	Initial surprise and interest in the AR/VR experience	"I was surprised," "I didn't know what to expect, but I liked it."

Table 4.2 Student feedback on understanding pollution concepts using VR/AR

Theme	*Description*	*Examples of student responses*
Plastic Pollution	Understanding the impact of plastic on marine life	"I learned how plastic hurts animals." "I learned how plastic harms fish." "I learned how plastic stays in the ocean."
Water Pollution	Learning about chemicals and trash affecting water quality	"I learned about dirty water." "I learned how chemicals hurt water." "I learned how oil spills harm the ocean."
Marine Life Impact	Exploring how pollution affects sea animals	"I saw how trash can kill fish." "I saw turtles eating plastic." "I saw how fishing nets can trap sea animals."
Human Impact	Connecting human actions to pollution	"I learned about the connection between human actions and pollution." "I saw how plastic bags float and look like jellyfish."
Pollution Effects	Gaining insight into broader environmental consequences	"I learned about how pollution affects ocean temperatures." "I saw how trash blocks sunlight and harms plants underwater."

students shared the long-term impacts of plastic waste, with responses such as, "I learned how plastic harms fish," and "I learned how plastic stays in the ocean." This immersive experience provided students with a deeper understanding of water pollution, offering insights into how chemicals and oil spills affect aquatic ecosystems. One student mentioned, "*I learned how oil spills harm the ocean.*"

Furthermore, this activity highlighted the impact of pollution on various marine animals, including fish, turtles, and birds. One student noted, "*I saw how trash can kill fish*," while another shared, "*I saw turtles eating plastic.*" These findings suggest that the AR and VR technologies make these complex issues more concrete and comprehensible.

VR/AR activity is advantageous for understanding pollution concepts because it provides an immersive and engaging way to explore these complex environmental issues. The possibilities provided by VR and AR to visualise pollution effects are key in facilitating and simplifying the understanding of these abstract concepts.

According to Table 4.3, the VR/AR-based pollution activity generally attracted positive responses from most students, as they found it fun and interesting. The immersive nature of the experience was prominent, as reflected in responses like, "*It was exciting to see things up close that I*

Table 4.3 Students' Engagement and Enjoyment when Interacting with a VR/AR Activity

Theme	*Description*	*Examples of student responses*
Fun and Interest	Understanding the impact of plastic on marine life	"It was fun! I liked seeing the ocean in 3D." "It was exciting to see things up close that I couldn't see in real life."
Visual Appeal	Learning about chemicals and trash affecting water quality	"I liked the pictures of the sea." "I loved seeing the underwater scenes."
Immersive Learning	Exploring how pollution affects sea animals	"It made everything feel real and important." "I enjoyed learning in a new way."
Divergent Preferences	Connecting human actions to pollution	"I think reading is better for me." "I don't think it showed enough information."

couldn't see in real life." The students remarked that visual richness and realism were especially impressive; they commented, *"I loved seeing the underwater scenes,"* and *"It made everything feel real and important."*

The activity also familiarised the students with an innovative way of learning. Some students found it fun and easy to follow, *"I enjoyed learning in a new way,"* while others remarked, *"It made learning exciting."* Such comments highlight the potential of VR/AR to create more engaging and effective educational content.

However, a handful of students preferred traditional learning methods, such as reading and discussion, as they were better suited to their learning styles. Statements like "*I think reading is better for me*" and "*I don't think it showed enough information to keep me interested*" point out possible improvements in the activity's design to address different learning preferences.

The VR/AR activity proved to be a practical activity that successfully enhanced students' engagement and enjoyment. It underscored the potential of immersive technologies to make learning more exciting and impactful, while emphasising the need for a more tailored approach to meet diverse student needs.

Table 4.4 shows that the students' environmental awareness was positively influenced through the VR/AR activity, as many students reported having more substantial commitments to practising green behaviours.

Table 4.4 Impact of VR/AR on environmental awareness

Theme	*Description*	*Examples of student responses*
Increased Awareness	Students demonstrated a greater understanding of the impact of pollution.	"Yes, it made me realise how small actions can cause big problems." "Yes, I now see how urgent it is to act against pollution."
Behavioural Motivation	Students were inspired to adopt eco-friendly habits.	"Yes, I want to recycle more now." "Yes, I want to stop using plastic."
Protecting Marine Life	Greater empathy for ocean ecosystems and animals.	"Yes, it made me care more about what happens to sea animals." "Yes, I want to protect the ocean now."
Call to Action	Desire to take an active role, such as participating in beach clean-ups or spreading awareness.	"Yes, I want to help clean beaches." "Yes, I want to spread awareness to my friends and family."

Students reported that the immersive experience deepened their understanding of pollution's consequences, as reflected in responses like, *"Yes, it made me realise how small actions can cause big problems."* Additionally, many students acknowledged the urgency of pollution control, saying, *"Yes, I now see how urgent it is to act against pollution."*

This activity also motivates behavioural changes. Several students expressed intentions to reduce plastic consumption or recycle more with statements like, *"Yes, I want to stop using plastic,"* and *"Yes, I want to recycle more now."* Furthermore, the emotional link created through the activity increased one's concern for marine life, as shown in responses such as, *"Yes, it made me care more about what happens to sea animals."*

Some students were encouraged to carry out direct interventions beyond personal habits, such as organising beach clean-ups and spreading awareness of environmental issues to others. For instance, one participant said, *"Yes, I want to spread awareness to my friends and family."*

In conclusion, the VR/AR learning activity increased environmental awareness among most students and developed a sense of responsibility and action. The immersive format helped familiarise students with

Table 4.5 Challenges in using VR and AR technologies during the activity

Theme	*Description*	*Examples of student responses*
Hardware Challenges	Issues with the VR headset's fit, weight, or adjustments.	"Yes, the headset felt heavy." "Yes, it was hard to adjust the headset to fit properly."
Control Difficulties	Problems navigating or using the controls.	"Yes, the controls were not very intuitive." "Yes, it was hard to control the headset."
Clarity of Display	Concerns about blurry or unclear visuals.	"A little, the images sometimes felt blurry." "Yes, the screen was not clear."
Ease of Use	Positive experiences with ease of use.	"No, everything worked smoothly for me." "No, it was very easy to follow."
Initial Learning Curve	Minor challenges at the beginning, later resolved.	"Not really, but I needed help at the beginning." "A little; it was hard to focus sometimes."

complex ecological issues, although future studies could focus on assessing different prior knowledge levels for maximum impact.

According to Table 4.5, the responses highlight several challenges students face when using VR/AR technology. Common hardware-related problems include a slightly heavy headset that is difficult to adjust. For instance, one student mentioned, "*Yes, the headset felt heavy*," while another stated, "*Yes, it was hard to adjust the headset to fit properly*."

Another frequently cited issue was trouble with control, mainly visual buttons and navigation. Responses like "*Yes, the controls were not very intuitive*" suggest that user-friendly controls could improve the overall experience.

Some students reported experiencing issues with visual clarity, claiming to see screens as either blurry or unclear. Comments like "*A little, the images sometimes felt blurry*" and "*Yes, the screen was not clear*" indicate a need for better visual optimisation.

Despite these difficulties, many students found the experience straightforward once they were guided by these processes. Positive feedback included statements such as "*No, it was very easy to follow*" and "*Not really, but I needed help at the beginning*."

4 Discussion

This study examines how AR and VR can be effectively used for EE. The research focused on studying the opportunities presented by the two technologies in enriching the learning experience and, problematically, their application in classrooms. The findings strongly suggest that AR and VR technologies will revolutionise EE by offering immersive, hands-on experiences that align with the experiential learning theory. These technologies actively engage students, fostering a deep understanding of complex ecological concepts and encouraging reflection, which are key elements in sustaining motivation and developing critical awareness. Furthermore, this transformative process supports shifts in students' perspectives and behaviours, consistent with the transformative learning theory, leading to greater environmental responsibility. This aligns directly with several Sustainable Development Goals (SDGs), particularly SDG 4 (Quality Education) by promoting innovative pedagogy, SDG 13 (Climate Action) by fostering environmental awareness, and SDG 15 (Life on Land) by inspiring attitudes and actions that support ecosystem conservation.

Despite their potential, integrating AR and VR into education presents several significant challenges. The high costs of hardware, software, and maintenance create significant financial barriers for many institutions. Additionally, the need for regular upkeep of software and specialised equipment further strains budgets, making widespread adoption difficult (Cao & Jian, 2023).

In terms of pedagogy, these technologies tend to transform the traditional learning processes. However, successfully applying these requires a defined framework to ensure that they augment education instead of duplicating the traditional ones (Medvedieva & Yamkovenko, 2024). A significant barrier posed by the complexity of AR and VR technologies is that teachers often cannot manage their use because of a lack of experience or training (Cao & Jian, 2023). Many institutions fail to provide adequate professional development and follow-up support, making it challenging for teachers to integrate these technologies into their teaching practices effectively.

Several actionable strategies must be implemented to overcome these challenges. The most vigorous approach is to introduce adequate professional development programmes for teachers. These programs should focus on specific technical skills and teaching methods, thereby giving students the confidence they need. Educators can effectively utilise AR/VR technologies in their EE curricula. The training should orient teachers with the technology, develop lesson plans aligned with the curriculum, and use experiential inquiry learning methods facilitated by AR and VR. Another point to note is that building partnerships with technology providers can lead to cost-effective, user-friendly tools for the school, thereby addressing both its financial and technological needs. Sharing resources, such as open-access AR/VR content, can reduce costs while enhancing the immersive experience.

From a student's perspective, the impact of AR and VR on understanding environmental issues such as pollution has been enormous. These technologies sparked curiosity and a desire for the immersive experience of being within environmental content, as evidenced by their favourable emotional reactions to it as a novel experience and their sense of being more "present" within pollution scenarios.

Focusing specifically on plastic and water pollution, immersive AR and VR experiences have been shown to deepen students' understanding of the environmental impact of pollution. These technologies also create an emotional connection that fosters a greater sense of caring for marine life. These findings suggest that immersive learning can trigger both cognitive and emotional responses, encouraging students to adopt more environmentally friendly behaviours, aligning with research on emotion-driven learning in immersive settings (Ladykova et al., 2024; Ying et al., 2024). In terms of behaviour, many students expressed a strong desire to adopt eco-friendly habits such as reducing plastic use and recycling. This highlights the potential of AR and VR to not only raise awareness, but also inspire real, meaningful actions (Ying et al., 2024).

Students voiced several concerns about using AR and VR in education, including comfort, visual clarity, and complex controls. They suggested making hardware more ergonomic, adding more interactive content, and simplifying controls to make these technologies more accessible and engaging for a broader range of learners. These insights highlight key technical barriers that must be addressed to create smoother and more effective learning experiences with AR and VR.

5 Conclusion

This study demonstrates how AR and VR technologies can transform EE by creating immersive, hands-on experiences that align with experiential learning theory's emphasis on active engagement and transformative learning theory's focus on shifting perspectives and behaviours. These technologies not only increase student motivation and engagement but also facilitate a deeper understanding of complex environmental concepts while fostering environmentally responsible attitudes and behaviours.

Despite these promising outcomes, challenges such as high implementation costs, technical limitations, and insufficient teacher training remain significant barriers to its widespread adoption. Addressing these challenges requires coordinated efforts: investing in professional development programs for educators, fostering partnerships between schools and technology providers, and advocating policies that promote equitable access to immersive learning tools.

Looking forward, AR and VR hold the potential to reshape the entire educational system, making it more interactive, student-centred, and aligned with the SDGs, particularly SDG 4 (Quality Education), SDG 13 (Climate Action), and SDG 15 (Life on Land). Future research should explore the long-term impacts of AR and VR on learners' environmental behaviours, assess their scalability across diverse educational contexts, and investigate their role in fostering interdisciplinary approaches to sustainability education. By continuing to innovate and invest in these tools, educators and policymakers can cultivate environmentally conscious and technologically empowered citizens.

References

Al-Ansi, A. M., Jaboob, M., Garad, A., & Al-Ansi, A. (2023). Analyzing augmented reality (AR) and virtual reality (VR) recent development in education. *Social Sciences & Humanities Open, 8*(1), 100532. https://doi.org/10.1016/J.SSAHO.2023.100532

Ardyansyah, A., & Rahayu, S. (2023). Development and implementation of augmented reality-based card game learning media with environmental literacy for improving students' understanding of Carbon compounds. *Orbital, 15*(2), 118–126. https://doi.org/10.17807/orbital.v15i2.17617

Beetul, A. V., Rajabalee, Y. B., & Santally, M. I. (2023). Augmented, virtual and mixed realities and their potential in teaching and learning: A systematic literature review. *2023 IST-Africa Conference, IST-Africa 2023.* https://doi.org/10.23919/IST-Africa60249.2023.10187875

Cao, F., & Jian, Y. (2023). *The role of AI and virtual reality in fostering environmental awareness and activism in college students.* https://doi.org/10.21203/rs.3.rs-3217197/v1

Cho, Y., & Park, K. S. (2023). Designing immersive virtual reality simulation for environmental science education. *Electronics (Switzerland), 12*(2). https://doi.org/10.3390/electronics12020315

Elmqaddem, N. (2019). Augmented reality and virtual reality in education. Myth or reality? *International Journal of Emerging Technologies in Learning, 14*(3). https://doi.org/10.3991/ijet.v14i03.9289

Imran, M., Almusharraf, N., & Abdellatif, M. S. (2024). Education for a Sustainable Future: The Impact of Environmental Education on Shaping Sustainable Values and Attitudes among Students. *International Journal of Engineering Pedagogy, 14*(6), 155–171. https://doi.org/10.3991/ijep.v14i6.48659

Kavanagh, S., Luxton-Reilly, A., Wuensche, B., & Plimmer, B. (2017). A systematic review of virtual reality in education. *Themes in Science and Technology Education, 10*(2), 85–119.

Kuriakou, G. (2023). Transformative learning in formal schools. *International Journal for Innovation Education and Research, 11*(11), 48–52. https://doi.org/10.31686/ijier.vol11.iss11.4176

Kurniya Rachmawati, A., & Tornando, H. (2024). *Challenges in consumer adaptation to immersive technologies: A comprehensive literature review.*

Ladykova, T. I., Sokolova, E. I., Grebenshchikova, L. Y., Sakhieva, R. G., Lapidus, N. I., & Chereshneva, Y. V. (2024). Augmented reality in environmental education: A systematic review. *Eurasia Journal of Mathematics, Science and Technology Education, 20*(8) https://doi.org/10.29333/ejmste/14914

Lege, R., & Bonner, E. (2020). Virtual reality in education: The promise, progress, and challenge. *JALT CALL Journal, 16*(3), 167–180. https://doi.org/10.29140/jaltcall.v16n3.388

Mahavidyalaya Champadanga, R. (2022). Education and environment HARIPADA DHARA assistant professor in education. *International Journal of Creative Research Thoughts, 10.* www.ijcrt.orgwww.ijcrt.orgc150

Medvedieva, M., & Yamkovenko, V. (2024). Overview of applications with AR and VR technologies in educational activities. *Věda a Perspektivy, 2*(33) https://doi.org/10.52058/2695-1592-2024-2(33)-168-177

Morris, T. H. (2020). Experiential learning – A systematic review and revision of Kolb's model. In *Interactive learning environments* (Vol. 28, Issue 8, pp. 1064–1077). Routledge. https://doi.org/10.1080/10494820.2019.1570279

Muniandy, N. (2024). *Institutions and schools: Overcoming hurdles in integrating Augmented Reality (AR) into stem education, IIP Series, 3*(3).

Ou, K. L., Chu, S. T., & Tarng, W. (2021). Development of a virtual wetland ecological system using VR 360° panoramic technology for environmental education. *Land, 10*(8). https://doi.org/10.3390/land10080829

Ozturk, E., & Akcay, G. (2023). The Eurasia Proceedings of Educational & Social Sciences (EPESS) The Eurasia Proceedings of Educational Can Environmental Education Supported by Augmented Reality (AR) Applications Improve the Environmental Awareness of Primary School Students? *& Social Sciences, 31*. www.isres.org

Porro, S., Spadoni, E., Bordegoni, M., & Carulli, M. (2022). Design of an intrinsically motivating AR experience for environmental awareness. *Proceedings of the Design Society, 2*, 1679–1688. https://doi.org/10.1017/pds.2022.170

Rodrigues, R., & Pombo, L. (2024). The potential of a mobile augmented reality game in education for sustainability: Report and analysis of an activity with the EduCITY app. *Sustainability, 16*(21), 9357. https://doi.org/10.3390/su16219357

Suarlin, S. (2023). Integrating environmental education to form environmental care characters in schools. *Advances in Community Services Research, 1*(2), 47–56. https://doi.org/10.60079/acsr.v1i2.335

Udaya Shankar, A., Tewari, V., Rahman, M., Mishra, A., & Bajaj, K. (2023). Impact of Virtual Reality (VR) and Augmented Reality (AR) in education. *TuijinJishu/Journal of Propulsion Technology, 44*(4).

Velempini, K. (2025). Assessing the role of environmental education practices towards the attainment of the 2030 sustainable development goals. *Sustainability (Switzerland), 17*(5). https://doi.org/10.3390/su17052043

Ying, T. W., Alias, N., & DeWitt, D. (2024). Sustainable environmental education using virtual reality: A module for improving environmental citizenship competences in secondary schools. *Eurasia Journal of Mathematics, Science and Technology Education, 20*(10), 1–15. https://doi.org/10.29333/ejmste/15177

CHAPTER 5

Green Guidance: Reconciling Individual Career Pathways and Sustainability Challenges. Exploratory Study with Guidance Counselors

Mohamed Haba and Amal Azeroual

1 Introduction

The current state of the world is marked by considerable and rapid progress on both economic and social fronts. Volatility, Uncertainty, Complexity, and Ambiguity, which are popularly referred to as VUCA, summarize these disruptions (Malaval, 2018). This scenario in turn adversely affects ecological equilibrium. Extreme forms of pollution, along with drought, and the destruction of the ecosystem have an enormous negative impact on the population's health and wellbeing and it will take more than noble sentiments to safeguard the planet. This is why educators must also rise to the occasion and help society transition to a sustainable future. The collective action spearheaded by the United Nations in pursuit of the Sustainable Development Goals (SDGs) provides the basic

M. Haba • A. Azeroual (✉)
Center for Guidance and Educational Planning, Rabat, Morocco
e-mail: mohamed.haba@usmba.ac.ma; amal.azeroual@um5r.ac.ma

K. Hirech (ed.), *Environmental Education and Sustainable Development*, https://doi.org/10.1007/978-3-032-05760-0_5

structure within which economic, social, and environmental efforts can be formulated.

Thus, education can be a powerful tool in increasing learner's ecological awareness and responsible environmental behavior. Students are aroused by the information services, campaigns, and education provided for them related to environmental issues. In schools, environmental education should not be solely focused on one subject but rather taught as an interdisciplinary and everyday issue that concerns all life and future existence on earth. Moreover, School and Career Guidance (SCG) as a subsystem of education has to engage in these efforts to preserve the environment. This problem, in substance, touches various disciplines; hence, it is a national obligation to work on progress in an integrated manner. It is accepted and can be understood that the educational achievements and career opportunities of learners should arouse their interests. However, every one of these routes should include the influence of ecological factors. It has to be emphasized that man's role should not be reduced to responding to society's challenges, but rather, solving ecological problems should be part of global planning. SCG thus is described as lever environmental control concerning SCG is directed toward enabling students to select sustainable strategies and understand the consequences of their professional choices (Bodoira & Rochat, 2021).

"Green Guidance," developed by Peter Plant in 2014, offers a practical approach to fully integrating ecological concerns into educational and career pathways (Plant, 2020). This concept emphasizes raising awareness among individuals about the environmental impacts of their educational and career choices, promoting paths aligned with sustainability principles. It also encourages career counselors to incorporate ecological considerations when supporting learners in developing their personal and professional projects. While career choices reflect an individual's self-commitment, shaped by their past experiences, present circumstances, and future aspirations (Haissat, 2016), SCG has traditionally been rooted in psychological frameworks. Professional practices in this field have consistently drawn upon psychology its methods and theoretical models, whether scientific, applied, differential, clinical, or pedagogical while maintaining ties to socio-economic realities (Boutet, 1999). However, it would be beneficial to question the relevance of individual choices through lenses beyond this "psychologization" of SCG. Indeed, an individual's educational and career trajectory is not only influenced by subjective identity processes but also shaped by mechanisms deeply tied to established norms and the immense, ongoing transformations of modern society (Guichard, 2004). In this context, socio-economic and, above all, environmental factors deserve particular attention in pursuing realistic and sustainable career choices.

While the Green Guidance approach offers an environmental perspective to align career choices with sustainability principles, it remains marginal and underexplored (Rochat, 2021), particularly in contexts like Morocco, where ecological challenges are especially acute. This limited adoption highlights the underutilization of the potential environmental contribution of SCG. This raises a critical question:

To what extent do Moroccan career counselors apply the Green Guidance approach to help learners align their individual choices with environmental needs?

The primary objective of this article is to address this question by pursuing the following goals:

- Highlight the awareness among career counselors regarding the importance of the Green Guidance approach.
- Exploring the potential for integrating Green Guidance into the SCG system in Morocco by identifying environmentally friendly professional practices of career counselors.
- Identifying obstacles that need to be overcome to better integrate the Green Guidance approach into counseling practices.
- Providing recommendations to assist career counselors in incorporating ecological dimensions into their professional practices.
- Through this exploration, the article aims to contribute to developing a more sustainable SCG system that aligns with the demands of sustainable development and prepares individuals to face future ecological challenges.

2 Related Works

2.1 *Career Choices as a Result of the Individual-Context Relationship*

To counter the limitations of the "psychologized" approaches of SCG, career decisions are increasingly seen as resulting from a combination of individual and environment interplay. Different explanations and theories have been proposed to add context to SCG. Among the theories are Vondracek's developmental contextual perspective, Young's contextual action theory as well as Patton's systems theory. These theories all explain how people make career choices while taking into account the context in which they live (Hooley & Rice, 2019).

According to the developmental contextual approach formulated by Vondracek and colleagues, vocational behavior results from interactions between individuals and the various social contexts in which they operate. This theory posits that career choices cannot be understood without accounting for cultural, social, and economic influences that shape individuals' opportunities and aspirations. It emphasizes that career development is a dynamic process evolving and influenced by external factors such as educational policies, sociocultural norms, and social networks (Vondracek & Porfeli, 2008).

Young's Contextual Action Theory also highlights the role of context in career development. This approach asserts that vocational actions are intrinsically linked to specific goals and projects within a given social framework. Young argues that vocational behavior is motivated by personal goals and the social meanings attached to these behaviors. This theory thus provides an integrative perspective, where individual actions are seen as being shaped by both internal factors (such as beliefs and emotions) and external factors (such as societal expectations) (Valach & Young, 2002).

In the same vein, the Systems Theory Framework of career development, developed by Patton and McMahon, views career development as a systemic process where various elements interact, including the individual, their environment, and the opportunities available (Patton & McMahon, 2014). From this perspective, the individual is an open system interacting with other systems (such as family, education, or the labor market), allowing for an exploration of how these interactions influence career decisions and how they evolve over time (Young & Mundy, 2023).

Finally, while it is commonly accepted that school and career choices are inherently individual, the theoretical approaches of contextual development, contextual action, and systems provide a useful framework for understanding how individual decisions are made within social, professional, and economic contexts. These choices thus become the result of a complex interaction between the individual and their environment (Guichard, 2013). By incorporating these different perspectives into the construction of career and educational choices, it becomes possible to understand better how individuals navigate their academic and professional paths while accounting for the varied contextual influences that shape their decisions (Gati et al., 2020).

These theories also emphasize the importance of adopting a holistic approach to SCG, one that considers not only individual aspirations but also the impact of social and cultural contexts. Furthermore, the environmental context should not be overlooked in this analysis, highlighting the need for guidance practices that integrate sustainability and ecological awareness into decision-making processes.

2.2 *The Environmental Context in Search of Sustainability*

In a world facing increasing environmental challenges, sustainability has become a central concern for both businesses and individuals. The alarming ecological situation calls for a profound transformation of human behavior at all levels, including in educational and professional spheres. The current environmental context is notably characterized by:

Global climate crises: Rising global temperatures, extreme weather events, and sea level increases represent major threats to ecological balance. These phenomena largely stem from greenhouse gas emissions linked to human activities, particularly in the industrial and energy sectors (Calvin et al., 2023).

Depletion of natural resources: The stress exerted on natural resources (water, soil, forests, biodiversity) by extractive industries and unsustainable consumption models undermines ecosystems' ability to regenerate. The Living Planet Index (WWF, 2022) highlights a significant decline in biodiversity due to unsustainable practices in sectors such as agriculture and fishing.

Widespread pollution: Pollution of air, soil, and water severely affects both human and ecological health. Plastics, toxic chemicals, and industrial discharges continue to contaminate global ecosystems, disproportionately impacting the most vulnerable populations.

It is important to highlight that environmental impacts are not evenly distributed. Poorer communities often bear the brunt of environmental degradation while having limited access to resilience solutions, thereby exacerbating social injustices (Menton et al., 2020).

In the face of these challenges, calls for a transition toward sustainable models are growing louder. This transition is not limited to public policies and businesses but also extends to individuals as economic actors and citizens. The SDGs, adopted by the United Nations, provide a critical framework to guide these efforts, emphasizing the interconnectedness of environmental sustainability, social equity, and economic prosperity.

2.3 *Green Guidance as the SCG's Response to the Environmental Situation*

Individuals' career choices play a pivotal role in the quest for environmental sustainability, as they directly influence economic structures and consumption practices. Several dimensions of this impact deserve exploration:

Ecological footprint of economic sectors: Economic sectors vary significantly in their environmental impact. For instance, the oil industry, air transport, and intensive agriculture generate high carbon emissions and significantly contribute to climate change. Conversely, sectors such as renewable energy, circular economy, and clean technologies foster sustainability.

Influence of career choices on consumption and production models: Professionals working in non-sustainable industries shape global production and consumption cycles. For example, a career in the fast-fashion industry supports resource exploitation and pollution, whereas involvement in sustainable industries promotes eco-responsible practices.

Importance of innovation in career choices: Individuals pursuing sustainability-focused careers, such as environmental engineering, waste management, or agro ecology, contribute directly to developing innovative solutions to reduce environmental impacts. This highlights the critical role of education and awareness in guiding career decisions.

Increased reliance on values in professional decision-making: Ethical and environmental criteria are increasingly integrated into career choices. A Deloitte survey (2022) revealed that young professionals prioritize companies with sustainable policies, prompting organizations to adopt more responsible practices. To make this fruitful connection between career choices and sustainability a reality, the education system is called upon to play a key role in raising awareness among students, the professionals of tomorrow, about environmental challenges. This goal can be achieved by integrating sustainability programs into school, university, and vocational curricula, helping students understand how their career choices can contribute to a more sustainable future (Carney, 2022). This approach enables learners to adopt proactive strategies to address environmental challenges and secure a sustainable future for future generations.

To better describe the integration of ecological concerns into career guidance processes, Peter Plant developed the concept of Green Guidance in 2014. This innovative approach emphasizes that school and career choices can no longer be driven solely by personal or socio-economic ambitions but must also account for environmental impacts. Within this framework, Green Guidance offers a holistic vision that links individual development to the preservation of ecosystems, while promoting career trajectories aligned with the goals of global sustainability (Plant, 2021). Concretely, Green Guidance provides a powerful framework for rethinking the relationship between career trajectories and environmental sustainability. It calls on career guidance and education professionals to raise awareness among individuals about the environmental consequences of their choices and to equip them with the tools needed to contribute positively to the ecological transition. By adopting a green perspective, career guidance becomes a key lever for promoting responsible behaviors and decisions (Chang & Mann, 2024). Its contributions can take several forms, such as:

Raising Awareness of Sustainable Careers: One of Green Guidance's fundamental missions is to introduce learners to sustainable economic sectors and professions (e.g., renewable energy, waste management, organic farming) or the development of eco-friendly products. By highlighting these opportunities, Green Guidance paves the way for career choices aligned with the objectives of reducing ecological footprints.

Integrating Environmental Considerations into Education and Training Pathways: Green Guidance emphasizes the importance of incorporating ecological dimensions from the earliest stages of planning educational and career pathways. This involves assisting learners in selecting training programs that foster the acquisition of sustainability-related skills (e.g., eco-design, resource management, green business practices). Through this approach, individuals will be better prepared to meet the growing demands of green job markets.

Promoting a Sustainability-Oriented Logic in Professional and Educational Decisions: Green Guidance encourages individuals to adopt a long-term perspective in their decision-making. This includes considering the social and environmental impacts of their choices and valuing professional practices that address sustainability challenges. In this sense, career counselors can support learners in evaluating how a position or industry contributes to reducing carbon emissions or preserving local ecosystems.

In conclusion, it is important to highlight that, unlike traditional career guidance approaches, which primarily focus on economic and professional success, Green Guidance offers an expanded definition of success by incorporating environmental impact alongside psychological and social well-being. This approach promotes a more balanced and sustainable conception of individual success. It should be noted that Green Guidance is far more than just an adaptation of career guidance practices to address ecological concerns; it represents a paradigm shift in how individuals perceive their relationship with the educational and professional worlds. For educational systems and career guidance professionals, adopting this perspective is an opportunity to prepare generations capable of addressing tomorrow's challenges.

3 Materials and Methods

3.1 Methods

Given that the Moroccan educational context is unfamiliar with the Green Guidance approach, the present study adopts an exploratory descriptive design. The exploratory aspect aims to characterize Green Guidance as the subject of study, making it more comprehensible and familiar. The information collected on the experiences of guidance counselors, who form the study population, is presented numerically and through descriptive graphs illustrating relationships between variables.

3.2 Materials and Survey Tool

The descriptive method, as a research approach, involves observing and describing the behavior, characteristics, or conditions of a specific population or phenomenon without altering the environment, meaning no variables are manipulated. Thus, the present study will focus on presenting and analyzing the collected data for statistical purposes. Data collection will be conducted through a field survey using a questionnaire comprising about ten questions distributed across four main axes.

This study employs a quantitative approach, relying on measurable data collected from a sample of participants using structured instruments. The responses will be coded and subjected to statistical analysis to identify trends, relationships, and patterns within the data. The choice of a quantitative method is particularly appropriate for obtaining objective and generalizable results, especially in an exploratory context where the phenomenon under investigation, Green Guidance, is still relatively unknown in the Moroccan educational landscape.

Axis 1: This first axis aims to gather personal and professional information about the respondents. It includes details such as the respondent's gender, years of experience as a guidance counselor, the subject they taught before attending the Center for Guidance and Educational Planning (COPE), and the type of district where they practice:

Axis 2: This section includes questions designed to assess guidance counselors' awareness of Green Guidance and document the respondents' professional practices related to SCG's environmental role.

Axis 3: This axis focuses on questions that reflect the obstacles hindering the integration of Green Guidance into the SCG.

Axis 4: The final section of the questionnaire is dedicated to the proposals from guidance counselors on measures that could be implemented in schools to integrate Green Guidance into their professional practices.

3.3 *Survey Population and Sample*

While the subject of this study could concern various educational stakeholders (guidance counselors, inspectors, administrative staff, students, etc.), given the adoption of the educational approach within the SCG framework in the Moroccan model, the focus was placed on the population that works directly with students. Thus, the target population of the survey consists of guidance counselors operating in school districts.

The study sample includes 207 education guidance counselors, of whom 144 are men, representing 69.6%, and 63 are women (30.4%). The majority of respondents (111 counselors) fall within the 35–45 age group, while 89 counselors are aged between 45 and 55 years (Table 5.1).

From a professional perspective, it is noteworthy that the vast majority of the samples have less than five years of experience as professional guidance counselors. This methodological choice is based on the assumption that newly trained guidance counselors may have had the opportunity to familiarize themselves with new approaches to guidance, including the Green Guidance approach:

Regarding the workplace, the data in Table 5.2 below show that 77 respondents, representing 37.2%, work in predominantly rural districts, while 74 guidance counselors, or 35.7%, are employed in schools located

Table 5.1 Distribution of the study sample by gender and seniority

Gender	*Seniority*			*Total*
	Less than 5 years	*5 to 10 years*	*11 years and more*	
Male	60.9%	0.5%	8.2%	69.6%
Female	27.5%	1.9%	1.0%	30.4%
Total	88.4%	2.4%	9.2%	100.0%

Source: Field survey

Table 5.2 Distribution of the sample by subject taught and district of practice

Subject taught	*Type of district*			*Total*
	Rural	*Urban*	*Other*	
Professional subjects	39	41	32	112
Non-professional subjects	38	33	24	95
Total	77	74	56	207

Source: Field survey

in urban districts. The remainder of the sample, comprising 56 guidance counselors, operates in peri-urban areas or in districts that include both rural and urban schools.

Another parameter highlighted is the subject taught before taking on the role of Guidance Counselor. In the study sample, 112 Guidance Counselors, representing 54.1% of respondents, taught subjects that can be considered as having an environmental component (Languages, Life and Earth Sciences, History-Geography, Islamic Education, etc.). This figure includes 71 respondents, or 34.29%, who were bilingual teachers in primary education, responsible for teaching all subjects, including those closely related to environmental topics. Additionally, 41 secondary school teachers taught subjects that can be described as "pro-environmental." The remaining 95 Guidance Counselors taught subjects that did not necessarily convey environmental content.

3.4 Data Analysis Techniques

The study adopts an exploratory approach, which necessitates the use of descriptive statistical techniques to analyze and interpret the data collected. These techniques serve to organize and simplify raw data, making it easier to present in a clear and accessible format for better understanding. Through this method, the study provides a comprehensive summary of the demographic characteristics, professional practices, and perspectives of guidance counselors concerning the central theme of the research.

4 Results

The findings from the analyses are presented in a manner that addresses the research objectives. First, the environmental role of SCG is clarified by exploring the awareness of guidance counselors and their professional practices. Next, the obstacles to be overcome for more effective integration of the "Green Guidance" approach into counseling practices are identified. Finally, recommendations are proposed to help guidance counselors incorporate the ecological dimension into their professional practices.

4.1 The Environmental Role of SCG Between Guidance Counselors' Awareness and Their Professional Practices

To gauge the level of awareness among guidance counselors regarding ecological issues, emphasis was placed on the subject they taught before joining the COPE. This approach is based on the idea that the subject taught serves as an indicator reflecting the nature of knowledge conveyed by the respondent in their professional practice.

Table 5.3 presents data related to the environmental backgrounds of the respondents. The underlying idea of environmental engagement originates from awareness, as it drives behaviors and life choices. The awareness of the surveyed guidance counselors may be the result of a range of preliminary activities. Based on the table's data, a significant proportion of respondents, exceeding 41%, have engaged either in school life activities related to environmental protection, civil activities through membership in associations working in this field, or scientific activities by participating in conferences addressing environmental issues. It is worth noting that teachers of subjects considered pro-environmental are more actively

Table 5.3 Proportion of Career Counselors involved in ecological actions based on the subject taught before the integration of the new framework

Subject taught	*Club facilitation*	*Association membership*	*Conference participation*
Professional subjects	58.2%	42.0%	41.9%
Non-professional subjects	41.8%	58.0%	58.1%

Source: Field survey

Table 5.4 Career Counselors' perceptions of the environmental role of the SCG: From idea to action

Seniority	*Awareness*	*Knowledge*	*Action organization*
Less than 5 years	96.2%	13.1%	10.9%
More than 5 years	91.7%	8.3%	20.8%

Source: Field survey

involved in the school sphere (58.2%), while other teachers are more present in the civil (58.0%) and scientific (58.1%) spheres.

If the engagement of Guidance Counselors in ecological actions before joining the COPE can be considered significant, as shown by the data in the previous table, this may explain their awareness of the environmental role of SCG. Indeed, more than 90% of respondents report being aware of this role. Seniority within the field does not create a notable difference among the respondents. However, the transition to action is lacking, as less than 14% of Guidance Counselors report being familiar with "Green Guidance." Approximately 10% of newly recruited counselors and 20% of others organize meaningful activities for learners to help them build personal projects that incorporate the environmental dimension (Table 5.4).

Despite the scarcity of actions related to the "Green Guidance" approach, as evidenced by the data in the previous table, some initiatives and guidance interventions with an environmental background, even implicitly, can be identified. The theme of "Information and Awareness" takes center stage. Thus, 145 Guidance Counselors, representing 70% of respondents, organize information sessions on pro-environmental careers. In contrast, 118 respondents work to raise awareness among learners about the importance of incorporating the environmental dimension into the process of building their personal projects. Other professional practices with an environmental focus involve only about 32% of the respondents (Fig. 5.1).

The data from Table 5.5 provide insights into the conviction of the surveyed Guidance Counselors regarding the importance and utility of integrating environmental issues into supporting students' personal projects. More than 60% of respondents, regardless of gender or the subjects they taught, believe that integrating environmental concerns into the guidance process would be highly beneficial for the effectiveness of the guidance and counseling system.

According to the respondents, this integration could enhance the quality of career choices related to the environment. It could also help learners build self-confidence as active contributors to their surroundings through the development of personal projects. These projects would take into account the rights of future generations, thereby jointly contributing to environmental protection and, in turn, to the sustainability of their professional futures.

4.2 *Factors Hindering the Integration of Green Orientation in Counseling Practice*

Figure 5.2 illustrates the opinions of Guidance Counselors regarding the obstacles the SCG system must overcome to better adopt Green Guidance. The primary obstacle, cited by 43.5% of respondents, is the insufficiency of educational offerings related to environmental fields. This is followed

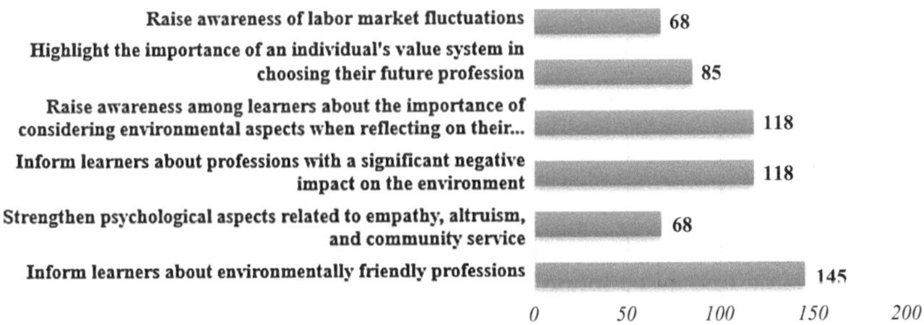

Fig. 5.1 Engagement of career advisors in pro-environmental professional practices. (Source: Field surveCy)

Table 5.5 Perceptions of respondents on the usefulness of integrating environmental issues into the process of supporting students' personal projects

		No advantage	*No opinion*	*Great advantage*	*Total*
Gender	Male	7.6%	30.6%	61.8%	100.0%
	Female	3.2%	27.0%	69.8%	100.0%
Subject taught	Professional subjects	7.1%	24.1%	68.8%	100.0%
	Non-professional subjects	5.3%	35.8%	58.9%	100.0%

Source: Field survey

by a lack of continuous training, noted by 40.6% of respondents, to enhance the skills of Guidance Counselors in new internationally adopted approaches. This challenge stems from the low priority given to integrating environmental issues into regional training plans for continuous education.

Additionally, the lack of available information on environment-related careers such as pedagogical resources tailored to sustainability and aligned with SCG programs was highlighted by 34.8% of respondents. Finally, approximately 29.5% of the surveyed counselors pointed out that the initial training provided at the COPE is insufficient, as it does not focus on recent advancements in the SCG domain nor offer tools to integrate environmental concerns into guidance practices.

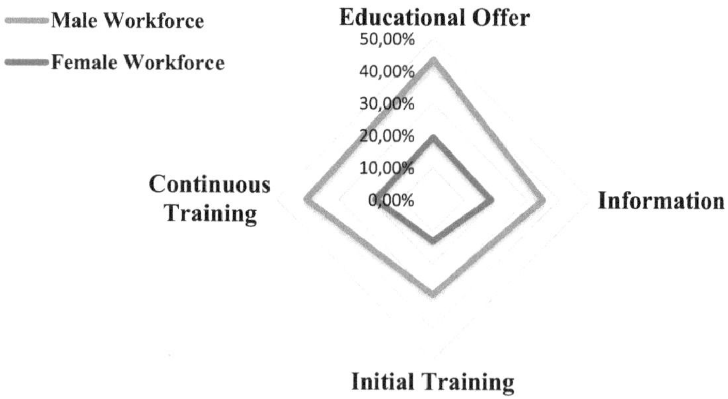

Fig. 5.2 Obstacles to overcome for better adoption of Green Orientation according to career advisors. (Source: Field survey)

Beyond these widely recognized obstacles, some respondents raised other issues, such as the high counselor-to-student ratio, which makes it challenging to plan orientation assistance activities effectively. They also pointed out the absence of an integrated school project addressing learner-centered concerns, such as environmental topics. Moreover, some respondents emphasized that the role of the Guidance Counselor is comprehensive and not limited to a specific professional domain, which leads to Green Guidance not being considered a priority in their work.

4.3 Recommendations from Guidance Counselors on Measures to Facilitate the Integration of Environmental Considerations into Career Guidance Practices

In the Moroccan SCG system, the school orientation district, which typically includes one to three schools, is regarded as a localized structure providing support and guidance to students in developing their personal projects. As such, the measures proposed by respondents for integrating Green Guidance into the SCG focus primarily on local-level interventions, i.e., within schools.

The data from Fig. 5.3 below reveals that respondents, regardless of gender, generally agree on the types of measures needed to promote Green Guidance locally. According to the chart, 183 respondents (88.40%) believe that establishing environmental clubs could serve as a practical tool for implementing Green Guidance. These school-based extracurricular entities highlight the professional dimension of their activities. Additionally, 158 respondents (76.32%) see the provision of informational materials about environment-related careers as a key initiative. This measure would require the development of guidance resources tailored to students' needs,

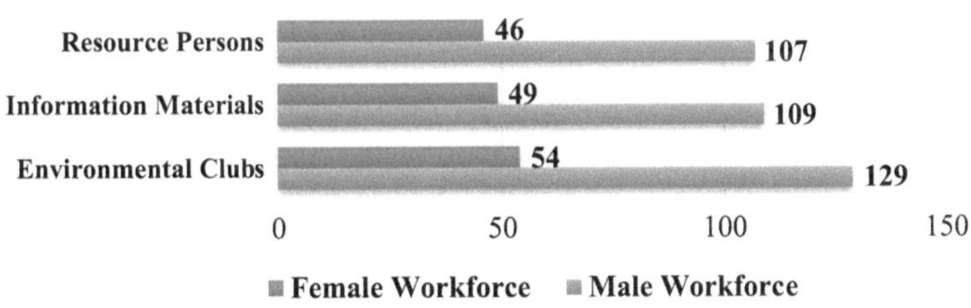

Fig. 5.3 Measures to be taken at the school level for the implementation of Green Orientation according to respondents. (Source: Field survey)

along with collaboration with specialized institutions and organizations to create relevant content.

The third measure, supported by 153 respondents (73.91%), involves opening the school to its surrounding community by hosting resource persons working in environment-related professions. These professionals would share their experiences and insights with students, thus reinforcing the environmental contribution of the SCG.

Other proposed measures, while significant, did not achieve unanimous agreement. These include:

- Incorporating the orientation dimension, including Green Guidance, into the school's integrated project. This could involve incentive policies to encourage schools to adopt this new approach and foster partnerships with their surrounding environment to pool sustainable initiatives.
- Highlighting environmental training programs and careers in information and orientation forums.
- Integrating environmental education into school curricula and teaching materials.

5 Discussion

The present research aimed to examine how Moroccan Guidance Counselors understand and implement the Green Guidance approach to help students reconcile their individual choices with environmental needs. Specifically, it sought to explore these educational actors' awareness and knowledge of this approach, to identify the professional practices they develop over time in this regard and to capture their perspectives on the challenges hindering its implementation, as well as the measures to be undertaken for the future. The originality of this study lies in its focus on the significance of innovation and the adoption of new practices in advancing the field of SCG.

In this context, the findings reveal two main observations. First, regardless of their environmental backgrounds, Guidance Counselors are generally aware of the importance of the Green Guidance approach in aligning students' individual and career choices with the environmental concerns of their communities, despite the recorded shortfall in concrete actions and professional practices. Second, beyond their awareness or even their commitment to this new approach, the participants highlight elements of the educational context that hinder its integration within the SCG, along with measures that could facilitate its adoption.

5.1 *Actors More Aware Than Committed to Green Guidance*

The interest shown by the Moroccan educational system in environmental issues highlights that most surveyed Guidance Counselors have some environmental background, whether through associative or school contributions to the collective effort of environmental protection. Consequently, they welcome the potential of the Green Guidance approach, acknowledging its transformative power as it embodies Nietzsche's principle of "self-transcendence through the pursuit of lofty goals and the means to achieve them" (Bernaud, 2016). Their openness to this approach is significant, regardless of gender, subject taught, professional seniority, or location of the practice.

However, the transition from awareness to action remains weak among the respondents. They are, therefore, more aware than committed, with their recorded pro-environmental professional practices generally confined to the level of information dissemination. This represents the preliminary stage in helping students construct their personal projects. This situation aligns with the perspective of Plan, who posits that Green Guidance could "ideally be proactive, thought-provoking, prompting reflection and action, but perhaps with a higher level of commitment." Without concrete commitment from actors, the potential benefits of the approach may not be fully realized.

With such a low level of tangible ecological engagement through SCG Systems, these field actors are still far from giving this theme the attention it deserves. While Western societies grapple with ecological issues under the mounting pressure of eco-anxiety "an anticipatory anxiety expressed as worry about the future and a chronic fear of environmental collapse" (Clayton, 2020), the current discourse in the Moroccan context remains centered on awareness. Moreover, there are difficulties in conceptualizing environmental commitment through the lens of social justice. Sustainability and social justice are interlinked, as career choices, no matter how individual, have broader societal implications (Irving & Malik, 2005). Green Guidance elevates these career choices, steering them toward a higher level of personal commitment, societal involvement, and meaning.

5.2 *Constraining Educational Context, Yet Open to Innovation in SCG*

In addition to the ambivalence between awareness and commitment among Guidance Counselors regarding Green Guidance, the constraints of the educational context hinder the valorization of this new approach.

While the precariousness of the educational offering in terms of pro-environmental training represents a significant barrier to any green approach innovation, it is essential to note that environmental issues and concerns about sustainable development transcend borders. Morocco, like other countries, must take steps to adapt its training system to environmental demands.

Efforts must also be directed toward training Guidance Counselors, whether through initial training at the COPE or through continuous training. Currently, the training model for Guidance Counselors remains rooted in psychology. This adherence explains some perspectives that view the Green Guidance approach as directive, as it challenges current practices and potentially the traditional Rogerian client-centered framework.

6 Limitations and Practical Implications of the Study

Although the results of the study can be considered satisfactory, it is important to acknowledge certain limitations encountered. First, the study was conducted with a sample of 207 Guidance Counselors, the majority of whom had less than five years of experience. This limits the ability to generalize the findings, particularly to other stakeholders involved in the field of SCG system, such as Orientation Inspectors or even educational actors within schools. Second, the research tool used as a questionnaire was designed primarily to explore field realities. However, this approach is susceptible to potential errors in its design and development. Thus, this work should be regarded as a preliminary attempt to probe a relatively underexplored area in the Moroccan educational system.

As for the practical utility of the results of the study, it should be noted that the Personal Student Project approach constitutes the cornerstone of the SCG system in Morocco. Consequently, the role of educational actors, including Guidance Counselors, is to support students through various tools and activities. In the context of collective efforts to protect the environment, the surveyed Guidance Counselors have expressed a favorable attitude toward integrating the "Green Guidance" approach into their professional activities. As career planning interventions are increasingly expected to encourage behaviors that foster student engagement (Fouad et al., 2016), this study reflects the commitment of Guidance Counselors to contribute concretely to this objective. By structuring career-planning efforts to promote environmentally responsible behaviors, Guidance Counselors can play a key role in aligning individual student goals with broader societal and environmental needs.

7 Conclusion

Despite its exploratory nature, this study has shed light on the degree of openness within the Moroccan educational context in general, and the field of SCG in particular, regarding the integration of new approaches to professional practice, such as Green Guidance. This approach aims to provide an orienting response to the challenges of ecological transition and the demands of sustainable development. While the majority of Career Guidance Counselors are convinced of the value and utility of integrating environmental issues into the process of supporting students' personal projects through the Green Guidance approach, it would be beneficial for the Moroccan SCG system to encourage joint efforts among all concerned educational stakeholders to overcome the obstacles hindering the success of this innovation.

Finally, it is important to highlight that while the study's results have demonstrated the potential and challenges of integrating Green Guidance into the SCG system, the success of this innovation depends on the implementation of various support measures. These include:

- Providing new work tools compatible with this approach for Career Guidance Counselors in school districts;
- Enhancing the training of actors involved in the SCG field to ensure better integration of the approach;
- Collaborating with stakeholders in the field of environmental protection to assist schools with expertise related to environmental themes.

Acknowledgment The authors express their sincere gratitude to Professor Abdelfettah Daoudy, an ESL teacher at Amideast and a researcher in the field of Applied Linguistics at Mohammed V University, for proofreading this article.

References

Bernaud, J. L. (2016). Le « sens de la vie » comme paradigme pour le conseil en orientation. *Psychologie Française, 61*(1), 61–72.

Bodoira, X., & Rochat, S. (2021). Une intervention basée sur les « besoins du monde » pour favoriser le sens de la vie et du travail dans l'insertion professionnelle de jeunes vulnérables. *Orientation Scolaire et Professionnelle, 50*(3), 451–467.

Boutet, M. (1999). Psychologie et orientation: une relation étroite au vécu éclaté. Le cas du conseiller d'orientation psychologue. *Travaux & documents. Psychologie et éducation, 12*, 81–96.

Calvin, K., et al. (2023). *IPCC, 2023: Climate change 2023: Synthesis report.* Contribution of Working Groups I, II and III to the Sixth Assessment Report of the Intergovernmental Panel on Climate Change [Core Writing Team, H. Lee and J. Romero (Eds.)]. IPCC, Geneva, Switzerland.

Carney, S. (2022). Reimagining our futures together: A new social contract for education. *Comparative Education, 58*(4), 568–569.

Chang, Y., & Mann, A. (2024). *Enhancing green career guidance systems for sustainable futures.* OECD Education Working Papers, No. 318, OECD Publishing.

Clayton, S. (2020). Climate anxiety: Psychological responses to climate change. *Journal of Anxiety Disorders, 74*, 102263.

Fouad, N. A., Ghosh, A., Chang, W. H., Figueiredo, C., & Bachhuber, T. (2016). Career exploration among college students. *Journal of College Student Development, 57*(4), 460–464.

Gati, I., Levin, N., & Landman-Tal, S. (2020). Decision-making models and career guidance. In *International handbook of career guidance* (pp. 115–145). Springer International Publishing.

Guichard, J. (2004). Se faire soi. *L'orientation scolaire et professionnelle, 33*(4), 499–533.

Guichard, J. (2013). *Career guidance, education, and dialogues for a fair and sustainable human development.* Inaugural conference of the UNESCO chair of Lifelong guidance and counselling, Wroclaw, Poland.

Haissat, S. (2016). *La notion d'identité personnelle en sociologie.* www.revue-interrogations.org

Hooley, T., & Rice, S. (2019). Ensuring quality in career guidance: A critical review. *British Journal of Guidance & Counselling, 47*(4), 472–486.

Irving, B. A., & Malik, B. (2005). *Critical reflections on career education and guidance: Promoting social justice within a global economy.* RoutledgeFalmer.

Malaval, C. (2018). *Le temps digital de l'entreprise. Médium, 54*(1), 139–148.

Menton, M., et al. (2020). Environmental justice and the SDGs: From synergies to gaps and contradictions. *Sustainability Science, 15*(6), 1621–1636.

Patton, W., & McMahon, M. (2014). *Career development and systems theory: Connecting theory and practice* (Vol. 2). Springer.

Plant, P. (2020). Paradigms under pressure: Green guidance. *Nordic Journal of Transitions, Careers and Guidance, 1*(1), 1–9.

Plant, P. (2021). Utopia revisited: Green guidance. *Journal of the National Institute for Career Education and Counselling, 47*(1), 53–58.

Rochat, S. (2021). Éco-orientation: quelles interventions pour quelles problématiques ? *L'Orientation Scolaire et Professionnelle, 50*, 565–590.

Valach, L., & Young, R. A. (2002). Contextual action theory in career counselling: Some misunderstood issues. *Canadian Journal of Counselling, 36*(2), 97–112.

Vondracek, F. W., & Porfeli, E. J. (2008). Social contexts for career guidance throughout the world. Developmental-contextual perspectives on career across the lifespan. In J. A. Athanasou & R. Van Esbroeck (Eds.), *International handbook of career guidance*. Springer.

WWF. (2022). *Living Planet Report 2022 – Building a nature-positive society* (R. E. A. Almond, M. Grooten, D. Juffe Bignoli, & T. Petersen, Eds.). WWF.

Young, R. A., & Mundy, A. K. S. (2023). Action and the active life: The contribution of contextual action theory to current career development challenges. *Australian Journal of Career Development, 32*(2), 85–93.

CHAPTER 6

Learning Science in Family Spaces: Engaging Maghrebi Mothers and Children for a Sustainable Future

Mounia Rachdi Errachdi

1 Introduction

Science education has traditionally been conceptualized as a process that unfolds primarily within formal school settings, where learning follows structured curricula and standardized assessments. However, this perspective overlooks the rich and diverse learning opportunities offered by non-formal educational spaces, which play a crucial role in the development of competencies and the construction of scientific knowledge, particularly in early childhood (Rogoff & Mejía-Arauz, 2022; Brooker et al., 2014). These spaces are especially relevant in multicultural contexts, where families bring unique cultural trajectories that enrich the educational experience (Esteban-Guitart & Moll, 2014; Hedges, 2022).

Non-formal educational spaces such as family support centers, community organizations, and science outreach programs offer flexible and

M. R. Errachdi (✉)
University of Vic – Central University of Catalonia (Uvic_UCC), Vic, Spain
e-mail: mounia.rachdi@uvic.cat

K. Hirech (ed.), *Environmental Education and Sustainable Development*, https://doi.org/10.1007/978-3-032-05760-0_6

culturally responsive environments for learning. Unlike formal institutions, these spaces are not bound by rigid structures, allowing them to adapt to the needs and cultural backgrounds of participating families. The Nana Space, the focus of this study, exemplifies such an environment. It is a family support center that has evolved into an inclusive educational setting where Maghrebi mothers and their children engage in science-related activities, particularly in the field of environmental education. Within this context, the study addresses two central research questions:

1. What role do Maghrebi mothers play in their children's science learning in non-formal educational spaces?
2. What barriers and facilitators shape their participation in these learning processes?

This research adopts a qualitative ethnographic-narrative methodology to highlight the lived experiences of participating mothers, illustrating how learning interactions are co-constructed and negotiated in culturally diverse settings (Riessman, 2008; Hammersley & Atkinson, 2019). The study is grounded in a robust theoretical framework that integrates perspectives from inclusive education, intercultural pedagogy, and situated learning, aiming to reveal how science education can become more accessible and socially meaningful for migrant families.

1.1 Educational Spaces and Cultural Contexts

Educational spaces are broadly classified as formal, non-formal, and informal (UNESCO, 2012; Rogoff, 2003):

- Formal spaces such as schools and universities are characterized by structured curricula, certified educators, and standardized assessments (Bernstein, 2000).
- Informal spaces, including homes and community interactions, provide spontaneous learning opportunities embedded in daily life and cultural routines.
- Non-formal spaces, like Nana Space, occupy an intermediary position: learning is intentionally organized but flexible, responding to the rhythms and interests of participants (Hedges, 2022).

In culturally diverse societies, non-formal spaces create bridges between home and institutional learning, fostering recognition of families' cultural resources. As González-Falcón (2021) and Ainscow and Booth (2020) suggest, inclusion in education requires more than removing physical or administrative barriers: it involves a cultural transformation that values the knowledge and experiences of all participants.

1.2 The Cultural Backpack and Migrant Families

A key concept underpinning this study is the "cultural backpack" the set of values, knowledge, and practices that each individual carries, shaping their approach to learning and education (Lozano-Pérez & Trinidad-Requena, 2019). For migrant families, this cultural backpack is marked by displacement, adaptation, and the negotiation of identity across multiple cultural contexts.

In the case of Maghrebi mothers, their participation in science activities is mediated by:

- barriers such as language limitations, initial mistrust of institutions, and the perception of science as a distant, specialized domain;
- facilitators including cultural mediation, the acknowledgment of prior knowledge, and flexible activities that resonate with their lived experiences (Baquedano-López et al., 2013; Revuelta, 2016; Brown et al., 2020).

Recognizing and valuing the cultural backpack of families is essential for designing inclusive pedagogical strategies that integrate environmental awareness without imposing assimilation into models misaligned with participants' identities.

1.3 Situated and Inclusive Science Learning in Early Childhood

Childhood is not merely a preparatory stage for adulthood but a period of active agency in which children construct meaning through interaction with their environment and significant adults (Rogoff & Mejía-Arauz, 2022; Lave & Wenger, 1991). Within Nana Space, children engage in hands-on science experiences observing, experimenting, and hypothesizing while strengthening their connection to environmental sustainability through practical activities such as recycling.

From a situated learning perspective, knowledge is co-constructed through participation in meaningful social practices. In Nana Space, learning occurs in dialogic processes that involve children, mothers, and educators, moving beyond the unidirectional transmission of knowledge. This aligns with the funds of knowledge framework (Esteban-Guitart & Moll, 2014), which emphasizes integrating the sociocultural experiences of families into educational practices.

Finally, this study contributes to the debate on inclusive and intercultural science education. Involving Maghrebi mothers in their children's scientific learning not only strengthens home-school connections but also transforms non-formal spaces into accessible, equitable, and socially relevant learning environments (Aguado & Sleeter, 2021; Hedges, 2022; Sterling, 2020). This approach aligns with the Sustainable Development Goals by fostering environmental literacy, social inclusion, and community-based educational innovation.

2 Methodology and Research Development

This research is framed within an interpretative and constructivist perspective, which enables an in-depth understanding of social phenomena from the standpoint of the participants themselves. This approach is particularly relevant for studies examining learning processes in culturally diverse contexts, as it captures the complexity of human interactions without reducing them to predefined categories (Creswell & Poth, 2018; Flick, 2018).

A qualitative methodology was adopted, following the principles of Bonilla-Castro (2005), prioritizing the understanding of specific cases over statistical generalizations, and emphasizing contextual richness and depth of individual experiences. This methodological approach aligns with the study's focus on the lived experiences of Maghrebi mothers in non-formal educational settings, where social meaning is co-constructed through interaction.

2.1 Research Design

The study employs a narrative ethnographic design, integrating the thick description characteristic of ethnography (Hammersley & Atkinson, 2019) with the interpretive depth of narrative inquiry (Riessman, 2008).

- Ethnography enables the documentation of the meanings participants attribute to their practices and contexts.
- Narrative analysis facilitates the construction of accounts that reflect participants' voices, experiences, and interpretations of science learning and environmental engagement.

This dual approach is ideal for exploring how Maghrebi mothers navigate their children's science education, as their participation is shaped by cultural heritage, migratory trajectories, and institutional relationships.

The research was conducted at Nana Space, a family support center in Catalonia, during six months of fieldwork, and involved:

- 10 Maghrebi mothers (aged 25–42)
- 12 children (aged 3–8)
- 3 educators/facilitators specialized in early childhood and environmental science activities

Participation was voluntary, and pseudonyms were used to ensure anonymity.

2.2 Research Objectives

The study was guided by four interrelated objectives:

1. Analyze the participation of Maghrebi mothers in their children's science education in non-formal spaces, with a focus on environmental science.
2. Identify the barriers and opportunities that shape family engagement, considering both individual and structural factors.
3. Explore pedagogical strategies that strengthen the inclusion of migrant families in community-based science learning.
4. Reflect on the role of educators as mediators in fostering environmental literacy and family participation.

These objectives aim to generate applicable knowledge for inclusive and sustainable educational practices in similar multicultural contexts.

2.3 Data Collection

Data collection combined multiple qualitative techniques to ensure triangulation (Patton, 2014; Flick, 2018):

1. Participant Observation – 10 structured sessions documenting interactions during science activities.
2. Semi-Structured Interviews – Conducted with mothers before and after the program to explore perceptions of science and participation.
3. Focus Groups – Facilitated collective reflection on challenges, opportunities, and inclusion strategies.
4. Field Journal & Visual Data – Reflexive notes, photographs, and videos captured context and non-verbal dynamics.

This multi-method approach allowed for a comprehensive and nuanced understanding of family participation and learning processes.

2.4 Data Analysis

Data were analyzed through thematic and narrative coding, following an iterative process (Braun & Clarke, 2019):

- Initial coding identified recurring patterns of participation, barriers, and opportunities.
- Axial coding connected individual experiences to broader cultural and pedagogical themes.
- Triangulation of observations, interviews, and focus groups strengthened validity.
- Member checking with participating mothers enhanced credibility and reflexivity.

This approach generated a contextualized interpretation of how science learning and environmental awareness are co-constructed in non-formal, culturally diverse spaces.

2.5 *Ethical Considerations*

The study followed the ethical standards of qualitative research (Tracy, 2020) and was approved by the Ethics Committee of the University of Vic – UCC (Code 241/2022).

- Informed consent was obtained from all participants.
- Confidentiality and anonymity were ensured.
- Reflexivity was practiced, recognizing the dual role of the researcher as both educator and observer.

This methodological design captures the complexity of family participation in science learning and aligns with the principles of inclusive and intercultural education, producing situated knowledge that contributes to sustainability education and environmental literacy in non-formal contexts.

3 Data Collection Strategies

The fieldwork was conducted at Nana Family Support Center, a non-formal learning environment where Maghrebi mothers and their children engage in science outreach and environmental education activities. Data collection took place over a six-month period and employed multiple complementary strategies to capture the depth and complexity of participants' experiences. This multi-method approach ensured methodological triangulation, enhancing the credibility and richness of the study (Creswell & Poth, 2018; Flick, 2018).

3.1 *Participant Observation*

Participant observation was the primary strategy for understanding the dynamics of science learning in the Nana space.

- A total of more than ten observation sessions were conducted, each lasting approximately 90–120 minutes.
- Observations documented interactions between mothers, children, and educators, focusing on:
  - mothers' levels of engagement in science activities;
  - mediation and scaffolding strategies used by educators;

- child–adult and peer interactions, including non-verbal communication and emotional expressions.

The researcher's active yet reflexive participation allowed the capture of spontaneous dynamics and provided a thick description of how environmental science learning unfolded in a flexible, non-structured context (DeWalt & DeWalt, 2011). Field notes included contextual details, descriptive narratives, and analytic reflections.

3.2 Semi-structured Interviews

To complement the observations, ten in-depth semi-structured interviews were conducted with Maghrebi mothers:

- Interviews explored perceptions of science and environmental education, roles in children's learning, and barriers to participation.
- Sessions lasted 45–60 minutes and were conducted in Spanish or with cultural mediation in Arabic when needed.
- Interviews were audio-recorded and transcribed verbatim to preserve the richness of participants' narratives, capturing tone, hesitation, and emotion.
- These interviews provided first-person accounts that revealed how cultural backgrounds and migratory experiences influence engagement with science education (Kvale & Brinkmann, 2021).

3.3 Focus Groups

Two focus groups, each with 6–8 participating mothers, were organized to promote collective reflection:

- Discussions encouraged the sharing of experiences, co-construction of knowledge, and identification of common barriers and facilitators in accessing science education.
- Focus groups lasted approximately 90 minutes and were audio-recorded with participants' consent.
- This group format revealed how social interaction and peer exchange shaped meaning-making processes, providing insight into shared cultural narratives (Morgan, 2019).

3.4 Field Journal and Reflexive Documentation

A field journal was maintained throughout the research to support reflexivity and analytic depth (Tracy, 2020):

- Daily entries documented observations, emergent themes, and reflections on the researcher's dual role as educator and ethnographer.
- Critical notes recorded emotional responses, instances of trust-building, and contextual factors affecting participation.
- Reflexive analysis helped identify how the researcher's positionality influenced both interactions and interpretations.
- Additionally, visual and audio documentation enhanced data collection:
- Photographs and short video clips captured non-verbal communication, gestures, and environmental interactions.
- WhatsApp communication with some participants (voluntary) supported follow-up reflections and participant feedback.
- This integration of multimodal evidence enriched the ethnographic narrative and supported the triangulation of observations, interviews, and focus groups.

3.5 Triangulation and Credibility

Combining participant observation, interviews, focus groups, field journal entries, and visual/audio records allowed for methodological triangulation:

- Cross-verification among data sources increased the credibility and trustworthiness of the findings (Patton, 2014).
- The use of member checking with participating mothers ensured that interpretations were aligned with their lived experiences.

By capturing individual narratives, collective reflections, and visual context, this multi-method strategy produced a comprehensive understanding of how Maghrebi mothers engage in science and environmental learning within non-formal educational spaces.

4 Data Analysis

The data analysis followed a systematic thematic and narrative coding process, integrating inductive and deductive approaches to capture the complexity of participants' experiences. This process aligned with the study's interpretative perspective and the narrative ethnographic design, allowing findings to emerge directly from the participants' voices while being informed by the research objectives (Braun & Clarke, 2019; Riessman, 2008).

4.1 *Analytical Approach*

An inductive thematic analysis was conducted to identify recurring patterns and themes across the various data sources, while also using a deductive lens to ensure alignment with the four guiding research objectives. The process involved multiple iterative steps:

1. Data Familiarization – Repeated reading of observation notes, interview transcripts, focus group discussions, and field journal entries.
2. Initial Coding – Generation of open codes that captured key moments, experiences, and reflections related to family participation in science and environmental learning.
3. Categorization and Theme Development – Codes were grouped into broader categories connected to the study objectives.
4. Triangulation and Cross-Verification – Patterns were cross-checked across interviews, observations, focus groups, and visual records to enhance credibility.

The main analytical categories were the following:

- Mothers' Participation – Levels of engagement, motivations, and evolving roles in science activities.
- Barriers and Opportunities – Structural and cultural factors influencing inclusion in non-formal educational spaces.
- Role of Educators – Mediation strategies and their impact on building an inclusive learning environment.
- Perceptions of Science and Environmental Education – How mothers conceptualize science and its relevance to their children's learning.

4.2 *Triangulation and Reflexivity*

Triangulation was fundamental for ensuring the trustworthiness of the findings (Patton, 2014):

- Observations provided evidence of interaction dynamics and non-verbal communication.
- Interviews and focus groups revealed participants' subjective experiences and collective reflections.
- Field journal and visual records added contextual richness and supported reflexive interpretation.

The researcher maintained a reflexive stance, continuously examining how her positionality as both educator and ethnographer influenced data interpretation. Member checking with mothers further validated that the final narratives authentically represented their experiences.

4.3 *Ethical Considerations in Analysis*

The study was approved by the Ethics Committee of the University of Vic – Central University of Catalonia (Code 241/2022). Ethical practices during analysis included:

- informed consent for all interviews, observations, and visual recordings;
- anonymity and confidentiality, using pseudonyms in transcripts and reports;
- respectful representation of participants' voices to minimize interpretative bias and avoid imposing external assumptions.

This multi-layered, narrative-driven analytical process allowed the study to generate situated and culturally responsive knowledge, highlighting how Maghrebi mothers co-construct science learning and environmental literacy in non-formal educational contexts.

5 Methodological Contributions

The methodology adopted in this research has provided a deep and nuanced understanding of the complexities surrounding Maghrebi mothers' participation in non-formal learning spaces, particularly in the field of science and environmental education. By combining ethnography and narrative inquiry, the study captured the lived experiences, voices, and reflections of participants, offering a holistic view of their engagement and the contextual factors shaping it.

5.1 Value of the Ethnographic-Narrative Approach

The ethnographic-narrative design allowed the research to:

- identify barriers such as linguistic challenges, institutional mistrust, and the perception of science as distant or specialized;
- highlight opportunities including cultural mediation, flexible activity planning, and the recognition of families' cultural backgrounds;
- capture multi-layered realities, integrating individual stories with collective reflections to better understand social dynamics in non-formal learning contexts (Riessman, 2008; Hammersley & Atkinson, 2019).

This methodological perspective foregrounds participants' agency and cultural knowledge, offering a culturally responsive framework for analyzing inclusion in science learning.

5.2 Multi-method Data Collection and Triangulation

The use of diverse and complementary data collection techniques participant observation, semi-structured interviews, focus groups, reflexive field journaling, and visual documentation provided a comprehensive perspective of the phenomenon under study. This multi-method strategy facilitated:

- Triangulation of data sources, increasing the credibility and depth of analysis (Creswell & Poth, 2018; Patton, 2014).
- The identification of convergences and divergences between individual experiences and collective narratives.

- The integration of reflexivity, acknowledging the researcher's dual role as educator and ethnographer in the interpretation of findings.

By leveraging triangulation and reflexive practices, the methodology ensured that the voices of the participants were represented accurately and respectfully, minimizing interpretative bias.

5.3 *Implications for Inclusive Science Education*

The methodological contributions of this study extend beyond its immediate context:

- It demonstrates how narrative ethnography can generate situated, culturally sensitive knowledge in multicultural, non-formal educational spaces.
- It provides a framework for designing inclusive pedagogical strategies that encourage family engagement in science and environmental education.
- It offers actionable insights for educators and policymakers aiming to promote equity, participation, and sustainability in early childhood and community-based science learning.

By emphasizing the integration of cultural diversity and environmental literacy, this methodological approach supports the creation of more equitable, context-responsive educational practices aligned with the principles of inclusive education and the Sustainable Development Goals.

6 Research Findings

This chapter presents the findings of the research conducted at the Nana Family Support Center, a non-formal pedagogical space that has emerged as a meeting point between science education, environmental responsibility, and the cultural identity of migrant families. Drawing on a narrative ethnographic approach, the results are organized into key dimensions that highlight both the voices of the participants and the transformative dynamics of the space.

The findings reveal how Maghrebi mothers and their children engage with science and environmental education in ways that are deeply shaped by cultural identity, family practices, and the inclusive strategies implemented in the Nana space. They also uncover the challenges, opportunities, and pedagogical implications for promoting family participation in non-formal educational settings.

6.1 My Personal Journey: Reflections from the Research

My dual role as researcher and educator in the Nana space, combined with my identity as a Maghrebi woman, became central to the process of analysis and knowledge generation. This positionality enabled me to build trust with participants, facilitating their willingness to share experiences, while also prompting self-reflection on my own assumptions about science education in migrant communities.

From the beginning of the fieldwork, I observed that mothers related to me not only as a researcher but as someone who shared linguistic, cultural, and migratory experiences. This created a safe and empathetic environment, which allowed for rich narrative data collection. At the same time, my positionality required reflexivity, as I navigated the tension between being part of the community and maintaining analytic distance.

Through this process, I realized that science, for many of these mothers, is not solely academic knowledge; it is intertwined with daily life, family responsibilities, and intergenerational learning. Recognizing this insight was fundamental for understanding how to bridge home and scientific learning in culturally diverse contexts.

6.2 The Nana Space: A Meeting and Learning Place

The Nana space has evolved from a family support initiative into an inclusive, non-formal educational setting where families explore science and environmental issues through hands-on, culturally responsive activities.

Initially, many mothers were hesitant to actively participate in science activities, perceiving science as distant or inaccessible. Over time, trust-building, culturally sensitive mediation, and contextualized pedagogy transformed the dynamics:

- Mothers began to actively engage in experiments and environmental activities.
- Activities connected science to everyday life: food, health, recycling, gardening, and environmental care.

- The space fostered a sense of belonging, encouraging mothers to see themselves as co-learners alongside their children.

This transformation illustrates how non-formal educational environments can bridge cultural and institutional gaps, serving as entry points for migrant families into broader educational and environmental engagement.

6.3 The Mothers: Voices and Experiences

Mothers were active agents in the learning process, and their narratives revealed diverse trajectories of engagement with science education:

- Initial perceptions: Science was associated with formal schooling and expert knowledge, perceived as external to their cultural backgrounds.
- Evolving participation: Through the Nana activities, mothers discovered that science is embedded in daily life and can be a tool for supporting their children's learning.
- Empowerment: Participation fostered confidence and self-efficacy, with mothers expressing a newfound ability to guide and accompany their children in scientific exploration.
- Despite these advances, several barriers were consistently identified:
- Language limitations hindered comprehension of some scientific concepts.
- Time constraints due to work and family responsibilities limited attendance.
- Residual perceptions of inaccessibility persisted, reflecting the distance between formal science and cultural experience.

To address these barriers, strategies were implemented, including bilingual materials, flexible scheduling, and culturally relevant content, reinforcing the inclusivity of the space.

6.4 Children as Protagonists in Learning

Children demonstrated curiosity, creativity, and agency in the learning process from the first sessions. Their engagement manifested in:

- spontaneous questioning and exploration of materials and natural phenomena;

- hands-on experiments and environmental activities, such as recycling and gardening;
- peer and family learning, where children often explained concepts to their mothers, fostering bidirectional learning.

One of the most remarkable findings was how children became mediators of science within their families. They transferred knowledge acquired in Nana to the home context, transforming science into shared family practice and reinforcing environmental awareness.

Learning at Nana thus emerged as a collective and dialogic process, where knowledge flows not only from educator to child but among all participants, generating a community of practice that embodies the principles of inclusive and situated learning (Lave & Wenger, 1991; Rogoff & Mejía-Arauz, 2022).

6.5 *Transformative Potential of Non-formal Spaces*

The findings underscore that non-formal educational spaces like Nana can:

- promote science and environmental literacy in migrant communities;
- strengthen family-school connections by empowering parents to engage with their children's learning;
- support social integration and cultural recognition, fostering a sense of belonging for families in educational ecosystems.

By embracing intercultural, inclusive, and sustainability-oriented pedagogies, the Nana space demonstrates the transformative potential of non-formal learning environments in advancing environmental responsibility and social equity.

References

Aguado, T., & Sleeter, C. (2021). *Intercultural education: Developing equity and social justice in diverse societies*. Routledge.

Ainscow, M. (2015). The index for inclusion. In *Struggles for equity in education* (pp. 95–105). Routledge.

Ainscow, M., & Booth, T. (2020). *The index for inclusion: Developing learning and participation in schools*. Routledge.

Auerbach, S. (2007). From moral supporters to struggling advocates: Reconceptualizing parent roles in education through the experience of working-class families of color. *Urban Education, 42*(3), 250–283. https://doi.org/10.1177/0042085907300433

Baquedano-López, P., Alexander, R. A., & Hernandez, S. J. (2013). Equity issues in parental and community involvement in schools: What teacher educators need to know. *Review of Research in Education, 37*(1), 149–182. https://doi.org/10.3102/0091732X12459718

Bernstein, B. (2000). *Pedagogy, symbolic control, and identity.* Rowman & Littlefield.

Bonilla-Castro, E. (2005). *Más allá del dilema de los métodos: La investigación en ciencias sociales.* Editorial Norma.

Braun, V., & Clarke, V. (2019). Reflecting on reflexive thematic analysis. *Qualitative Research in Sport, Exercise and Health, 11*(4), 589–597. https://doi.org/10.1080/2159676X.2019.1628806

Brooker, L., Blaise, M., & Edwards, S. (2014). *The Sage handbook of play and learning in early childhood.* SAGE.

Brown, A., Roche, J., & Hurley, M. (2020). Engaging migrant and refugee communities in non-formal science learning spaces. *Journal of Science Communication, 19*(4), R01. https://doi.org/10.22323/2.19040601

Creswell, J. W., & Poth, C. N. (2018). *Qualitative inquiry and research design: Choosing among five approaches* (4th ed.). SAGE.

Denzin, N. K., Lincoln, Y. S., MacLure, M., Otterstad, A. M., Torrance, H., Cannella, G. S., Koro, M., & McTier, T. (2017). Critical qualitative methodologies: Reconceptualizations and emergent construction. *International Review of Qualitative Research, 10*(4), 482–498.

DeWalt, K. M., & DeWalt, B. R. (2011). *Participant observation: A guide for fieldworkers* (2nd ed.).

Edwards, P. A. (1995). Empowering low-income mothers and fathers to share books with young children. *The Reading Teacher, 48*(7), 558–564.

Esteban-Guitart, M., & Moll, L. C. (2014). Funds of identity: A new concept based on the funds of knowledge approach. *Culture & Psychology, 20*(1), 31–48. https://doi.org/10.1177/1354067X13515934

Flick, U. (2018). *An introduction to qualitative research* (6th ed.). SAGE.

González-Falcón, I. (2021). Culturally responsive teaching in diverse classrooms. *Journal of Intercultural Education, 32*(4), 451–467. https://doi.org/10.1080/14675986.2021.1931071

Guralp, G., McHugh, M., & Hayes, S. (2024). Non-formal science education: Moving towards more inclusive pedagogies for diverse classrooms. *CEPS Journal, 14*(1), 105–142. https://doi.org/10.26529/cepsj.1707

Hammersley, M., & Atkinson, P. (2019). *Ethnography: Principles in practice* (4th ed.). Routledge.

Hedges, H. (2022). *Children's learning in informal and non-formal contexts.* Routledge.

Kvale, S., & Brinkmann, S. (2021). *InterViews: Learning the craft of qualitative research interviewing* (4th ed.). SAGE.

Lave, J., & Wenger, E. (1991). *Situated learning: Legitimate peripheral participation.* Cambridge University Press.

Lozano-Pérez, C., & Trinidad-Requena, A. (2019). La mochila cultural: Identidad y aprendizaje en contextos migrantes. *Revista de Educación Intercultural, 21*(2), 45–60.

Manrique, A. L., Dirani, E. A., Frere, A. F., Moreira, G. E., & Arezes, P. M. (2019). Teachers' perceptions on inclusion in basic school. *International Journal of Educational Management, 33*(2), 409–419. https://doi.org/10.1108/IJEM-02-2018-0058

Morgan, D. L. (2019). *Basic and advanced focus groups.* SAGE.

Morin, G., & Marwah, S. (2024). Engaging in emotional and cultural work: Migrant mothers' experiences of school-based sexuality education. *Health Education Journal, 83*(1), 45–58. https://doi.org/10.1177/00178969241281271

Patton, M. Q. (2014). *Qualitative research & evaluation methods* (4th ed.). SAGE.

Revuelta, P. (2016). La construcción de la confianza en contextos educativos interculturales. *Revista de Ciencias Sociales, 32*(1), 95–112.

Riessman, C. K. (2008). *Narrative methods for the human sciences.* SAGE.

Rogoff, B. (2003). *The cultural nature of human development.* Oxford University Press.

Rogoff, B., & Mejía-Arauz, R. (2022). *Learning together: Children and families in cultural context.* Oxford University Press.

Sterling, S. (2020). Transformative learning for sustainable futures. *Sustainability, 12*(6), 2490. https://doi.org/10.3390/su12062490

Tracy, S. J. (2020). *Qualitative research methods: Collecting evidence, crafting analysis, communicating impact* (2nd ed.). Wiley.

UNESCO. (2012). *International Standard Classification of Education: ISCED 2011.* UNESCO Publishing.

PART II

Environmental Challenges and Green Education Roles

CHAPTER 7

Environmental Education and Climate Change Awareness: Toward Sustainable Change

Kamal Hirech, *Mohammed Rhiat*, *Rachid Hjaji*, *Abdelkader Hammaoui*, *and Hicham Berbar*

K. Hirech (✉) • M. Rhiat
Higher School of Education and Training, Mohammed First University, Oujda, Morocco
e-mail: k.hirech@ump.ac.ma; mohammed.rhiat@ump.ac.ma

R. Hjaji
Mohammed First University, Oujda, Morocco

A. Hammaoui
Higher School of Technology, Mohammed First University, Oujda, Morocco
e-mail: abdelkader.hammaoui@ump.ac.ma

H. Berbar
Normal Superior School of Tétouan, Abdelmalek Essaâdi University, Tetouan, Morocco
e-mail: h.berbar@uae.ac.ma

K. Hirech (ed.), *Environmental Education and Sustainable Development*, https://doi.org/10.1007/978-3-032-05760-0_7

1 Introduction

In recent years, climate change has emerged as an unprecedented threat to ecological, economic, and social systems worldwide (Abbass et al., 2022). Rising average temperatures, melting glaciers, rising sea levels, and the intensification of extreme weather events such as hurricanes, droughts, and floods underline the seriousness of this environmental crisis (World Health Organization, UNEP United Nations Environment Programme, and World Organisation for Animal Health, 2022). These disruptions are mainly linked to greenhouse gas emissions, whether of natural origin or as a result of human activities, including the intensive use of fossil fuels, deforestation, and unsustainable production and consumption patterns (Dreyfus et al., 2022).

The effects of climate change go beyond environmental damage; they also pose significant socio-economic challenges, exacerbating inequalities, threatening food and water security, and increasing climate-induced migration (Williams et al., 2021). In this global context, and given the planetary urgency, collective and proactive action is essential to mitigate the effects of climate change and build resilient and sustainable societies (Mehan, 2023).

While political, technological, and economic responses are at the heart of mitigation and adaptation strategies, the role of education in combating climate change is often underestimated (Borde et al., 2022). Education plays an essential role in improving the understanding of the causes, consequences, and possible solutions to this phenomenon. It also fosters the adoption of responsible, environmentally friendly behavior and encourages civic commitment to preserving and protecting the environment (Kurup et al., 2021).

Environmental and climate education is not just about acquiring scientific knowledge about natural phenomena and their environmental impacts. It also aims to transform the attitudes, values, and practices of individuals and communities (Trott, 2022). By raising awareness from an early age, incorporating participatory pedagogical approaches, and strengthening problem-solving skills, education can become a powerful tool for mobilizing current and future generations (Shlobin et al., 2021). Initiatives such as school curricula, awareness campaigns, and community projects help to instill a culture of sustainability in everyday practices and mindsets (Oe et al., 2022).

Climate education is particularly important in regions that are highly vulnerable to climate change, such as North Africa and the MENA region. Communities in these regions are already facing impacts such as soil erosion, depletion of water resources, and forced displacement. Raising awareness of climate change can strengthen their resilience, providing them with tools to adapt to new climatic realities and promoting local solutions, which often draw on traditional knowledge and modern innovations (Leal Filho et al., 2022).

Globally, many international initiatives recognize the fundamental role of education in the fight against climate change. The 2015 Paris Agreement, through its Article 12, stresses the importance of improving public education, awareness, and participation in climate issues (Segger, 2016). Similarly, the United Nations Sustainable Development Goals (SDGs), in particular SDG 4 (Quality Education) and SDG 13 (Climate Action), emphasize the need to strengthen education for sustainable development (Nazar et al., 2018). However, implementing these commitments remains a challenge, particularly in developing countries where education systems often lack adequate resources and infrastructure (Browning et al., 2021).

Beyond institutional frameworks, climate education requires active collaboration between governments, non-governmental organizations, academic institutions, and the private sector (Dzvimbo et al., 2022). This synergy makes it possible to create relevant educational content, train qualified educators, and develop innovative methodologies adapted to local contexts. In addition, the use of technology offers immense potential for broadening access to information and enhancing learning capacities on climate change (Zhang et al., 2022). Online platforms, interactive applications, and digital communication campaigns make it possible to reach diverse audiences and deliver messages tailored to different age groups and levels of education (Kaplan & Haenlein, 2016).

This chapter aims to explore how education can become a catalyst for change to address climate challenges. It first examines the global context of climate change and the challenges it poses to individuals and communities. It then analyzes the role of education in transforming behavior and promoting sustainable solutions. Finally, it highlights exemplary educational initiatives and offers recommendations for strengthening their impact.

The aim is to demonstrate that education is not just about imparting knowledge, but is a driving force for inspiring collective action and building a more sustainable future. In a world increasingly faced with complex environmental challenges, investing in climate education is not only a strategic response, but also a moral obligation to future generations.

2 Fundamental Principles and Importance of Environmental Education

2.1 Importance of Environmental Education

Environmental education is based on the idea that understanding ecosystems, human interactions with the environment, and their impacts is essential to fostering sustainable and responsible behavior (Uralovich et al., 2023). It is built around key principles (Fig. 7.1):

- Providing clear scientific information about the causes, effects, and potential solutions to climate change.
- Helping individuals analyze complex environmental issues and design appropriate responses.
- Encouraging responsible attitudes, such as climate justice and intergenerational equity.

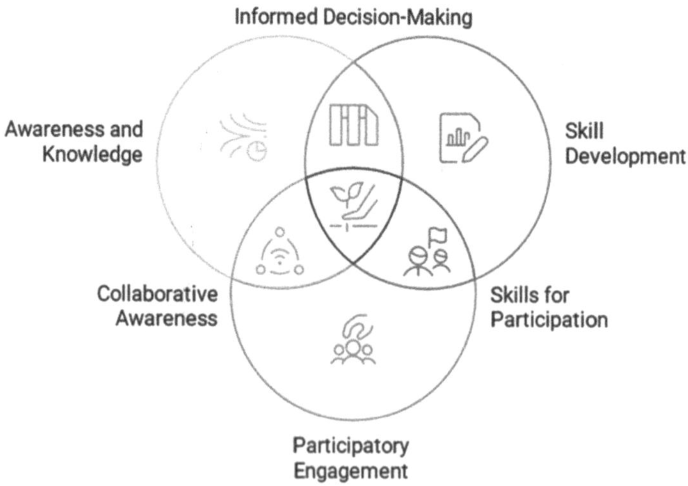

Fig. 7.1 Pathways to sustainable behavior

- Promoting cooperation between individuals and communities to drive collective environmental action.

These principles underline the crucial role of environmental education in meeting climate challenges and preserving ecosystems. It's not just about imparting knowledge, but also about shaping behavior and inspiring structural transformations toward sustainability (Yadav et al., 2022).

2.2 *Education as a Driver of Youth Mobilization*

Environmental education goes beyond the transfer of knowledge; it seeks to turn younger generations into active agents of change. Strengthening young people's leadership on climate issues is essential, as evidenced by global movements such as student climate strikes. By giving them a good education, they acquire the tools they need to understand the challenges and advocate solutions to political decision-makers (Sloam et al., 2022).

Encouraging action through concrete projects such as tree planting, waste reduction, or community awareness campaigns allows young people to put into practice what they have learned while seeing the real impact of their efforts.

In addition, environmental education helps to promote a culture of sustainability by integrating eco-friendly practices into schools, such as recycling and the use of renewable energies. These initiatives change social norms and influence behavior beyond the classroom. Developing critical thinking skills is also fundamental: young people need to learn how to analyze complex climate challenges, differentiate scientific fact from misinformation, and make a meaningful contribution to public debate.

3 The Climate Change and the Need for Awareness

3.1 *Impact of Climate Change on a Global and Local Scale*

Climate change represents a significant global threat that affects ecosystems, economies, and human societies. On a global scale (Fig. 7.2), the rapid rise in temperatures leads to concerning phenomena such as glacier melting, rising sea levels, and increasingly frequent natural disasters like hurricanes, floods, and prolonged droughts. These disturbances disrupt

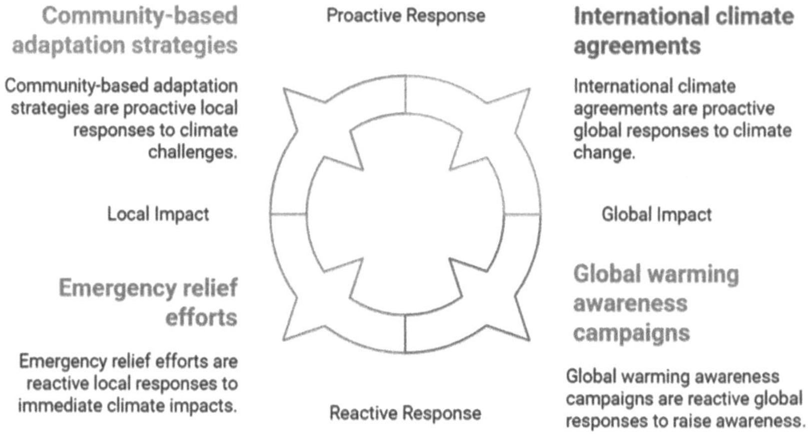

Fig. 7.2 Climate change impact and response matrix

ecosystems, reduce biodiversity, and threaten food and water security in many regions around the world (Jha & Dev, 2024).

At the local level, the impacts of climate change vary according to the geographical, social, and economic characteristics of the affected areas. Coastal regions, for example, suffer from soil erosion and the intrusion of saltwater into groundwater, threatening freshwater resources. In arid areas, prolonged drought exacerbates water scarcity, severely affecting agriculture and the populations that depend on it. Urban areas, on the other hand, face heatwaves intensified by the urban heat island effect, endangering the health of residents, especially the most vulnerable, such as children and the elderly.

These global and local impacts highlight the magnitude of the challenges posed by climate change and call for collective and coordinated actions to mitigate its effects and adapt to new environmental realities.

3.2 *Urgent Need for Collective Awareness and Immediate Action*

In the face of the magnitude of climate challenges, collective awareness and swift action are indispensable to mitigate damage and adapt to inevitable changes (Perdomo, 2024). Education plays a vital role in informing the population about the causes and consequences of climate change while encouraging more sustainable behaviors. Enhancing educational programs and awareness campaigns is vital to reaching a diverse audience.

At the same time, promoting locally adapted solutions provides effective responses to the specific challenges facing communities, while contributing to the reduction of greenhouse gas emissions. Citizen participation is also essential to pressure decision-makers (Yadav, 2025).

Governments and companies must also take responsibility for the environmental transition by adopting ambitious policies that align with global climate goals. This includes tangible commitments to reduce emissions, invest in clean technologies, and finance adaptation initiatives in the most vulnerable areas.

Climate change is an emergency that requires coordinated responses at both the local and global levels.

4 Integrative Approaches to Climate Change Education and Public Engagement

4.1 *Educational Initiatives for Climate Awareness*

Educational programs and awareness initiatives play a key role in combating climate change by transferring knowledge, developing skills, and encouraging teamwork. As examples:

- In Germany, educational programs for high school students include practical workshops on renewable energy. These workshops allow young people to build small models of solar panels and wind turbines, providing a tangible understanding of sustainable technologies (Amalu et al., 2023).
- In the Sub-Saharan Africa region, where local initiatives such as the Farmer Field Schools program teach farmers agricultural practices suitable for local climatic conditions, improving food security and reducing greenhouse gas emissions (Osumba et al., 2021).
- Furthermore, in Asian cities like Jakarta and Manila, non-governmental organizations organize awareness campaigns that combine artistic events and public discussions to raise awareness about the risks of flooding due to rising sea levels (Geldin, 2022).

These examples illustrate how educational programs, designed to suit different contexts, can mobilize diverse audiences, from students to local communities, to effectively address climate challenges.

4.2 *Complementary Approaches to Climate Education*

Environmental and climate change education relies on two complementary approaches: combining formal and non-formal education to maximize impact and reach a broad audience.

- Formal education integrates climate issues into school and university curricula, raising awareness among young people from an early age (Okada & Gray, 2023). For example, interdisciplinary courses allow students to study the impacts of CO2 emissions in science while exploring climate justice solutions in geography or economics. Laboratories and practical projects, such as air quality measurement or visits to solar farms, provide students with concrete experiences to apply their knowledge to real-world issues.
- At the same time, non-formal education complements these efforts by reaching audiences outside academic settings. Community workshops, organized in cultural centers or local associations, raise awareness among adults about climate issues and propose solutions adapted to their lifestyles (Becher, 2024). Awareness campaigns, conducted in public spaces or on social media, reach a diverse audience, fostering collective awareness.

Combining these two approaches creates a continuous, accessible, and engaging learning framework capable of sustainably mobilizing different communities around environmental issues.

4.3 *Technological and Media-Driven Approaches to Environmental Education*

Modern technologies and media play a vital role in expanding the scope of environmental awareness initiatives. These tools provide powerful means to educate, motivate, and connect individuals worldwide, enhancing collective mobilization in response to climate challenges.

- Mobile applications and online platforms stand out as key channels. Interactive tools, such as Earth School or Climate Action Now, allow users to learn the basics of climate change while engaging in tangible local actions. Additionally, massive open online courses (MOOCs), organized by institutions or universities, attract thousands of participants each year who wish to deepen their knowledge of climate change (Voudoukis & Pagiatakis, 2022).
- Social media plays a significant role in spreading environmental messages to younger generations. Viral campaigns on Instagram, Twitter, or TikTok, accompanied by popular hashtags, gather millions of people around common environmental issues, encouraging collective action (Shutaleva et al., 2021).
- Finally, immersive technological tools, such as virtual reality (VR), are revolutionizing climate education. By immersing participants in simulations of the consequences of climate change, such as rising sea levels or melting glaciers, these technologies make environmental issues tangible and unforgettable (Negi, 2024).

5 Empowering Teacher Training Institutes for Climate Change Education

Effective climate change mitigation hinges on robust and systematic education that disseminates knowledge and cultivates environmental awareness among future generations. Teacher training institutes and schools constitute pivotal actors in this endeavor, serving as strategic leverage points to embed climate and environmental issues into curricula and pedagogical practices. By preparing educators capable of effectively transmitting this knowledge, these institutions initiate a multiplier effect across society.

5.1 Methodological Framework for Sustainable Environmental Education

To effectively integrate environmental education into teacher training, a structured integration methodology is essential. This requires a systematic approach that goes beyond isolated initiatives to fit within a broader, cohesive framework. The key elements of this methodology include:

- Analyzing curricula helps identify existing environmental content and uncover gaps to coherently integrate climate change education across different subjects.
- The creation of interdisciplinary modules allows complex environmental themes to be addressed by connecting science, social studies, languages, and the humanities, while encouraging critical thinking and problem-solving.
- Adopting experiential pedagogy including hands-on activities such as field visits or ecological projects promotes active learning and strengthens environmental responsibility among teacher trainees.
- Collaborative learning among future teachers enhances pedagogical skills and facilitates the joint design of environmental education projects, while fostering cooperation and pedagogical innovation.
- Appropriate assessment methods such as portfolios, practical projects, or classroom observation effectively measure the ability of future teachers to convey environmental knowledge in an engaging way.
- Continuous professional development is essential to keep teachers' skills up to date by providing training on new pedagogical approaches, digital tools, and the latest advances in climate science.
- Establishing partnerships with external stakeholders (NGOs, researchers, local communities) enriches pedagogical practices and offers concrete opportunities for involvement in local environmental initiatives.

5.2 *Teacher Training as a Critical Catalyst*

Teacher training represents a cornerstone for the propagation of climate literacy. Embedding specialized modules on environmental protection and climate change within teacher education curricula equips future educators with the pedagogical tools necessary to foster awareness and understanding among their students. This structured knowledge transfer fosters a sustainable educational continuum that reaches diverse societal sectors.

5.3 *Innovative Environmental Education Practices at ESEF Oujda*

The Higher School of Education and Training (ESEF) in Oujda stands as a compelling example of proactive engagement in environmental education, despite the structural and curricular constraints that may exist. The institution has launched a range of impactful initiatives, including the establishment of environmental clubs, the integration of interdisciplinary projects centered on the sustainable management of natural resources, the facilitation of awareness-raising workshops on environmentally responsible practices, and the creation of innovative, locally relevant teaching materials. Central to these efforts is the emphasis on experiential learning, with trainee teachers actively participating in the design and implementation of educational activities. This approach not only deepens their theoretical understanding but also cultivates essential practical skills, empowering them to effectively engage and motivate their future students toward environmental responsibility.

5.4 *Advancing Pedagogical Methodologies for an Integrated Approach*

A significant challenge for teacher training institutes lies in fostering an integrated pedagogical framework that interlinks environmental sciences with complementary disciplines such as geography, biology, and social sciences. Employing active, student-centered methodologies including project-based learning, case studies, and experiential workshops enables teachers to develop the critical skills needed to address climate change and environmental topics effectively. Such approaches nurture deeper engagement and empower students to apply theoretical insights to real-world challenges.

In conclusion, teacher training institutes serve as essential agents of educational transformation, instrumental in raising environmental consciousness. The initiatives observed at ESEF Oujda underscore the imperative of a well-structured, experiential, and methodologically sound educational strategy to confront contemporary and future environmental and climate imperatives.

6 Challenges and Opportunities in Climate Change Education and Awareness

6.1 Persistent Challenges in Teacher Training

Integrating climate change education within teacher training institutions remains a significant challenge, especially in the context of Morocco. Despite increasing awareness of environmental issues, teacher education programs often struggle with structural and pedagogical barriers (Fig. 7.3):

- The scarcity of financial support and educational infrastructure limits access to modern, interactive tools necessary for effective environmental education.
- Many programs continue to rely on rigid disciplinary frameworks, impeding the integration of interdisciplinary content related to climate change and sustainability.
- Most teacher trainees are not offered dedicated modules on climate change, hindering their ability to design and deliver environmental awareness activities effectively.

Fig. 7.3 Enhancing climate education in Morocco

- Limited engagement of local stakeholders, including schools and municipalities, weakens the real-world application of environmental learning and limits long-term impact.

6.2 Opportunities and Initiatives for Advancing Climate Education

Despite persistent challenges, a number of promising opportunities are emerging to mainstream environmental and climate change education within teacher training programs. The Higher School of Education and Training in Oujda (ESEFO) has taken a pioneering role in this movement. In collaboration with the Hans Seidel Foundation and the Green Initiatives Center, ESEFO is implementing a comprehensive and forward-thinking strategy to embed climate education into the heart of teacher preparation:

- ESEFO is among the first teacher training institutions in the region to integrate digital technologies, interactive platforms, and interdisciplinary approaches into its climate education framework. This marks a breakthrough in Moroccan higher education. These tools not only enhance learner engagement but also foster more adaptive, inclusive, and forward-looking teaching practices, equipping future educators with the competencies to address the climate crisis effectively.
- Through partnerships with national and international actors such as the Hans Seidel Foundation and the Green Initiatives Center, ESEFO benefits from technical expertise, co-development of educational resources, and access to global knowledge-sharing networks. These collaborations reinforce the institutional capacity to deliver high-quality, globally informed environmental education.
- ESEFO actively empowers students through experiential learning, encouraging them to lead local environmental projects, participate in eco-initiatives, and join climate-focused student clubs. This hands-on involvement connects theory with practice and instills a culture of environmental responsibility among future educators.
- As part of their training, future teachers design and implement environmental education projects in primary and secondary schools. This outreach has a dual benefit: it cultivates environmental awareness and responsibility in younger students while giving trainee teachers real-world experience. The cascading impact of this initiative

strengthens the link between higher education institutions and local communities.

- ESEFO is also committed to sustaining momentum at the university level through the establishment of climate education clubs, digital platforms, and public awareness campaigns. These initiatives create ongoing spaces for reflection, dialogue, and innovation, both on campus and in the wider educational ecosystem.

7 Conclusion

Environmental education represents a powerful catalyst in the global response to climate change. By equipping current and future generations with the knowledge, values, and competencies needed to understand and address environmental challenges, it fosters a culture of sustainability and collective responsibility. This education empowers individuals not only to become more conscious of their ecological footprint, but also to engage actively in the transition toward greener, more resilient societies.

In this transformative process, teacher training institutions occupy a strategic position. Through interdisciplinary curricula, experiential learning, eco-pedagogical innovation, and the integration of digital tools, these institutions build dynamic and inclusive pathways for knowledge transfer that extend far beyond the classroom, reaching local communities, schools, and public institutions.

Moreover, ESEF's pilot initiatives demonstrate that well-designed training programs can generate a tangible impact at multiple levels. Future teachers not only deepen their own ecological literacy, but also pass it on to students in primary and secondary schools through school-based climate projects, workshops, and awareness campaigns. This cascading effect enhances civic engagement and encourages the adoption of sustainable behaviors from a young age, laying the groundwork for a more environmentally conscious generation.

However, to meet the urgency and scale of climate challenges, it is essential to go beyond isolated initiatives. Environmental and climate education must be integrated into the core of educational systems, from curriculum design and teacher professional development to institutional governance and educational policy. This transformation requires sustained commitment and collaboration among educators, policymakers, academic institutions, civil society organizations, and international partners.

References

Abbass, K., Qasim, M. Z., Song, H., Murshed, M., Mahmood, H., & Younis, I. (2022). A review of the global climate change impacts, adaptation, and sustainable mitigation measures. *Environmental Science and Pollution Research, 29*(28), 42539–42559.

Amalu, E. H., Short, M., Chong, P. L., Hughes, D. J., Adebayo, D. S., Tchuenbou-Magaia, F., Lähde, P., Kukka, M., Polyzou, O., Oikonomou, T. I., Karytsas, C., Gebremedhin, A., Ossian, C., & Ekere, N. N. (2023). Critical skills needs and challenges for STEM/STEAM graduates increased employability and entrepreneurship in the solar energy sector. *Renewable and Sustainable Energy Reviews, 187*, 113776.

Becher, K. M. (2024). *Non-formal sustainability education: Case study of environmental NGOs in Iceland.* Doctoral dissertation.

Borde, B., Léna, P., & Lescarmontier, L. (2022). Education as a strategy for climate change mitigation and adaptation. In *Handbook of climate change mitigation and adaptation* (pp. 3089–3113). Springer International Publishing.

Browning, S., Beymer-Farris, B., & Seay, J. R. (2021). Addressing the challenges associated with plastic waste disposal and management in developing countries. *Current Opinion in Chemical Engineering, 32*, 100682.

Dreyfus, G. B., Xu, Y., Shindell, D. T., Zaelke, D., & Ramanathan, V. (2022). Mitigating climate disruption in time: A self-consistent approach for avoiding both near-term and long-term global warming. *Proceedings of the National Academy of Sciences, 119*(22), e2123536119.

Dzvimbo, M. A., Mashizha, T. M., Zhanda, K., & Mawonde, A. (2022). Promoting sustainable development goals: Role of higher education institutions in climate and disaster management in Zimbabwe. *Jàmbá-Journal of Disaster Risk Studies, 14*(1), 1206.

Geldin, S. (2022). *Rethinking the heavy hand of flood prevention in informal settlements: An investigation of retreat-related practices in the Philippines.* Doctoral dissertation, University of Pennsylvania.

Jha, M. K., & Dev, M. (2024). Impacts of climate change. In *Smart internet of things for environment and healthcare* (pp. 139–159). Cham.

Kaplan, A. M., & Haenlein, M. (2016). Higher education and the digital revolution: About MOOCs, SPOCs, social media, and the Cookie Monster. *Business Horizons, 59*(4), 441–450.

Kurup, P. M., Levinson, R., & Li, X. (2021). Informed-decision regarding global warming and climate change among high school students in the United Kingdom. *Canadian Journal of Science, Mathematics and Technology Education, 21*, 166–185.

Leal Filho, W., Barbir, J., Gwenzi, J., Ayal, D., Simpson, N. P., Adeleke, L., Tilahun, B., Chirisa, I., Gbedemah, S. F., Nzengya, D. M., Sharifi, A.,

Theodory, T., & Yaffa, S. (2022). The role of indigenous knowledge in climate change adaptation in Africa. *Environmental Science & Policy, 136*, 250–260.

Mehan, A. (2023). Re-theorizing the collective action to address the climate change challenges: Towards resilient and inclusive agenda. *Canadian Journal of Regional Science, 46*(1), 8–15.

Nazar, R., Chaudhry, I. S., Ali, S., & Faheem, M. (2018). Role of quality education for sustainable development goals (SDGS). *International Journal of Social Sciences, 4*(2), 486–501.

Negi, S. K. (2024). Exploring the impact of virtual reality and augmented reality technologies in sustainability education on green energy and sustainability behavioral change: A qualitative analysis. *Procedia Computer Science, 236*, 550–557.

Oe, H., Yamaoka, Y., & Ochiai, H. (2022). A qualitative assessment of community learning initiatives for environmental awareness and behaviour change: Applying UNESCO education for sustainable development (ESD) framework. *International Journal of Environmental Research and Public Health, 19*(6), 3528.

Okada, A., & Gray, P. (2023). A climate change and sustainability education movement: Networks, open schooling, and the 'CARE-KNOW-DO' framework. *Sustainability, 15*(3), 2356.

Osumba, J. J., Recha, J. W., & Oroma, G. W. (2021). Transforming agricultural extension service delivery through innovative bottom–up climate-resilient agribusiness farmer field schools. *Sustainability, 13*(7), 3938.

Perdomo, J. C. (2024). The struggle to anticipate the future: Political uses, technological approaches, and moral debates on climate change. *Environmental Reviews, 33*, 1–16.

Segger, M. C. C. (2016). Advancing the Paris Agreement on climate change for sustainable development. *Cambridge International Law Journal, 5*(2), 202–237.

Shlobin, N. A., Kanmounye, U. S., Ozair, A., de Koning, R., Zolo, Y., Zivkovic, I., Niquen-Jimenez, M., Affana, C. K., Jogo, L., Abongha, G. B., Garba, D. L., & Rosseau, G. (2021). Educating the next generation of global neurosurgeons: Competencies, skills, and resources for medical students interested in global neurosurgery. *World Neurosurgery, 155*, 150–159.

Shutaleva, A., Martyushev, N., Nikonova, Z., Savchenko, I., Abramova, S., Lubimova, V., & Novgorodtseva, A. (2021). Environmental behavior of youth and sustainable development. *Sustainability, 14*(1), 250.

Sloam, J., Pickard, S., & Henn, M. (2022). Young people and environmental activism: The transformation of democratic politics. *Journal of Youth Studies, 25*(6), 683–691.

Trott, C. D. (2022). Climate change education for transformation: Exploring the affective and attitudinal dimensions of children's learning and action. *Environmental Education Research, 28*(7), 1023–1042.

Uralovich, K. S., Toshmamatovich, T. U., Kubayevich, K. F., Sapaev, I. B., Saylaubaevna, S. S., Beknazarova, Z. F., & Khurramov, A. (2023). A primary factor in sustainable development and environmental sustainability is environmental education. *Caspian Journal of Environmental Sciences, 21*(4), 965–975.

Voudoukis, N., & Pagiatakis, G. (2022). Massive open online courses (MOOCs): Practices, trends, and challenges for the higher education. *European Journal of Education and Pedagogy, 3*(3), 288–295.

Williams, P. C., Bartlett, A. W., Howard-Jones, A., McMullan, B., Khatami, A., Britton, P. N., & Marais, B. J. (2021). Impact of climate change and biodiversity collapse on the global emergence and spread of infectious diseases. *Journal of Paediatrics and Child Health, 57*(11), 1811–1818.

World Health Organization, UNEP United Nations Environment Programme, & World Organisation for Animal Health. (2022). *One health joint plan of action (2022–2026): Working together for the health of humans, animals, plants and the environment*. World Health Organization.

Yadav, S. (2025). Education for sustainable awareness with integrating eco awareness into educational curricula: Strategies and challenges. In *Exploring pillars of sustainability for modern age improvements* (pp. 103–122). IGI Global Scientific Publishing.

Yadav, S. K., Banerjee, A., Jhariya, M. K., Meena, R. S., Raj, A., Khan, N., Kumar, S., & Sheoran, S. (2022). Environmental education for sustainable development. In *Natural resources conservation and advances for sustainability* (pp. 415–431). Elsevier.

Zhang, C., Khan, I., Dagar, V., Saeed, A., & Zafar, M. W. (2022). Environmental impact of information and communication technology: Unveiling the role of education in developing countries. *Technological Forecasting and Social Change, 178*, 121570.

CHAPTER 8

Challenges and Potential Improvements in Implementing Environmental Education amid Climate Change

Rosemary Shikangalah, Shoopala Uugulu, and Eveline Omagano Anyolo

1 Introduction

Industrial development in Namibia has increased significantly, leading to a higher risk of land degradation and pollution. Through environmental education (EE) and Education for Sustainable Development (ESD), the

R. Shikangalah (✉)
Department of Environmental Science, University of Namibia, Windhoek, Namibia
e-mail: rshikangalah@unam.na

S. Uugulu
Department of Geo-Sciences, University of Namibia, Keetmanshoop, Namibia
e-mail: suugulu@unam.na

E. O. Anyolo
Department of Intermediate and Vocational Education, University of Namibia, Ongwediva, Namibia
e-mail: eanyolo@unam.na

K. Hirech (ed.), *Environmental Education and Sustainable Development*, https://doi.org/10.1007/978-3-032-05760-0_8

country has been moving towards producing an environmentally conscious citizenry with positive attitudes and values for sustainability (Kanyimba et al., 2014; Anyolo et al., 2018). These people can, therefore, advocate for and develop environmental remedial and mitigation solutions. Pushing towards the need and the implementation of EE and ESD to produce "a critical mass of citizens who are not just informed and trained, but who are above all capable of using their achievements to bring about the economic, social, cultural and political changes required for sustainable development" and to address sustainability challenges, EE and ESD were recognised in Article 95 of the Namibian Constitution, the Green Plan of 1992, the Ministry of Education, Arts and Culture (MEAC) curriculum of 2015, and Ministry of Education Namibian Education for Sustainable Development Strategy of 2009 (Association for the Development of Education in Africa, 2012; Ruppel, 2022; Dzikowski, 2024; Scharfbillig et al., 2025). After Namibia gained independence in 1990, the education system underwent reform, and one of the changes was the introduction and implementation of an environmental education (EE) curriculum (Scharfbillig et al., 2025). The fundamental goal of EE is to shape the minds and skills of the young and promote the importance of sustainable use of natural resources (Malua & Olvitt, 2024; Nuumbosho, 2024; Kuusalu et al., 2025). However, implementing environmental education faces numerous challenges, primarily due to Namibia's ongoing climate change crisis (Reid et al., 2021; Saari et al., 2024). This study aims to provide an overview of educators' challenges in implementing EE under current climate change conditions and identify potential strategies that could be employed to strive for climate resilience, thereby improving the curriculum's implementation to achieve the greater goal of present and future resource sustainability in Namibia. This chapter is divided into seven sections. These sections include an introduction, the current climate and future conditions, the evolution of environmental education, methodology, challenges in implementing EE and ESD, potential opportunities and solutions for improvement, and a conclusion.

2 Current Conditions and Future Predictions of Climate Change in Namibia

Namibia is ranked second in aridity after the Sahara Desert and is one of the driest countries in Southern Africa, frequently facing erratic rainfall and prolonged droughts. The sequence of Standardised Precipitation Index (SPI-3) from mid-October 2023 to mid-March 2024 (Fig. 8.1) shows that the drought has been persistent, expanding and intensifying quickly over southern Africa regions (Toreti et al., 2024). For Namibia, around half of the country broadly experienced even lower precipitation compared to the rest of the regions, classified as drier than normal conditions (−1.5 to −2) during the years 2023 to 2024, except for October and

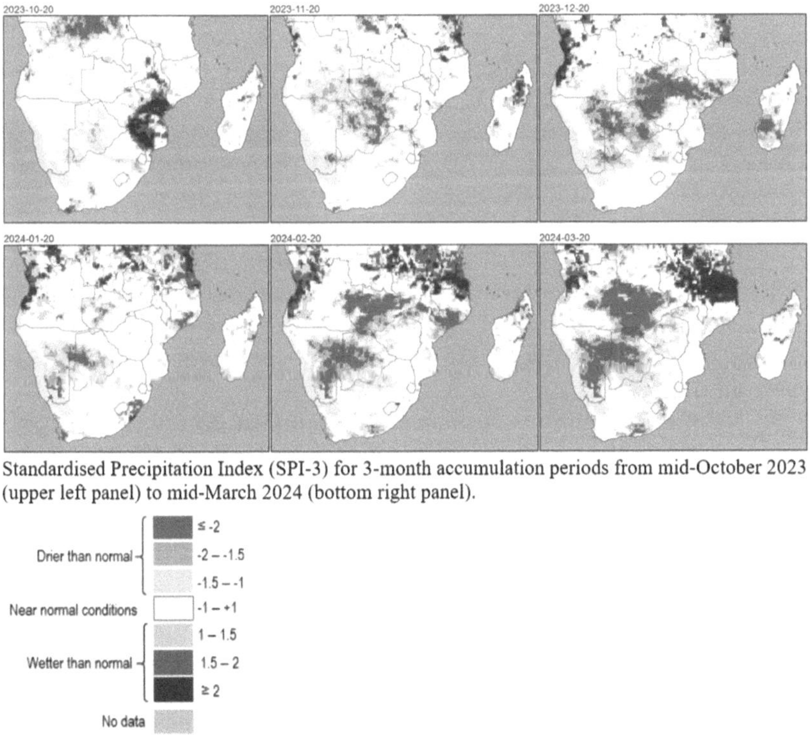

Fig. 8.1 Precipitation of Namibia, based on the 1991–2020 period (Toreti et al., 2024)

November, which show near-normal conditions −1 to 1 (Fig. 8.1). According to the country's National Determined Contributions of 2023, rainfall has decreased by 7–14% since the baseline period of 1981–2018. The temperature for the region generally shows above-average temperatures (0.5 °C and 1.5 °C) from October 2023 to March 2024, compared to the baseline 1991–2020 (Toreti et al., 2024). However, for Namibia, more than half of the country falls between 1 °C and 2.5 °C (Toreti et al., 2024). Toreti et al. (2024) also showed the evolution of NDVI anomalies from mid-October 2023 to mid-March 2024, indicating a progressive increase of vegetation stress in the country, starting from mid-October 2023 to mid-December 2023, and how the vegetation is in severely bad conditions by mid-March 2024, despite some recovery in January.

The mean annual precipitation is 269.2 mm, and mean annual temperatures range between 14.3 °C and 24.2 °C, with some areas reaching daily maximum temperatures of 40 °C (World Bank Group, 2021). Figure 8.2 shows the deviation of rainfall from long-term mean from 1990 to 2022 (Hipondoka & Hamunyela, 2024). A clear picture of frequent droughts can be observed where half of the country experienced at least 12 years of below-average rainfall (Fig. 8.2). Consequently, national disasters were declared during the years 1992/1993, 1995/1996, 2012/2013, 2013/2014, 2015/2016, and 2018/2019 (Nakanyete et al., 2020). The 2012/2013 and 2018/2019 droughts are perceived as particularly bad and widespread (Shikangalah, 2020).

Standardised Precipitation Index (SPI-3) for 3-month accumulation periods from mid-October 2023 (upper left panel) to mid-March 2024 (bottom right panel).

In Namibia, the impacts of climate change on natural resource productivity exacerbate the country's environmental sustainability challenges and add to the array of existing problems, including unsustainable economic development, energy management issues, land degradation, soil erosion, deforestation, and pollution (NaDEET, 2019; Murray & Ruppel, 2025).

3 Evolution of Environmental Education and Education for Sustainable Development

Environmental education (EE) and Education for Sustainable Development (ESD) serve as transformative frameworks addressing global sustainability challenges (Scharfbillig et al., 2025; Velempini, 2025). These frameworks

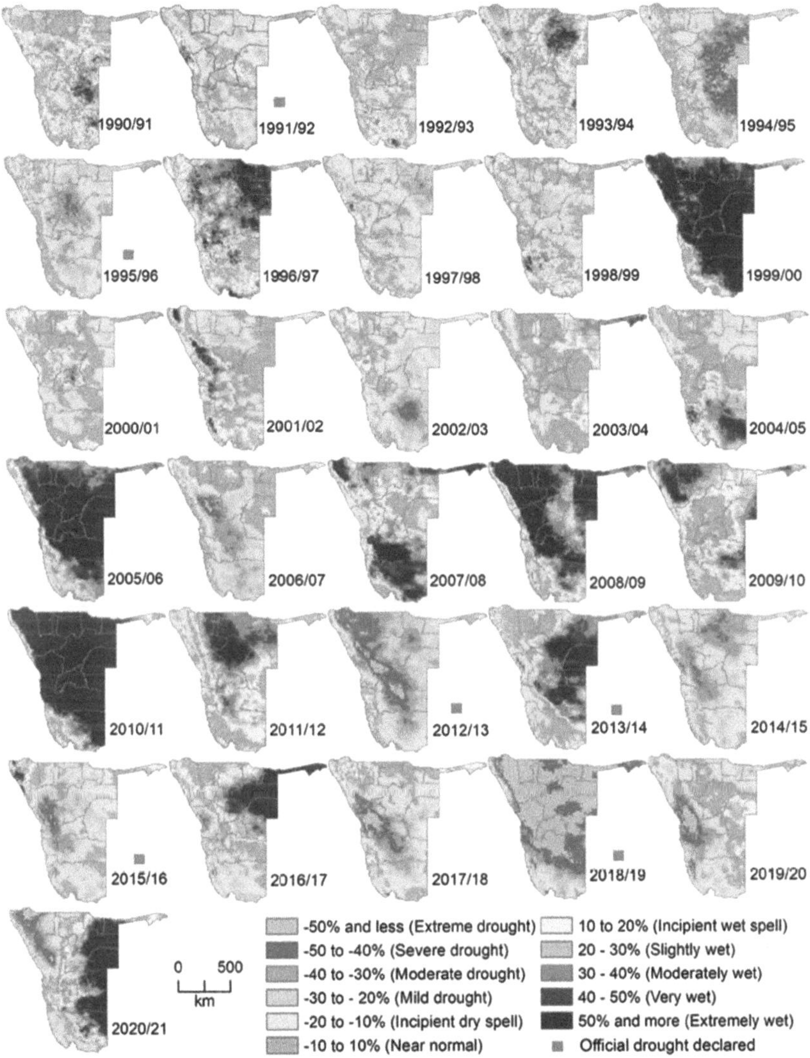

Fig. 8.2 Deviation of rainfall from long-term mean from 1990 to 2022 (Hipondoka & Hamunyela, 2024)

emphasise the transformative role of education in fostering climate action, biodiversity conservation, and sustainable resource use (UNESCO, 2018). Since independence, the Namibian government has enacted several key regulations, laws, and policies to accelerate the journey towards sustainable development. These include the Namibia Vision 2030 (Government of the Republic of Namibia [GRN], 2004), the Harambee Prosperity Plan (GRN, 2016a), and the Fifth National Development Plan (GRN, 2016b). These provisions are guided by Article 95 of the National Constitution (Ministry of Information and Broadcasting, 2007), which is the country's supreme law (Kalumbu et al., 2025). Article 95 (I) of the Constitution states that the government must actively promote and maintain the welfare of the people by adopting policies aimed at preserving ecosystems and biological diversity and utilising natural resources sustainably for the benefit of all Namibians, both present and future (Ruppel, 2022; Dzikowski, 2024).

The concept of sustainable development in Namibia is shaped by the outcomes of several international agreements concerning the environment and sustainability that the country backs, including Agenda 21 from the 1992 Earth Summit, the Biological Diversity Convention (1992), Combating Desertification from 1994, the Basel Convention from 1999, the Johannesburg Plan of Implementation from 2002, the 'Future We Want' declaration from the Rio+20 Conference in 2012, and most recently, the Sustainable Development Goals adopted in 2015 (Ruppel-Schlichting, 2022; Nuumbosho, 2024; Murray & Ruppel, 2025). These documents collectively emphasise and recognise the key role of sustainable development in harmonising socio-economic growth and social progress needs with natural resource availability and conservation (NaDEET, 2019). Furthermore, as a result of the United Nations General Assembly, through its Resolution 57/254, which declared 2005–2014 as the Decade of Education for Sustainable Development, a new framework for EE and (ESD) was proposed as "Education for Sustainable Development: Towards achieving the Sustainable Development Goals (ESD for 2030)" (NaDEET, 2019). Namibia is a signatory to many UN conventions and protocols; therefore, the country is obligated to observe and mainstream the EE strategy as part of the agreed-upon objectives of the Sustainable Development Goals and the Multilateral Environmental Agreements (Lotz-Sisitka, 2002; Fröhlich, 2006; NaDEET, 2019).

Although the development of this policy was initiated in the early 1990s, it was not until the adoption of the Namibian Constitution that the first EE conference was convened in 1991 (Scharfbillig et al., 2025). The goals of the policy are to (1) establish and implement quality EE and ESD programmes, education and training systems across all sectors, including formal education, vocational and technical education, and informal and non-formal education processes; (2) improve synergies and partnerships across all sectors to mainstream EE and ESD programmes and initiatives, including the development of sector-specific policies; (3) improve research and innovation outputs on existing EE and ESD programmes and initiatives; (4) mobilise increased funding and resource allocations for EE and ESD programmes and initiatives; and (5) ensure recognition of the central role of EE and ESD in national sustainable development plans (NaDEET, 2019).

According to the above strategies, the EE and ESD had to be integrated into the existing structures and subjects of all educational phases (Şahin, 2022; Malua & Olvitt, 2024; Saari et al., 2024; Kuusalu et al., 2025). Several projects were established and played key roles in integrating EE and ESD into the Namibian school curriculum. These included the development of the Namibia Environmental Education Network (NEEN) (Kanyimba, 2002, p. 20; Tshiningayamwe, 2012, p. 3; Sheya, 2014); the Life Sciences Project (1991–2000); the Enviroteach Project (1992–1999) of the Desert Research Foundation of Namibia; the Forestry Awareness and Tree Planting Project (1996–2002); Support Environmental Education issues in Namibia (SEEN) by National Institute for Educational Development (NIED) (NaDEET, 2019); and Environmental Awareness Project in 2017 under the Hanns Seidel Foundation (HSF), which is to continue and finalise the policy process (NaDEET, 2019; Iipinge & De Sousa, 2024; Koen & De Souza, 2025).

Even though ESD was established before the adoption of the Sustainable Development Goals (SDGs) (Hjorth Warlenius, 2022), ESD has become a crucial instrument in the current SDG agenda to address economic, social, and environmental issues and reorient education towards a more sustainable future (Stein et al., 2022; Velempini, 2025). Adopted in 2015, the SDGs call on all countries worldwide to ensure inclusive and quality education, and to take urgent action to combat climate change and its impacts, among other goals (Malua & Olvitt, 2024). Hence, ESD is seen as a means to achieve SDGs in Namibia, particularly those that are directly connected to the environment, education, and communities such as SDG 4 (Quality Education), SDG 11 (Sustainable Communities), SDG 12

(Responsible Consumption and Production), SDG 13 (Climate Action), and SDG 15 (Life on Land) (Saari et al., 2024; Kuusalu et al., 2025).

4 Methodology

The study employed a qualitative research approach that relied on a desktop literature review to explore the challenges faced in implementing EE and ESD in the context of climate change in Namibia. The approach was used to analyse annual reports, policy documents, reviews, and peer-reviewed journal articles that directly deal with EE and ESD and their implementation. The study used a purposive sampling to select five EE centres based on their prominence in EE/ESD and vulnerability to climate impacts. The key EE centres play significant roles in fostering environmental awareness and sustainable practices, which are also actively engaged in promoting EE and ESD through practical and hands-on learning experiences in Namibia. Thematic analysis was used to analyse several documents, identify challenges (e.g., resource scarcity), and propose solutions (e.g., fog harvesting). This triangulation strengthened the validity of the study and outcomes (Meydan & Akkaş, 2024).

5 Challenges in Implementing EE and ESD

Various socio-cultural and economic factors influence the implementation of EE and ESD in Namibia. Practices such as hunting or land use for grazing can sometimes lead to resistance to change, even if they contribute to environmental degradation and hinder efforts to promote sustainable land management through community-based natural resource management (CBNRM) (Namibian Association of CBNRM Support Organisations [NACSO], 2018). Additionally, there is a lack of awareness and understanding of the importance of environmental conservation and sustainable practices, particularly in rural areas where communities are often unaware of the long-term consequences of activities such as deforestation for firewood, leading to continued unsustainable practices (Shiimi et al., 2020). In addition, efforts to introduce sustainable fishing practices, such as in the Zambezi region, were also met with resistance from local fishermen, who viewed the new regulations as an infringement of their traditional livelihoods (NACSO, 2019). Another socio-cultural barrier is the language gap (Kanyimba et al., 2014), as well as access to materials in school.

It was reported that only 20% of rural schools had access to EE materials, compared to 70% of urban schools (MEAC, 2019).

Economic factors, such as poverty and resource constraints, unemployment, and tourism pressure, are some aspects that negatively impact the implementation of EE and ESD. Many communities prioritise immediate economic survival over long-term environmental sustainability. Despite awareness of their environmental impacts, poverty drives communities to practices such as overfishing and logging activities (Ministry of Environment and Tourism [MET], 2011). High unemployment rates force many individuals to prioritise short-term economic gains over long-term environmental sustainability, as individuals rely on these activities for income despite their harmful effects (Chamber of Mines of Namibia, 2021). Increased tourist activities have led to habitat disruption, highlighting the need for sustainable tourism practices (MET, 2021).

Although EE and ESD face many challenges in their implementation and teaching processes initially, one particular persistent issue is resource constraints. The curriculum requires hands-on experiences, field trips, and requisites such as equipment and materials; the absence of these results in a paucity of practical engagement (Msezane & Mudau, 2014; Sikhosana, 2025). Fien (1998) explained how education through the environment is a pedagogical technique that uses the environment as a resource for learning, adding reality and practical exercises to learning and providing learners with an appreciation of the environment through direct contact with it, leading to the acquisition of a great deal of knowledge and understanding as a result of first-hand experiences (Anyolo et al., 2018). The most undermining factor in this regard is the impact of climate change on the necessary resources, as these activities essentially require direct access to and exploitation of natural resources. In addition, drought, floods, diseases, and poverty affect many regions in Namibia, leaving teachers and students unable to attend school, which consequently hinders the implementation of the curriculum.

The Cheetah Conservation Fund (CCF) is one of the centres aligned with the implementation of EE and ESD. The centre offers a range of activities that are suitable for different visitors. For environmental education, activities include overnight educational camping in natural environments that many learning groups make use of more often, guided nature walks that lead to the observation of natural predator-prey interactions or human feeding experiences; wildlife tracking and data collection for research-oriented students, primarily from higher education institutes;

and also demonstrations of farming-related activities such as cattle husbandry, herd and veld management, disease and vaccination programmes that involves local farmers (CCF Annual Report, 2023). The centre has had up to 300,000 student participants since 1994, with up to 56 schools, a total of 17,209 students, and 352 teachers visiting in 2023 alone (CCF Annual Report, 2023). Environmental resources that underpin a wide range of ecological and socio-economic activities include the grasslands, which in parts of the centre adjoin livestock farms and agricultural lands while simultaneously functioning as natural hunting grounds and wildlife habitats. Additional resources includes arborel system, diverse founal species such as cheetah, antelopes, oryx, kudu, red hartebeest, and warthog, some of which are subjected to regulated hunting activities. Others are such as soil, freshwater, fibrewoods/bush blocks, and livestock (CCF Annual Report, 2023), which are increasingly vulnerable to the impacts of climate change.

The Desert Research Foundation of Namibia (DRFN) is another centre that is dedicated to promoting EE and ESD through training, research, and consultation through different relevant sectors in Namibia (DRFN, 2025). Activities directed towards water management and conservation, including practices that utilise renewable resources such as solar and wind projects and green hydrogen, are at the forefront. Approaches also aim to encourage reforestation, sustainable agriculture, and the preservation of ecosystems, while reducing land degradation. Such activities require resources such as freshwater systems, including surface water and groundwater systems, vegetation cover, land, crops, and livestock. These resources are easily impacted by climate change, whether through drought, limited rainfall amounts, high temperatures, increased evaporation, or flooding in many parts of Namibia.

The Giraffe Conservation Foundation (GCF)—Khomas Environmental Education Programme (KEEP) operates the most extensive environmental education programme in Namibia. This programme offers an interactive and hands-on experience within a bush-based environmental education initiative, designed to connect and reconnect students with nature through field-based environmental education (GCF.org, 2025). This entails activities such as bushwalks to observe and learn about local flora and fauna, surveying of giraffes and other wildlife, tracking and footprint identification of wildlife, monitoring of freshwater quality for the animals, and interactive lessons on ecosystems and food chains in the field (GCF Annual Reports 2023/2024). This requires natural resources such as vegetation, watercourses, animal habitat quality and availability, and wildlife species, as well as soil, which are vital for understanding the ecological needs of giraffes and identifying key areas for conservation efforts, including those for animals like giraffes and lions. According to the GCF Annual Report (2023/2024), the primary threats to conservation include habitat loss, fragmentation, and degradation, as well as poaching, human-wildlife conflict, and disease. Climate change and its impacts (drought) on grazing resources and degradation of the environment are stated as one of the problems (NW Namibia Impact Report, 2022). To date, the programme has had over 21,500 students and 1150 teachers participating (GCF.org, 2025).

The Gobabeb Namib Research Institute is another centre at the heart of environmental education. The surrounding environment comprises three distinct but fragile ecosystems: the Namib (the Sand Sea to the south, the Gravel Plains to the north, and the riparian woodlands) of the ephemeral Kuiseb River (Gobabeb.org, 2025). The centre offers training programmes that focus on participatory methods and hands-on approaches in various areas, including understanding desertification and different aspects of ecology, learning arid land management approaches, and providing opportunities to research the biota and abiotic components of the environment (Gobabeb.org, 2025). This entails dune and desert ecology research, field-based studies on desert biodiversity and adaptations, restoration ecology projects for degraded desert land, weather and climate data collection using environmental sensors and sustainable technology approaches (Gobabeb.org, 2025), therefore, required resources such as dunes, wind, desert plants, insects, reptiles, wildlife, & soil among others. The frequent and severe droughts in Namibia, particularly in this hyper-arid region, significantly impact water supply, vegetation growth, and wildlife populations. Wildlife is forced to migrate to new areas in search of water, leading to the disappearance of traditional animal tracks (Gobabeb.org, 2025).

The Namib Desert Environmental Education Trust (NaDEET) is another centre that works towards EE and ESD. According to the NaDEET annual report of 2023, the centre reached up to 75,000 learners and teachers through Teach For ESD activities. The centre seeks to teach Namibians sustainable living methods through hands-on, practical training courses in sustainable cooking technology and water and energy-saving methods (NaDEET, 2025). This includes hands-on sustainable living demonstrations (e.g., solar cooking and water conservation exercises), night walks to study nocturnal desert animals, biodiversity monitoring and habitat protection projects, planting of trees and recycling activities. While many activities aim to understand climate change and its impacts, like many centres for EE and ESD, other activities still require resources. For example, at NaDEET, the programme "Time to Identify" involves identifying and describing selected plants and animals commonly seen at NaDEET in the Namib Rand Nature Reserve. Resources needed include natural desert habitat, nocturnal species, water sources, plants, animals, and conservation areas. All these are vulnerable to the impacts of climate change. For instance, increased dust and sand particles from desertification and/or land degradation can reduce the efficiency of solar panels, posing challenges for renewable energy projects. According to the study by Mani and Pillai (2010), dust accumulation on solar panels can reduce their efficiency by up to 50%, depending on the type and amount of dust, reducing their performance (Saidan et al., 2016).

At Higher Education Institutions (HEIs), educators are faced with numerous challenges in teaching aspects of sustainability and choosing approaches to teaching ESD (Burgener & Barth, 2018). According to Rieckmann 2019), one of the challenges is the lack of understanding of the meaning, context, and implementation of ESD. Although educators must consider an interdisciplinary approach in preparing their work, they consider it challenging due to a shortage of instructional resources on ESD issues (Richter-Beuschel & Bögeholz, 2019). The study conducted in Namibia by Kanyimba et al. (2014) concluded that the following needs to be improved for the ESD to be successful in Namibia: there is a need for Namibian HEIs to lead the agenda for ESD, training should include the lecturers and administrative officials of Namibian HEIs, administrative and other officials should be convinced about the need to provide funding for ESD, lecturers must have resources that can be used to teach ESD concepts to their students, the importance of ESD for the present and future generation needs to be demonstrated to ensure understanding by students and the larger community, and ESD must be linked to the core function of teaching and research.

Namibian schools are also faced with challenges in implementing EE and ESD. Issues of learner motivation, limited timetable allocation that does not account for the need for field trips, a lack of teaching and learning materials, unclear syllabi, and inadequate training are identified as major barriers faced by teachers in implementing ESD (Anyolo, 2015). The lack of awareness and learners' involvement concur with the findings of Velazquez et al. (2005) and Majumdar (2007), the lack of time as a barrier is also found by Summers et al. (2005) and Simasiku (2012), and the lack of training is also supported by Corney (2006) and Kimaryo (2011). In the face of climate change, many Namibian schools often face closures due to climate-related floods and droughts, disrupting education. For instance, in 2009, 38 schools in Omusati Region closed due to floods (Kambo, 2009), a pattern that repeated in 2011 for over 100 schools (United Nations International Children's Emergency Fund [UNICEF], 2011; Ashipala & Simasiku, 2020); in 2017, another 102 schools (Xinhua, 2017) and 11 schools are affected of which 09 closed due to rising water levels in the Omusati region (Kaapanda, 2025). Droughts similarly forced several schools to close. Around 15 schools were closed in 2019 (Namibia Vulnerability Assessment Committee [NVAC], 2019), while recent cases in 2024 schools, such as IK Tjimuhiva Combined School (Omusati Region), Ndama Junior Primary School (Kavango East Region), and Uis schools (Erongo Region), are some of the schools reported to have been closed due to water scarcity (Tafirenyika, 2024). These recurring events are intensifying absenteeism and dropouts, directly undermining educational stability. Often, social and economic aspects are exacerbated by floods and droughts, resulting in poverty that affects approximately 40% of vulnerable households and directly undermines educational stability (World Bank, 2022).

6 Potential Opportunities and Solutions for Improvement

In Namibian schools, teachers are familiar with the concepts of EE and ESD, although their perceptions are limited to the ecological aspects acknowledged. Educators can enhance learning outcomes by integrating sustainable development aspects (environmental, social, and economic) into their teaching through a variety of methods such as participatory approaches, filedwork and experiental learning activities (Anyolo, 2015). The fieldwork method involves utilising school grounds and communities (Jeronen et al., 2017) to enable learners to interact with the environment and have a hands-on experience. Similarly, the classroom observation revealed aspects such as water scarcity, environmental destruction and its associated impacts, poverty, and unsustainable farming practices as part of teachers' lesson presentations. Although teachers reported that they use participatory strategies to integrate ESD into their teaching, this was not evident in the classroom observation, as all teachers primarily used lecture and question-and-answer methods only during their teaching (Anyolo, 2015), which encourages routinised learning. However, Käpylä and Wahlström (2000) believe that to change learners' attitudes and behaviour towards sustainable development, they need to be exposed to sustainability issues directly and act to solve problems.

In terms of environmental centres in Namibia, various centres have specific aims but are also aligned with serving the implementation of EE and ESD. The CCF could, for instance, partner with local farmers to practise silvopasture and agroforestry, which integrates drought-resistant trees into both grazing and agricultural-based fields within their system. This could serve as a win for both sides, farmers and the CCF, in terms of minimising habitat loss while improving soil and providing shade for wildlife and livestock. This system has been successfully practised in Kenya by the World Agroforestry Centre (ICRAF) and has been reported to have increased drought resilience and improved fodder availability (ICRAF, 2020).

The DRFN, in addition to their current practices of renewable resources and green hydrogen, could explore the use of large-scale rainwater harvesting systems and/or artificial groundwater recharge techniques to combat water scarcity. This could involve guiding local communities to assist in building rainwater harvesting structures and recharge pits where appropriate. Namibia already has knowledge of the pockets of groundwater and the use of artificial recharge from which DRFN could tap guidance (Sarma & Xu, 2017; Matengu, 2020; Uugulu, 2022). This will not only give ample time for the ephemeral rivers to revive whenever little rainfall is received, but it will also ensure sustainable water availability in drought environments.

The KEEP, together with the local people, could venture into establishing a wildlife corridor to connect fragmented habitats, allowing giraffes and other species to migrate in response to climate change, as seen in the Kavango-Zambezi Transfrontier Conservation Area (KAZA) corridor (Lupiya, 2019; Rosen et al., 2024).

The Gobabeb Namibia Research Institute and NaDEET could explore harvesting fog water, as the fog is one of the elements they experience constantly. It has been successfully practised in Chile through the Atacama Desert Fog collection project (United Nations Environment Programme [UNEP], 2018). Another alternative is the use of desalination. Currently, desalination is practised minimally and primarily for mining activities in

Namibia. However, the practice has the potential to provide clean water for numerous other activities in arid coastal areas. The Sahara Forest Project has successfully demonstrated solar desalination in desert environments (Sahara Forest Project, 2020).

7 Conclusion

Namibia, like many other countries, is working towards sustainable development. The country took the path of using environmental education (EE) and Education for Sustainable Development (ESD) in accomplishing sustainable development and management goals, using a robust policy framework that includes Article 95 of the Constitution, the Green Plan of 1992, and the National EE and ESD Strategy, to promote sustainable development and environmental stewardship. In pursuit of the practical realisation of the SDGs, the country ensured curriculum transformation at all levels of education to respond to the country's developmental objectives and those of the world. While the country has made significant progress, it has not been without challenges. However, the country faces a persistent challenge of climate change impacts such as droughts, floods, and rising temperatures, which threaten and hinder the implementation of EE and ESD programmes. Recurring droughts, floods, and extreme weather events disrupt much of the efforts put in place to promote hands-on learning and sustainable practices by schools and several centres, limiting access to essential resources. While strengthening climate-resilient infrastructure is vital, securing government and international funding to support and maintain EE and ESD activities is particularly important for the country's future development. Centres and schools are encouraged to seek improved initiatives that could be more climate resilient.

References

Anyolo, E. O. (2015). *Implementing Education for Sustainable Development in Namibia: A case of three senior secondary schools.* Doctoral dissertation, University of Namibia. https://erepo.uef.fi/server/api/core/bitstreams/b2d10c50-3087-49fd-92d2-021f48091b1d/content

Anyolo, E. O., Kärkkäinen, S., & Keinonen, T. (2018). Implementing Education for Sustainable Development in Namibia: School teachers' perceptions and teaching practices. *Journal of Teacher Education for Sustainability, 20*(1), 61–78.

Ashipala, V., & Simasiku, O. (2020, March 5). Over 100 schools closed due to floods. *ReliefWeb*. https://reliefweb.int/report/namibia/over-100-schools-closed-due-floods

Association for the Development of Education in Africa. (2012). *Sub-theme 1: Common core skills for lifelong learning*.

Burgener, L., & Barth, M. (2018). Sustainability competencies in teacher education: Making teacher education count in everyday school practice. *Journal of Cleaner Production, 174*, 821–826.

Chamber of Mines of Namibia. (2021). *Informal mining in Namibia: Challenges and opportunities*. Author.

Cheetah Conservation Fund (CCF). (2023). *Annual report 2023*. https://cheetah.org/learn/resource-library/

Corney, G. (2006). Education for Sustainable Development: An empirical study of the tensions and challenges faced by geography student teachers. *International Research in Geographical and Environmental Education, 15*(3), 224–240.

Desert Research Foundation of Namibia (DRFN). (2025). *Vision, mission and core values*. https://drfn.org.na/about-us/vision-mission-and-core-values

Dzikowski, M. (2024). *The political ecology of the human wildlife conflict in the Kunene region in Namibia*. Doctoral dissertation, Ghent University.

Fien, J. (1998). *Environmental education: A pathway to sustainability*. Deakin University.

Fröhlich, G. (2006). *Namibian environmental education certificate: Module 2 (Study guide)*. Center for Open and Lifelong Learning, Polytechnic of Namibia.

Giraffe Conservation Foundation (GCF). (2024). *Annual report 2023/24*. https://giraffeconservation.org/wp-content/uploads/2024/11/2023-24-GCF-Annual-Report.pdf

Giraffe Conservation Foundation (GCF). (2025). *Khomas environmental education program (KEEP)*. https://giraffeconservation.org/programs/khomas-environmental-education-programme-keep/

Gobabeb Namib Research Institute. (2025). *Welcome to Gobabeb!* Retrieved February 13, 2025, from https://gobabeb.org/

Government of the Republic of Namibia (GRN). (2004). *Namibia Vision 2030: Policy framework for long-term national development*. Office of the President.

Government of the Republic of Namibia (GRN). (2016a). *Harambee Prosperity Plan 2016/17–2019/20: Namibian Government's action plan towards prosperity for all.*

Government of the Republic of Namibia (GRN). (2016b). *Namibia's 5th National Development Plan (NDP5): Working together towards greater prosperity for all 2017/18–2020/21.*

Hipondoka, M., & Hamunyela, E. (2024). *Risks profile of natural hazards and selected diseases in Namibia*. Risk Profile Project.

Hjorth Warlenius, R. (2022). Learning for life: ESD, ecopedagogy and the new spirit of capitalism. *Journal of Environmental Education, 53*, 141–153. https://doi.org/10.1080/00958964.2022.2070102

Iipinge, W. N., & De Sousa, L. O. (2024). Stakeholder involvement in reforming a Namibian Physical Science syllabus through climate change environmental education. In L. O. de Sousa & M. P. Koen (Eds.), *Collaborative community partnerships across education settings* (NWU Community-based Educational Research Series) (Vol. 1, pp. 73–100). AVARSITY Books. https://doi.org/10.4102/aosis.2024.BK432.04

Jeronen, E., Palmberg, I., & Yli-Panula, E. (2017). Teaching methods in biology education and sustainability education including outdoor education for promoting sustainability: A literature review. *Educational Sciences, 7*(1), 1.

Kaapanda, V. (2025, February 19). Flood hits Omusati. *New Era*. https://neweralive.na/floods-hit-omusati

Kalumbu, M. G., Tomas, M., & Proponent, E. S. (2025). *Environmental scoping assessment (ESA) study for the establishment of mining activities of base and rare metals and industrial minerals on mining claims: 70010, 70011 & 70012 Omaruru District, Erongo Region.*

Kambo, K. (2009, February 24). Namibia: Thirty-seven schools closed. *ReliefWeb*. https://reliefweb.int/report/namibia/namibia-thirty-seven-schools-closed

Kanyimba, A. T. (2002). *Towards the incorporation of environmental education in the Namibian secondary school curriculum*. Master's thesis, University of South Africa.

Kanyimba, A. T., Richter, B. W., & Raath, S. P. (2014). The effectiveness of environmental Education for Sustainable Development in Namibia. *Southern African Journal of Environmental Education, 30*, 68–82.

Käpylä, M., & Wahlström, R. (2000). An environmental education program for teacher trainers in Finland. *The Journal of Environmental Education, 31*(2), 31–37.

Kimaryo, L. A. (2011). *Integrating environmental education in primary school education in Tanzania: Teachers' perceptions and teaching practices*. Åbo Akademi University Press.

Koen, M. P., & De Souza, L. O. (Eds.). (2025). *Collaborative community partnerships across education settings*. AOSIS.

Kuusalu, S. R., Maijala, M., Maunumäki, M., Maunula, M., & Shiweda, M. A. (2025). Ecological and economic sustainability in Namibian and Finnish pre-service language teachers' reflections. In *Generating sustainable futures through teacher education: Globally rethinking higher education* (pp. 223–239). Singapore.

Lotz-Sisitka, H. (2002). Curriculum patterning in environmental education: a review of developments in formal education in South Africa. *Environmental Education, Ethics and Action in Southern Africa. Pretoria. Environmental Education Association of Southern Africa/Human Sciences Research Council*, 97–120.

Lupiya, A. (2019). *The Kavango-Zambezi Conservation Area (KAZA) and Its Dynamics in Zambia.*

Majumdar, S. (2007). Integrating sustainable development in TVET curriculum. In *11th UNESCO–APEID International Conference on "Reinventing Higher Education: Toward Participatory and Sustainable Development"* Bangkok, Thailand

Malua, A., & Olvitt, L. (2024). The emergence of Education for Sustainable Development in Namibia's senior primary English language curriculum: A morphogenetic case study. *Journal of Education for Sustainable Development*, 09734082241282891.

Mani, M., & Pillai, R. (2010). Impact of dust on solar photovoltaic (PV) performance: Research status, challenges, and recommendations. *Renewable and Sustainable Energy Reviews, 14*(9), 3124–3131. https://doi.org/10.1016/j.rser.2010.07.065

Matengu, B. M. (2020). *Groundwater assessment and sustainable management of the coastal alluvial aquifers in Namib Desert, Namibia: Omdel Aquifer as a case study.*

Meydan, C. H., & Akkaş, H. (2024). The role of triangulation in qualitative research: Converging perspectives. In *Principles of conducting qualitative research in multicultural settings* (pp. 98–129). IGI Global.

Ministry of Education, Arts and Culture (MEAC). (2019). *National report on environmental education in Namibian schools.* Government of Namibia.

Ministry of Environment and Tourism (MET). (2011). *National policy on climate change for Namibia.*

Ministry of Environment and Tourism (MET). (2021). *National policy on climate change for Namibia.* Republic of Namibia.

Ministry of Information and Broadcasting. (2007). *The Constitution of the Republic of Namibia.* Government of the Republic of Namibia.

Msezane, S. B., & Mudau, A. V. (2014). Reconnoitering the stimulus of environmental education in reducing improper solid waste disposal: A case of St. Marcia School in the Mkhondo Village in Mpumalanga in South Africa. *Journal of Human Ecology, 48*(3), 367–374. https://doi.org/10.1080/09709274.2014.11906806

Murray, R., & Ruppel, O. C. (2025, March). Country report for Namibia. In *Legal pathways to sustainable soil management in Africa* (pp. 467–548). Nomos Verlagsgesellschaft mbH & Co. KG.

Nakanyete, N.F., Shikangalah, R.N. & Vatuva. A. (2020). Drought as a disaster in the Namibian context. *International Journal of Science and Research*, 9, 377-386. https://www.academia.edu/42913880/Drought_as_a_Disaster_in_the_Namibian_Context

Namibia Vulnerability Assessment Committee (NVAC). (2019). *Drought impact report*. https://www.npc.gov.na

Namibian Association of CBNRM Support Organizations (NACSO). (2018). *Annual report on natural resource management in Namibia*.

Namibian Association of CBNRM Support Organizations (NACSO). (2019). *Community-based natural resource management in the Zambezi Region*.

Nuumbosho, J. (2024). *Promoting Sustainable Development Goal 4 through ICT integration into Geography advanced subsidiary curriculum in selected secondary schools, Ohangwena region*. Doctoral dissertation, University of Namibia.

Reid, A., Dillon, J., Ardoin, N., & Ferreira, J. A. (2021). Scientists' warnings and the need to reimagine, recreate, and restore environmental education. *Environmental Education Research, 27*(6), 783–795. https://doi.org/10.1080/13504622.2021.1937577

Richter-Beuschel, L., & Bögeholz, S. (2019). Student teachers' knowledge to enable problem-solving for sustainable development. *Sustainability, 12*(1), 79.

Rieckmann, M. (2019). Education for Sustainable Development in teacher education: An international perspective. *Environmental Education Research, 5*, 33–48.

Rosen, L. E. L. E., Amuthenu, N. S., Atkinson, S. A., Babayani, N. D., Elago, S. A. T., Hikufe, E., ... & Osofsky, S. A. (2024). Veterinary Fences in the KAZA TFCA: *Assessment of Livestock Disease Risks of Potential Removal of Specific Fence Sections, with an Emphasis on the Botswana-Namibia Border*. AHEAD Programme, Cornell University on behalf of the KAZA Animal Health Sub Working Group.

Ruppel, O. C. (2022). Environmental law in Namibia: An overview. In *Environmental law and policy in Namibia* (pp. 75–108). Nomos Verlagsgesellschaft mbH & Co. KG.

Ruppel-Schlichting, K. (2022, April). Introducing environmental law. In *Environmental law and policy in Namibia* (pp. 49–64). Nomos Verlagsgesellschaft mbH & Co. KG.

Saari, M. H., Poulton-Busler, R., & Vladimirova, A. (2024). Does sustainability really start with teachers? Reflections on integrating environmental education in pre-service teacher education in Namibia and Finland. *The Journal of Environmental Education, 55*(6), 494–508.

Sahara Forest Project (SFP). (2020). *Solar desalination in Jordan*. https://www.saharaforestproject.com

Şahin, E. (2022). Building primary schools as a model of sustainable communities: Hints for teachers. In *Education for Sustainable Development in primary and secondary schools: Pedagogical and practical approaches for teachers* (pp. 167–177). Springer International Publishing.

Saidan, M., Albaali, A. G., Alasis, E., & Kaldellis, J. K. (2016). Experimental study on the effect of dust accumulation on solar panel performance in the Middle East. *Energy Procedia, 92*, 499–505. https://doi.org/10.1016/j.egypro.2016.07.064

Sarma, D., and Xu, Y. (2017). The recharge process in alluvial strip aquifers in arid Namibia and implications for artificial recharge. *Hydrogeology Journal, 25*(1), 123

Scharfbillig, B., van Rooyen, G., Ndjavera, P., & Noichl, T. (2025). *Advancing Education for Sustainable Development: A German-Namibian Educator Exchange Program*. BoD–Books on Demand.

Sheya, E. (2014). *Indigenous knowledge and environmental education: A case study of selected schools in Namibia*. Doctoral dissertation, Stellenbosch University.

Shiimi, T., Kanyimba, A., & Hamunyela, M. (2020). Community perceptions of deforestation in the Oshikoto Region. *Journal of Namibian Studies, 27*, 45–60.

Shikangalah, R. N. (2020). The 2019 drought in Namibia: An overview. *Journal of Namibian Studies, 27*, 37–58. https://namibian-studies.com/index.php/JNS/article/view/198

Sikhosana, L. (2025). Examining the integration of environmental education into pedagogy by educators: Environmental education into pedagogy. *International Journal of Curriculum and Instruction, 17*(1), 27–45.

Simasiku, F. S. (2012). *Investigating opportunities for the development of action competence through fieldwork in Namibia senior secondary school geography curriculum*. Master's dissertation, Rhodes University.

Stein, S., Andreotti, V., Suša, R., Ahenakew, C., & Čajková, T. (2022). From "Education for Sustainable Development" to "education for the end of the world as we know it". *Educational Philosophy Theory, 54*, 274–287. https://doi.org/10.1080/00131857.2020.1835646

Summers, M., Childs, A., & Corney, G. (2005). Education for Sustainable Development in initial teacher training: Issues for interdisciplinary collaboration. *Environmental Education Research, 11*(5), 623–647.

Tafirenyika, T. (2024, April 14). Uis schools close over water shortage. *The Namibian*. https://www.namibian.com.na/uis-schools-close-over-water-shortage/

The Namib Desert Environmental Education Trust (NaDEET). (2019). *National environmental education and Education for Sustainable Development policy 2019*. https://nadeet.org/namib-desert-environmental-education-trust-nadeet

The Namib Desert Environmental Education Trust (NaDEET). (2025). *OUR surroundings*. https://nadeet.org/our-surroundings

Toreti, A., Bavera, D., Acosta Navarro, J., Acquafresca, L., Asega, C., Barbosa, P., et al. (2024). *Drought in Southern Africa April 2024*. Publications Office of the European Union. https://doi.org/10.2760/960341

Tshiningayamwe, S. A. N. (2012). *Implementation of environmental learning in the NSSC biology curriculum: A case study of two schools in Windhoek, Khomas region, Namibia*. Master's dissertation, Rhodes University. https://commons.ru.ac.za/vital/access/manager/Repository/vital%3A1564

UNESCO. (2018). United Nations educational, scientific and cultural organization. *Policy brief, Education for Sustainable Development and SDGs*. https://en.unesco.org/sites/default/files/gap_pn1_-_esd_and_the_sdgs_policy_brief_6_page_version.pdf

United Nations Environment Programme (UNEP). (2018). *Fog Harvesting in Chile*. https://www.unep.org

United Nations International Children's Emergency Fund (UNICEF). (2011). *Namibia Floods: Impact on Education*. https://www.unicef.org

Uugulu, S. (2022). *Estimation of groundwater recharge along a precipitation gradient for Savannah aquifers in Namibia with special emphasis on the impact of vegetation* (Doctoral dissertation), University of Namibia, Windhoek.

Velazguez, L., Munguia, N., & Sanchez, M. (2005). Deterring sustainability in higher education institutions: An appraisal of the factors which influence sustainability in higher education institution. *International Journal of Sustainability in Higher Education*, 6(4), 383–391.

Velempini, K. (2025). Assessing the role of environmental education practices towards the attainment of the 2030 sustainable development goals. *Sustainability, 17*(5), 2043.

World Agroforestry Centre (ICRAF). (2020). *Agroforestry in Kenya*. https://www.worldagroforestry.org

World Bank Group. (2021). *Climate Change Knowledge Portal*: Namibia. https://climateknowledgeportal.worldbank.org/

World Bank Group. (2022). *Namibia Climate Vulnerability Assessment*. https://www.worldbank.org

Xinhua, M. (2017, April 5). Flood affects more than 27000 students in Namibia. *Xinhuanet*. http://www.xinhuanet.com//english/2017-04/05/c_136182718.htm

CHAPTER 9

The Role of Schools in Educating Learners on Waste Management During Health Crises: The Case of the COVID-19 Pandemic

Omar Ouhsine, *Abdellah Ouigmane*, *and Mohamed Berkani*

1 Introduction

The COVID-19 pandemic, caused by the SARS-CoV-2 virus, has spread rapidly from Wuhan, China, to the entire globe. Declared a public health emergency of international concern by the WHO on 30 January 2020, the disease has caused significant loss of life since its outbreak in Wuhan on 31 December 2019. As the disease is mainly transmitted between humans, governments have implemented measures such as quarantining infected people and general containment, in line with the recommendations of the WHO's International Health Regulations (2005), to limit its spread and protect public health (WHO, 2020; Zhu et al., 2020; Gorbalenya et al., 2020; Lescure et al., 2020; Huang et al., 2020; Hassani & Sandali, 2020; Li et al., 2020; Azouaghe et al., 2022).

O. Ouhsine (✉) • A. Ouigmane • M. Berkani
Faculty of Sciences and Technology, University of Sultan Moulay Slimane, Beni Mellal, Morocco
e-mail: omar.ouhsine@usms.ma

K. Hirech (ed.), *Environmental Education and Sustainable Development*, https://doi.org/10.1007/978-3-032-05760-0_9

The introduction of containment and quarantine measures in developing countries such as Morocco in response to the COVID-19 pandemic has had a profound impact on the lifestyle and socio-economic context of the citizens of these wastes (Ouhsine et al., 2020; Ouigmane et al., 2021).

In addition, waste management during a pandemic becomes a major concern, because of the impact that waste can have on public health and the environment (Ouhsine et al., 2020; Ouigmane et al., 2021).

Several studies have been carried out in Morocco, notably by Rassam et al. (2012), Ouigmane et al., (2017, 2018), and Azougarh et al. (2019), focusing on the quantitative and qualitative characterization of waste in relation to changes in citizens' lifestyles in different urban socio-economic contexts. However, due to the novelty of this issue, none of this research deals with waste management during the pandemic (Rassam et al., 2012; Ouigmane et al., 2017; Azougarh et al., 2019), with the exception of those by Ouhsine et al. (2020) and Ouigmane et al. (2021), which looked at waste management in Morocco during the COVID-19 pandemic.

Educational establishments, such as schools and universities, play a key role in raising awareness and encouraging the adoption of good waste management practices, particularly in times of crisis. The COVID-19 pandemic is seen as a representative example of a global health crisis.

The aim of this study is to analyze, through a literature review, the impact of the pandemic on waste management, the measures put in place to deal with it, and the results obtained to date. It also highlights the learners' awareness of the importance of rational waste management in a pandemic context. Finally, it offers some thoughts on the future of waste management and suggests skills and learning content to help improve waste management during health crises.

2 Materials and Methods

This study adopts a methodology based on an in-depth review of the literature. It is based on the collection and analysis of scientific articles, reports, official documents, and other relevant sources dealing with the impact of the COVID-19 pandemic on waste management. To this end, an exhaustive search was conducted in various academic databases, including PubMed, Scopus, Web of Science, Google Scholar, and ResearchGate. Keywords used included 'COVID-19', 'pandemic', 'waste management', 'medical waste', 'environmental impact', and other related terms. The articles selected were examined in detail and summarized in order to identify the main findings and trends in the existing literature on the subject. This methodological approach allows for a rigorous, evidence-based analysis, highlighting the impact of the pandemic on waste management and formulating recommendations to address these challenges.

In addition, a questionnaire was distributed to secondary school students in order to better understand their awareness of waste management during and after the pandemic. Based on recent references in environmental education, this questionnaire included various questions covering several key aspects: awareness of the effects of waste on health during the pandemic, sorting habits, adoption of eco-friendly alternatives, good practices implemented during the health crisis, and personal commitment to protecting the environment. Other elements, such as reducing plastic use, reusing objects, composting organic waste, participating in cleanup campaigns, and promoting eco-responsible actions, were also addressed. Responses were collected using a simple scale allowing participants to express their level of agreement (Likert scale). The entire process was conducted in full compliance with anonymity and participant consent, in an ethical and participatory manner. The primary aim of this survey was to collect concrete data to help identify more effective educational approaches to encourage environmentally friendly behavior.

Finally, the aim of this work is to propose appropriate learning content to reinforce the knowledge and skills needed for rational waste management during a health crisis.

2.1 Reliability of the Scale

To ensure the reliability of the scale used in this survey, a statistical test based on Cronbach's alpha was performed. This indicator allows us to verify the extent to which the different questions in the same questionnaire accurately measure the same concept. The closer the value obtained is to 1, the more reliable the scale is considered to be. In our case, the calculated alpha is 0.783, which indicates a good level of consistency between the items. This value falls within a range generally considered acceptable for social science research, as highlighted in the work of Taber (2018), who considers that an alpha between 0.65 and 0.77 at a 95% confidence level reflects satisfactory internal consistency.

3 Results AND Discussion

3.1 *Impact of the Pandemic on Waste Management*

The COVID-19 pandemic has had a significant impact on waste management worldwide. The containment and quarantine measures implemented to curb the spread of the virus have led to changes in consumption and production habits, affecting the quantity, composition, and management of waste. The increased demand for disposable products such as masks, gloves, thermometers, disinfectants, cleaning supplies, and food packaging has resulted in a surge in medical and plastic waste, some of which is classified as infectious waste (Ouhsine et al., 2020; Sarkodie and Owusu, 2021; Van Fan et al., 2021; Olawade et al., 2023).

Furthermore, the closure of recycling centers and waste treatment facilities has made waste management and proper disposal more challenging. This situation has also heightened awareness of the importance of responsible waste management, emphasizing the need for sustainable practices to mitigate the environmental and health impacts of this global crisis (Sarkodie & Owusu, 2021).

The main potential sources of contamination from COVID-19 waste include household waste, quarantine centers, hospitals, and temporary care facilities, all of which increase the risk of virus transmission (Blenkharn, 2018; Ghung and Brooks, 2019; Musson et al., 2007; Sasu et al., 2011).

It is important to note that the coronavirus (SARS-CoV-2) has a variable lifespan depending on the surface: approximately 3 hours in aerosols, 4 hours on copper, 24 hours on cardboard, two to three days on stainless steel, and up to 3 days on plastics. Consequently, transmission of the virus through contact with contaminated surfaces or interaction between individuals represents a major source of infection. For instance, refuse collectors handling potentially infected waste risk contracting the virus and spreading it to those around them, potentially accelerating its transmission (Chin et al., 2020; Nghiem et al., 2020; Tamin et al., 2020; Olawade et al., 2023).

However, this study will focus on household waste, which poses a contamination risk within homes and necessitates changes in individual behavior. To minimize these risks, such waste should be stored in hermetically sealed double bags for a specific period before collection.

Following this storage phase, the waste is collected, transported, and then treated for final disposal.

3.2 *Waste Management Processes in the Event of a Pandemic*

The analysis of the article by Hantoko et al. (2021) shows that each waste management process during a pandemic must be dealt with in a specific way in order to ensure effective management and protect public health (Fig. 9.1). However, in our study, we will focus on waste production at household level and on the individual actions to be taken to limit the risks of contamination, both for the individual and for those around him. The aim is to encourage individual actions to help slow the spread of the pandemic.

The key steps to effective waste management during a pandemic are summarized below. Waste generation varies according to origin: households generate mainly organic waste, while quarantined areas produce mixed waste. Temporary hospitals generate organic, recyclable waste similar to conventional hospital waste, while medical establishments generate hazardous waste, specifically identified as related to COVID-19. Waste separation and storage also differ: households sort organic and hospital waste, while confined areas suspend all sorting 72 hours before collection. Temporary hospitals, on the other hand, separate different categories of waste, while medical facilities must label hazardous waste and disinfect collection vehicles. With regard to collection and transportation, the number of containers and frequency of collection are increased, except in

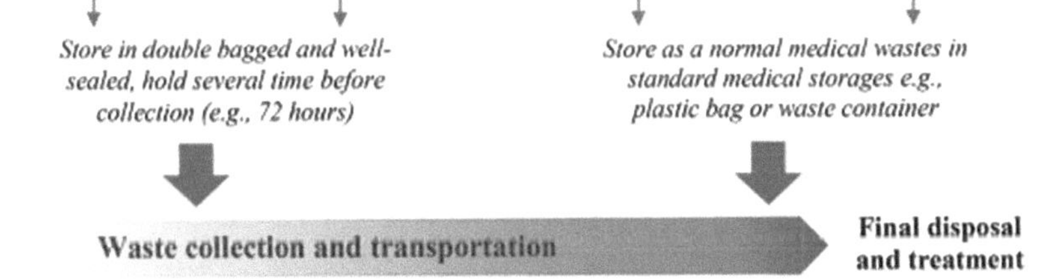

Fig. 9.1 The potential sources of contaminated waste during the COVID-19 pandemic. (Source: Hantoko et al., 2021)

quarantine, where priority is given to waste associated with COVID-19. Finally, treatment and disposal methods are adapted to each type of waste: composting or recycling (limited activities), anaerobic digestion, secure landfill for mixed waste, steam sterilization or emergency incineration for temporary hospital waste, and priority incineration for hazardous waste, with preference given to automated facilities. All these measures are designed to reduce health and environmental risks during pandemic crises.

3.3 *Best Practices for Managing Household Waste During the COVID-19 Pandemic*

Best practices for managing household waste during the COVID-19 pandemic are based on several essential recommendations. The safe management of medical and related waste, such as masks, gloves, and other protective equipment, is paramount. This waste must be used responsibly, disposed of in securely sealed plastic bags, and kept separate from household waste (Ndiaye, 2020; Beraud et al., 2022; Jyotirmayee et al., 2024).

Reducing the use of single-use products, particularly disposable plastics, is also crucial to limiting the increase in such waste during a pandemic (Ndiaye, 2020; Beraud et al., 2022; Česonienė et al., 2022; Olawade et al., 2023). In this regard, the use of reusable personal protective equipment (PPE) is encouraged whenever possible (Centers for Disease Control and Prevention (U.S.) (2020); WHO, 2020). Additionally, opting for recyclable or biodegradable materials helps to minimize environmental impact (WHO, 2020).

It is also essential to follow specific health precautions when handling homecare waste, such as syringes or bandages, to prevent any risk of contamination (Oubachir & Oukaci, 2020; Sonia, 2021). Furthermore, investing in innovative waste treatment technologies, such as autoclaves, allows for environmentally friendly waste management without systematically resorting to incineration.

The adoption of responsible practices is essential to protect both public health and the environment, both during and after the pandemic (Table 9.1).

Table 9.1 Responsible practices to protect health and the environment

Recommendation	*Details*	*Reference*
Safe management of medical and related waste	Responsible use, disposed in sealed plastic bags, disposed separately from household waste	Ndiaye (2020) ; Beraud et al. (2022)
Reduction of single-use items	Limitation of disposable plastics	Ndiaye (2020) ; Beraud et al. (2022) ; Česonienė et al. (2022)
Use of reusable personal protective equipment (PPE)	Encouragement of the use of reusable PPE	WHO (2020)
Encourage the use of recyclable or biodegradable materials	Reduction of environmental impact	WHO (2020)
Adherence to specific health precautions for managing home care waste	Safe management of syringes, bandages, etc.	Oubachir and Oukaci (2020) ; Sonia (2021)
Investment in innovative waste treatment technologies	Use of household autoclaves for eco-friendly waste treatment, avoiding incineration	Current study

3.4 The Role of Schools in Waste Management Education in the Event of a Pandemic

To address pandemics, schools play a crucial role in waste management education by raising awareness of environmental issues and promoting sustainable practices such as recycling, reuse, and waste reduction. They must adapt their teaching methods and implement appropriate sanitary measures to ensure everyone's safety while actively engaging the community (Centers for Disease Control and Prevention (U.S.) (2020)).

Furthermore, close collaboration with local authorities and specialized waste management companies is essential to ensuring effective waste management and protecting public health, even during pandemics.

3.5 Inclusion of Pandemic Themes in Life and Earth Sciences Lessons in Morocco: An Analysis of School Curricula

The Moroccan school curriculum for life and earth sciences (L&E) is designed to cover various aspects of natural phenomena related to fauna, flora, the environment, biodiversity, and ecology, as well as the interactions between living organisms and their environment. It also aims to develop a scientific and civic culture among learners, encouraging them to adopt a critical and reflective approach.

Regarding the pandemic, it is worth noting that the L&E curriculum includes learning content related to public health and disease prevention. Classroom discussions have addressed topics such as understanding viruses, their transmission, preventive measures, and the importance of vaccination (PG, 2007; PG, 2009). However, the pandemic remains a relatively new phenomenon in this context, highlighting the need to incorporate education on health and environmental behavior during health crises into school curricula.

Updating curricula to include contemporary global issues, such as pandemics, is therefore essential. This would not only provide students with a deeper understanding of these phenomena but also equip them with the knowledge and skills necessary to address such challenges in the future.

3.6 Assessing Learners' Awareness of Pandemic Waste Management Using AI

To assess learners' awareness of waste management during a pandemic, AI was used to develop a set of questions designed to directly and clearly evaluate their level of awareness and commitment to environmental practices in such a context (Table 9.2).

The results of this survey show that, although learners have strong environmental intentions, they still struggle to translate these into concrete actions. The example of waste sorting is revealing: although 91% of respondents say they want to adopt environmentally friendly practices, 73% say they do not sort their waste at home. This type of discrepancy, often observed in environmental education, is shed light on by the transformative learning approach (Mezirow, 2000), which reminds us that change rarely comes about through simple top-down information. It is critical experiences and personal questioning that ultimately trigger real awareness. More recently, Macintyre et al., (2024) have emphasized the role of local and participatory contexts in the emergence of such sustainable learning.

Furthermore, a large majority of participants (70%) expressed a lack of clear information on waste management during the pandemic. This suggests that current educational programs do not sufficiently meet students' needs in terms of everyday ecology. To remedy this, it would be useful to

Table 9.2 Eco-gestures for waste management at home during the pandemic

Aspect to be measured	*Questions*	*The terms of rewards*	*%*
Awareness of the health impact of waste during the pandemic	To what extent are you aware of the health impact of your waste during the pandemic?	1: Not at all aware	0
		2: Slightly aware	18
		3: Somewhat aware	60
		4: Aware	12
		5: Very aware	10
Waste sorting	How often do you sort your waste at home during the pandemic?	1: Never	73
		2: Rarely	15
		3: Sometimes	8
		4: Often	4
		5: Always	0
Ecological alternatives	To what extent are you willing to adopt ecological solutions to reduce your waste during the pandemic?	1: Not at all willing	0
		2: Slightly willing	3
		3: Somewhat willing	6
		4: Willing	51
		5: Very willing	40
Best waste management practices during the pandemic	To what extent are you informed about good waste management practices during the pandemic?	1: Never informed	1
		2: Slightly informed	70
		3: Moderately informed	20
		4: Informed	9
		5: Very informed	0
Ecological commitment during the pandemic	To what extent do you participate in ecological initiatives aimed at reducing waste during the pandemic?	1: Never engaged	62
		2: Slightly engaged	13
		3: Moderately engaged	0
		4: Engaged	5
		5: Very engaged	20

(*continued*)

Table 9.2 (continued)

Aspect to be measured	*Questions*	*The terms of rewards*	*%*
Reduce single-use plastics	Are you willing to reduce your use of single-use plastics (straws, bags, bottles, etc.)?	1: Not at all willing	2
		2: Slightly willing	18
		3: Somewhat willing	43
		4: Willing	25
		5: Very willing	12
Reuse of objects	To what extent are you willing to reuse items (fabric bags, bottles, containers, etc.)?	1: Not at all willing	2
		2: Slightly willing	18
		3: Somewhat willing	43
		4: Willing	24
		5: Very willing	13
Composting organic waste	Are you willing to compost your organic waste at home or at school?	1: Not at all willing	12
		2: Slightly willing	18
		3: Somewhat willing	47
		4: Willing	20
		5: Very willing	3
Participation in cleanup initiatives	Are you willing to participate in environmental cleanup activities in your community?	1: Not at all willing	0
		2: Slightly willing	5
		3: Somewhat willing	13
		4: Willing	30
		5: Very willing	52
Promotion of eco-friendly actions	To what extent are you willing to encourage others to adopt environmentally friendly waste management practices?	1: Not at all willing	2
		2: Slightly willing	5
		3: Somewhat willing	13
		4: Willing	31
		5: Very willing	49

incorporate more knowledge rooted in local practices, as proposed by Dei (2000) and Battiste (2005), who advocate for the recognition of indigenous knowledge in formal education. This type of approach allows students to reconnect with more concrete practices, often derived from their own environment. Battiste (2019) also points out that this knowledge, which has long been neglected, offers a much more balanced relationship with living things, based on collective responsibility rather than exploitation.

In the same vein, the results reveal that 43% of respondents want to reduce their use of plastic, the same percentage are considering reusing objects, and nearly half are interested in composting. These actions, even if they remain at the intention stage, show a growing awareness of the impact of our lifestyles. Here, posthumanism offers an original interpretation: rather than placing humans at the center of everything, it invites us to think of education as a space for cohabitation with all living beings (Braidotti, 2013; Pascoe, 2024). This perspective provides an interesting ethical basis for rethinking everyday actions as forms of mutual respect for our environment.

Finally, community spirit is clearly evident in the data: 82% of learners say they are willing to participate in cleanup activities, and 80% want to raise awareness among those around them about more responsible practices. This enthusiasm is promising, but it needs to be supported. Criticism from decolonial movements rightly points out that many educational initiatives remain locked into rigid models that are often far removed from the social realities of students (Escobar, 2008; Mignolo, 2011; Gómez-Barris, 2018). If we want young people to take action, it is essential to offer them spaces for participation where their knowledge, experiences, and territories are truly taken into account.

To move in this direction, several concrete steps can be considered. Intensifying awareness campaigns in schools, clearly explaining the benefits of sorting and composting, would be a good starting point. Practical activities, such as workshops run by school environmental clubs, can make these actions more accessible and concrete. It would also be beneficial to integrate waste management education into school curricula, taking into account exceptional circumstances such as pandemics. Organizing field trips to waste sorting or treatment centers would give students first-hand experience of the ecological chain in which they participate. Finally, encouraging participation in collective actions, such as neighborhood cleanups or local awareness events, would strengthen the sense of belonging and shared responsibility. In other words, for intentions to become sustainable habits, conditions must be created so that everyone feels involved in the transition.

3.7 *Using Artificial Intelligence (AI) to Generate Learning Content for Waste Management During the Pandemic*

The COVID-19 pandemic affected the entire world, including Morocco, and led to the emergence of various environmental challenges. One notable impact was the significant increase in certain types of waste, particularly personal protective equipment (PPE) such as masks and gloves, as well as medical waste like syringes and thermometers. Proper management of this waste is crucial to preventing the spread of the virus and minimizing its environmental impact (Ouhsine et al., 2020; Sarkodie & Owusu, 2021; Jyotirmayee et al., 2024).

3.7.1 *Theme: Waste Management During a Pandemic*

Learning Skills

Among the learning skills that can be suggested to teachers are, first, understanding the impact of the pandemic on waste production and management; second, the ability to identify the types of waste generated during the pandemic; and finally, learning good environmental practices such as sorting and recycling waste specific to the pandemic.

Lesson 1: Waste Classification During a Pandemic

Guidelines for Lesson Development

The session should focus on two main aspects: medical and similar waste, including masks, gloves, and other personal protective equipment (PPE); and household waste, whose increase is linked to the higher consumption of disinfectant products, single-use plastics, and food packaging.

The learning session sheet (Table 9.3) will provide participants with a comprehensive understanding of waste classification during a pandemic. It will emphasize the distinction between medical and similar waste and household waste while presenting the best management practices tailored to each category.

Table 9.3 Learning session sheet on waste classification

Aspect	*Description*
Objective	Understand the different categories of waste generated during a pandemic, focusing on medical and similar waste, as well as household waste.
Duration of the activity	To be determined.
Number of learners	To be determined.
Guidelines	The session should be based on two main characteristics of waste classification during a pandemic: medical and similar waste, and household waste.
Medical and similar waste	Medical and similar waste includes masks, gloves, personal protective equipment (PPE), and any other materials used in medical care. These wastes require specific handling and disposal to prevent the spread of diseases.
Household waste	This includes packaging from increased use of disinfectants, single-use plastics, and cans. These wastes increase during a pandemic and require proper management to minimize environmental impact.
Learning tasks	1. Identify the different types of medical and household waste. 2. Discuss appropriate management methods for each waste category. 3. Case study on the environmental consequences of improper waste management.
Expected outcomes	– Understand the different types of waste generated during a pandemic. – Learn the appropriate management methods for each waste category. – Raise awareness about the environmental consequences of improper waste management.

Lesson 2: Understanding Environmental Risks and Impacts

To structure this session effectively, it is essential to select learning tasks that highlight several key aspects.

- Health Impact: Infectious waste can contribute to the spread of the virus if not properly treated.
- Environmental Impact: Uncontrolled incineration, waterway pollution, and the increase in plastic waste are major concerns. It is also crucial to consider air and water pollution, as well as the potential damage to natural ecosystems caused by the increased use of disinfectant chemicals.

Taken together, these elements provide a comprehensive understanding of the risks and environmental consequences of inadequate waste management during a health crisis.

Aim of the Learning Session
To raise awareness of the health risks associated with infectious waste, explain the environmental impact of inadequate waste management and promote eco-friendly waste management practices (Table 9.4).

Lesson 3: Best Waste Management Practices During a Pandemic
This lesson sheet (Table 9.5) will equip participants with the knowledge and skills necessary to implement effective waste management practices, including selective sorting, safe disposal, and recycling. These measures contribute to responsible and sustainable waste management.

4 Conclusion

Effective waste management during a pandemic is not only crucial for minimizing the risk of contamination but also for safeguarding public health, protecting the environment, and fostering eco-responsible behaviors. In educational settings, this can be strengthened by integrating dedicated modules into the curriculum, organizing hands-on workshops, educational field trips, and launching targeted awareness campaigns to deepen understanding.

Active engagement from learners is essential for success. This can be achieved by encouraging participation in environmental clubs, providing staff training, installing infrastructure such as waste sorting bins and disinfection stations, and fostering collaboration with local organizations to create a community-wide impact.

This chapter underscores the critical health risks linked to improper waste disposal during a pandemic, emphasizes the necessity for efficient waste management practices, and outlines essential competencies for developing impactful educational content. By working together, all stakeholders can mitigate the harmful effects of waste, protect both human health and ecosystems, and play a vital role in controlling the spread of pandemics. Such collective action will ensure a more sustainable, resilient future for all.

Table 9.4 Proposed learning tasks based on environmental risks and impacts

Aspect	*Description*	*Learning tasks*	*Expected outcomes*
Health impact	– Infectious waste can spread the virus if not properly treated.	– Analyze real case studies. – Discuss the consequences.	– Understanding the link between waste management and the spread of diseases.
Air pollution	– Increased use of disinfectant chemicals and uncontrolled incineration of waste.	– Debate the effects of incineration. – Disinfection workshops.	– Knowledge of the negative effects of certain practices on the environment.
Water pollution	– Contamination of waterways due to poor waste management and disinfectant chemicals.	– Case studies on river pollution. – Discussion of solutions.	– Adoption of more sustainable and environmentally friendly waste management strategies.
Plastic pollution	– Increased plastic pollution related to the higher use of single-use products.	– Plastic reduction projects.	– Adoption of strategies to reduce the use of single-use plastics.
Natural ecosystems	– Potential damage to natural ecosystems caused by intensive disinfection practices.	– Discussions on alternatives.	– Awareness of potential damage to ecosystems and promotion of environmentally friendly practices.

Table 9.5 Sorting of waste learning session sheet

Aspect	*Description*
Objective	Understand and apply best waste management practices, focusing on selective sorting, safe disposal, and recycling.
Project duration	To be determined.
Target audience	Environmental learners, healthcare professionals, waste managers, and anyone involved in waste management.
Guidelines	The session will focus on three best waste management practices: selective sorting, safe disposal, and recycling.
Selective sorting	Separation of infectious waste from recyclable waste to minimize contamination risk and facilitate the proper treatment of each type of waste.
Safe disposal	Use of special containers designed for infectious waste to ensure safe disposal and reduce the risk of disease transmission.
Recycling and reuse	Encourage recycling of non-infectious materials and reuse of single-use items by the same person, thus reducing waste production and promoting sustainable resource management.
Learning tasks	1. Waste sorting simulation. 2. Visit to a waste treatment facility to observe safe disposal procedures. 3. Discussions on the benefits of recycling and reusing materials.
Expected outcomes	– Understanding of best waste management practices. – Application of selective sorting and safe disposal techniques in real-life situations. – Increased awareness of the importance of recycling and reusing materials.

References

Azouaghe, S., Boua, M., Khadoui, A., El Bardouni, A., & Belhaj, A. (2022). Comportements d'autoprotection face à la Covid-19 à l'aune des caractéristiques sociodémographiques et des moyens de communication. *Revue de Recherches Francophones en Sciences de l'Information et de la Communication.* https://hal.science/hal-03589826/

Azougarh, Y., Abbaz, M., Hafid, N., Anfar, Z., el haouti, R., AIT El Fakir, A., Benafqir, M., Ez-zahery, M., & el Alem, N. (2019). The use of natural materials for the treatment of leachates of the agadir dump. *Moroccan Journal of Chemistry, 7*(1), 132–140. https://doi.org/10.48317/IMIST.PRSM/MORJCHEM-V7I1.15243

Battiste, M. (2005). Indigenous knowledge: Foundations for first nations. *WINHEC: International Journal of Indigenous Education Scholarship, 1*(1), 1–17.

Battiste, M. (2019). *Decolonizing education: Nourishing the learning spirit.* University of British Columbia Press.

Beraud, H., Durand, M., Oates, R., & Gbedande, F. (2022). Fondamentaux hygiénistes ou résilience environnementale ? La gestion des déchets en crise Covid. Développement durable et territoires. *Économie, géographie, politique, droit, sociologie, 13*(2). https://doi.org/10.4000/developpementdurable.21154

Blenkharn, J. I. (2018). Clinical wastes in the community: Local authority management of clinical wastes from domestic premises. *Public Health., 122*(5), 526–531. https://www.sciencedirect.com/science/article/abs/pii/S0033350607002752

Braidotti, R. (2013). Posthuman humanities. *European Educational Research Journal, 12*(1), 1–19.

Centers for Disease Control and Prevention (U.S.). (2020). Établissements scolaires: enjeux opÉrationnels des mesures d'attÉnuation des risques de COVID-19 dans les rÉgions dÉfavorisÉes du monde. (cdc.gov). https://stacks.cdc.gov/view/cdc/95997

Česonienė, L., Simanavičiūtė, R., & Šileikienė, D. (2022, November). Analysis of hazardous plastic waste generated during the COVID-19 pandemic in Lithuania. In *International conference on Mediterranean Geosciences Union* (pp. 3–5). Springer Nature Switzerland. https://doi.org/10.1007/978-3-031-48754-5_1

Chin, A. W. H., Chu, J. T. S., Perera, M. R. A., Hui, K. P. Y., Yen, H. L., Chan, M. C. W., Peiris, M., & Poon, L. L. M. (2020). Stability of SARS-CoV-2 in different environmental conditions. *Lancet., 1*(1), e10. https://doi.org/10.1016/S2666-5247(20)30003-3

Dei, G. J. S. (2000). Rethinking the role of Indigenous knowledges in the academy. *International Journal of Inclusive Education, 4*(2), 111–132. https://doi.org/10.1080/136031100284849

Escobar, A. (2008). Territories of difference: Place, movements, life, redes. In *Territories of difference.* Duke University Press.

Ghung, S., & Brooks, B. W. (2019). Identifying household pharmaceutical waste characteristics and population behaviors in one of the most densely populated global cities. *Resources, Conservation & Recycling, 140*, 267–277. https://doi.org/10.1016/j.resconrec.2018.09.024

Gómez-Barris, M. (2018). *The extractive zone: Social ecologies and decolonial perspectives.* Duke University Press.

Gorbalenya, A. E., Baker, S. C., Baric, R. S., de Groot, R. J., Drosten, C., Gulyaeva, A. A., Haagmans, B. L., lauber, C., leontovich, A. M., Neuman, B. W., Penzar, D., Perlman, S., Poon, L. L. M., Samborskiy, D., Sidorov, I. A., Sola, I., & Zeibuhr, J. (2020). The species severe acute respiratory syndrome- related coronavirus: Classifying 2019-nCoV and naming it SARS-CoV-2. *Nature Microbiology, 5*, 536–544. https://doi.org/10.1038/s41564-020-0695-z

Hantoko, D., Li, X., Pariatamby, A., Yoshikawa, K., Horttanainen, M., & Yan, M. (2021). Challenges and practices on waste management and disposal during COVID-19 pandemic. *Journal of Environmental Management, 286*, 112140. https://doi.org/10.1016/j.jenvman.2021.112140

Hassani, R. T. J., & Sandali, O. (2020). Le nouveau Coronavirus Covid-19: quels risques ophtalmiques? *Journal Francais D'Ophtalmologie, 43*(4), 291. https://doi.org/10.1016/j.jfo.2020.02.001

Huang, C., Wang, Y., Li, X., Ren, L., Zhao, J., Hu, Y., Zhang, P., Fan, G., Xu, G., Gu, X., Cheng, Z., Yu, T., Xia, J., Wei, Y., Wu, W., Xie, X., Yin, W., Li, H., Liu, M., Xiao, Y., Goa, H., Guo, L., Xie, J., Wang, G., Jiang, R., Gao, Z., Jin, Q., Wang, J., & Cao, B. (2020). Clinical features of patients infected with 2019 novel coronavirus in Wuhan, China. *Lancet., 395*(10223), 497–506. https://doi.org/10.1016/S0140-6736(20)30183-5

Jyotirmayee, B., Samal, I. P., Jena, S., & Mahalik, G. (2024). Solid waste management during and after the COVID-19 outbreak: Possibilities, difficulties and approaches. In A. P. Das & S. Mishra (Eds.), *Impact of COVID-19 waste on environmental pollution and its sustainable management. Environmental science and engineering*. Springer. https://doi.org/10.1007/978-3-031-50840-0_16

Lescure, F. X., Bouadma, L., Nguyen, D., Parisey, M., Wicky, P. H., Behillil, S., et al. (2020). Clinical and virological data of the first cases of COVID-19 in Europe: A case series. *The Lancet Infectious Diseases, 20*(6), 697–706. https://doi.org/10.1016/S1473-3099(20)30200-0

Li, Q., Guan, X., Wu, P., Wang, X., Zhou, L., Tong, Y., Ren, R., Leung, K. S. M., Lau, E. H. Y., Wong, J. Y., Xing, X., Xiang, N., Wu, Y., Li, C., Chen, Q., Li, D., Liu, T., Zhao, J., Liu, M., Tu, W., Chen, C., Jin, L., Yang, R., Wang, Q., Zhou, S., Wang, R., Liu, H., Luo, Y., Liu, Y., Shao, G., Li, H., Tao, Z., Yang, Y., Deng, Z., Liu, B., Ma, Z., Zhang, Y., Shi, G., Lam, T. T. Y., Wu, J. T., Goa, G. F., Cowling, J. B., Yang, B., Leung, G. M., & Feng, Z. (2020). Early transmission dynamics in Wuhan, China, of novel coronavirus-infected pneumonia. *The New England Journal of Medicine, 382*, 1199–1207. https://doi.org/10.1056/NEJMoa2001316

Macintyre, T., de Souza, D. T., & Wals, A. E. J. (2024). A regenerative decolonization perspective on ESD from Latin America. *Compare: A Journal of Comparative and International Education, 54*(5), 821–838.

Mezirow, J. (2000). *Learning as transformation: Critical perspectives on a theory in progress* (The Jossey-Bass Higher and Adult Education Series). Jossey-Bass Publishers.

Mignolo, W. D. (2011). *The darker side of Western modernity: Global futures, decolonial options.* Duke University Press.

Musson, S. E., Townsend, T., Seaburg, K., & Mousa, J. (2007). A continuous collection system for household pharmaceutical wastes: A pilot project. *Journal of the Air & Waste Management Association, 57*(7), 828–835. https://www.tandfonline.com/doi/abs/10.3155/1047-3289.57.7.828; https://doi.org/10.3155/1047-3289.57.7.828

Ndiaye, M. C. (2020). Analyse de l'impact du COVID-19 dans la gestion des déchets solides. https://doi.org/10.13140/RG.2.2.31746.91843

Nghiem, L. D., Morgan, B., Donner, E., & Short, M. D. (2020). The COVID-19 pandemic: Considerations for the waste and wastewater services sector. *Case Study: Chemical and Environmental Engineering*, 1, 100006. https://doi.org/10.1016/j.cscee.2020.100006

Olawade, D. B., Wada, O. Z., Ore, O. T., David-Olawade, A. C., Esan, D. T., Egbewole, B. I., & Ling, J. (2023). Trends of solid waste generation during COVID-19 pandemic: A review. *Waste Management Bulletin, 1*(4), 93–103. https://doi.org/10.1016/j.wmb.2023.10.002

Oubachir, T., & Oukaci, K. (2020). Contribution à l'évaluation de la gestion des déchets d'activités de soins durant les périodes avant et pendant la COVID-19: cas de l'EPSP d'Iferhounene. Doctoral dissertation, Université Mouloud Mammeri. https://dspace.ummto.dz/handle/ummto/13959

Ouhsine, O., Ouigmane, A., Layati, E., Aba, B., Isaifan, R. J., & Berkani, M. (2020). Impact of COVID-19 on the qualitative and quantitative aspect of household solid waste. *Global Journal of Environmental Science and Management, 6*(Special Issue), 41–52. https://doi.org/10.22034/GJESM.2019.06.SI.05

Ouigmane, A., Boudouch, O., Hasib, A., & Berkani, M. (2018). Management of municipal solid waste in Morocco: The size effect in the distribution of combustible components and evaluation of the fuel fractions. *Handbook of Environmental Materials Management, 1–13.* https://doi.org/10.1007/978-3319-58538-3_82-1

Ouigmane, A., Boudouch, O., Hasib, A., Berkani, M., Aadraoui, M., & Dhairi, E. (2017). The size effect in the distribution of combustible components in the municipal solid waste produced in the summertime. Case of the city of Beni Mellal Morocco. *Journal of Materials and Environmental Sciences, 8*, 2729–2737. https://www.jmaterenvironsci.com/Document/vol8/vol8_N8/312-JMES-Ouigmane.pdf

Ouigmane, A., Boudouch, O., Hasib, A., Ouhsine, O., Layati, E., Isaifan, R. J., et al. (2021). Effect of COVID-19 on the generation of waste in Marrakech, Morocco. *Journal of Health and Pollution, 11*(30), 210606. https://doi.org/10.5696/2156-9614-11.30.210606

Pascoe, J. (2024). The art of being posthuman: Who are we in the 21st century? By Francesca Ferrando, Polity Press, 2023, 250 pp., USD24. 95 (paperback), ISBN: 9781509548965.

PG. (2007). Pedagogical guidelines for life and science studies, Qualifying high school, Morocco.

PG. (2009). Pedagogical guidelines for life and science studies, Collegial high school, Morocco.

Rassam, A., Bourkhiss, B., Chaouch, A., el Watik, L., Chaouki, H., & Ghannami, M. (2012). Caractérisation de lixiviats des décharges contrôlées au Maroc et solutions de traitement : cas de lixiviats de la ville d'Al Hoceima. *Environnement Science Lib, 4*(120204).

Sarkodie, S. A., & Owusu, P. A. (2021). Impact of COVID-19 pandemic on waste management. *Environment, Development and Sustainability, 23*(5), 7951–7960. https://doi.org/10.1007/s10668-020-00956-y

Sasu, S., Kümmerer, K., & Kranert, M. (2011). Assessment of pharmaceutical waste management at selected hospitals and homes in Ghana. *Waste Manag. Res., 30*(6), 625–630. https://doi.org/10.1177/0734242X11423286

Sonia, B. (2021). La gestion des déchets d'activité de soins à risques infectieux en période de Covid-19 au CHU de Tizi-Ouzou. *Revue Namaa pour l'économie et le commerce* 355–344 ,(01)*5* ,___. https://asjp.cerist.dz/en/article/160826

Taber, K. S. (2018). The use of Cronbach's alpha when developing and reporting research instruments in science education. *Research in Science Education, 48*(6), 1273–1296. https://doi.org/10.1007/s11165-016-9602-2

Tamin, A., Harcourt, J. L., Thornburg, N. J., Gerber, S. I., Lloyd-Smith, J. O., de Wit, E., & Munster, V. J. (2020). Aerosol and surface stability of SARS-CoV-2 as compared with SARS-CoV-1. *The New England Journal of Medicine, 382*(16), 1564–1567. https://doi.org/10.1056/NEJMc2004973

Van Fan, Y., Jiang, P., Hemzal, M., & Klemeš, J. J. (2021). An update of COVID-19 influence on waste management. *Science of the Total Environment, 754*, 142014. https://doi.org/10.1016/j.scitotenv.2020.142014

WHO. (2020). Rational use of personal protective equipment (PPE) against coronavirus disease 2019 (COVID-19) and considerations for severe shortages: Provisional guidance, April 6, 2020 (No. WHO/2019-nCov/IPC_PPE_use/2020.3). World Health Organization. WHO Reference Number: WHO/2019-nCoV/IPC_PPE_use/2020.4.

Zhu, N., Zhang, D., Wang, W., Li, X., Yang, B., Song, J., Zhao, X., Huang, B., Shi, W., Lu, R., Niu, P., Zhan, F., Ma, X., Wang, D., Xu, W., Wu, G., Gao, G. F., & Tan, W. (2020). A novel coronavirus from patients with pneumonia in China, 2019. *The New England Journal of Medicine, 382*(8), 727–733. https://doi.org/10.1056/NEJMoa2001017

CHAPTER 10

Environmental Ethics and Behavioral Change: Legal Rule vs. Education

Tarbalouti Essaid, *El Aidi Abir*, *and Ouali Abderrazak*

1 Introduction

Respect for the environment remains a central concern in contemporary economic debates, primarily due to the persistent absence of effective, inclusive, and truly sustainable solutions. Over the past decades, significant efforts have been made to formalize environmental protection through regulatory frameworks and the integration of ecological concerns into development strategies. These efforts culminated in the recognition of environmental protection as a fundamental principle of international law Facchini & Falque (2022), as enshrined in Principle 4 of the 1992 Rio Declaration: "*In order to achieve sustainable development, environmental protection must constitute an integral part of the development process and cannot be considered in isolation.*" United Nations. (1992). This principle reflects a growing consensus on the need to harmonize economic, social, and environmental priorities to ensure equitable and lasting progress.

T. Essaid • E. A. Abir, (✉) • O. Abderrazak
FSJES, Cadi Ayyad University, Marrakech, Morocco
e-mail: e.tarbalouti@uca.ac.ma; a.elaidi.ced@uca.ac.ma; A.ouali@uca.ac.ma

K. Hirech (ed.), *Environmental Education and Sustainable Development*, https://doi.org/10.1007/978-3-032-05760-0_10

In this context, the Sustainable Development Goals (SDGs) adopted by the United Nations in 2015 offer a holistic and ambitious global agenda. These 17 interconnected goals serve as a roadmap for tackling major global challenges such as poverty, inequality, climate change, and environmental degradation. While they are universal in scope, their success depends on the ability of societies to translate them into concrete, context-specific strategies that respect the balance between growth, equity, and ecological responsibility.

Despite institutional milestones such as the establishment of the United Nations Environment Programme (UNEP) in 1972, many economists, including Nobel laureate Joseph Stiglitz, argue that these efforts have not yet led to a fully operational environmental ethic. The proliferation of fragmented treaties and disconnected environmental bodies has created overlaps and inefficiencies, often compromising the coherence and effectiveness of public action.

Government responses to environmental issues have largely relied on the traditional tools of regulation and taxation. However, in countries like France, taxation remains marginal, and regulation dominates environmental policymaking. These regulatory approaches are based on the assumption that market failures are the sole origin of ecological degradation Agoulon et al. (2016) and that state intervention is the only viable corrective mechanism. Yet recent evidence shows that overly rigid environmental rules can backfire triggering opportunistic behavior, regulatory evasion, or perverse incentives that deepen rather than mitigate ecological damage.

Contrary to popular belief, the state is not always a protector of the environment. In many cases, public intervention has either aggravated environmental harms or proved incapable of addressing them effectively. This paradox, well documented in the literature, calls for a rethinking of current governance models and a search for more adaptive, participatory, and preventive strategies.

In this light, education emerges as a pivotal lever in fostering environmental responsibility. Rather than focusing solely on coercive tools, a more sustainable solution lies in building ecological awareness through formal and informal learning. Environmental education though often mentioned only indirectly in policy documents is central to nurturing long-term behavioral change and empowering citizens Dijksterhuis et al. (2005). Several recent studies, including Alaamri et al. (2024), confirm that education can transform attitudes and cultivate a durable environmental ethic, making it a compelling alternative to regulation.

This chapter contributes to this perspective by exploring the limitations of regulation and advocating for the transformative role of education as a vehicle for sustainable environmental ethics. It aligns specifically with SDG 4 (quality education) and SDG 13 (climate action), while also contributing to the broader mission of integrating environmental sustainability into public policy and governance frameworks.

The chapter is structured into four sections. The first section presents the analytical methodology used, which relies on a comparative theoretical analysis of regulatory and educational approaches. The next three sections constitute the development. Section 2, titled "Environmental Regulation and Its Limitations," explores classical and behavioral economic perspectives, focusing on issues such as transaction costs and the unintended consequences of rigid regulation. Section 3, "A Sustainable Alternative: Education as a Driver of Environmental Ethics," argues for the superior long-term benefits of education and presents empirical examples supporting this thesis. Section 4, "Toward an Integrated Model for Sustainable Environmental Ethics," proposes a hybrid framework combining education and adaptive regulation to maximize environmental policy effectiveness. The chapter concludes with key policy recommendations aimed at fostering an environmentally conscious society through education-based governance.

2 Methodology Adopted in the Analysis

This research adopts a comparative theoretical approach that examines the limitations of strict environmental regulations by highlighting their costs, effectiveness, and impact on environmental ethics. The aim is to demonstrate that an alternative based on education and awareness-raising may be more effective in achieving genuine environmental respect and sustainable ethics.

2.1 Methodological Framework and Rationale for Choice

Our analysis is based on a critical review of the literature on the effectiveness of environmental regulations and the resulting behavioral dynamics. This approach makes it possible to examine the results of current policies and identify any shortcomings. Within this framework, we will analyze regulations along four key dimensions: cost, effectiveness, ethics and environmental compliance, and opportunistic behavior.

2.2 *Comparative Approach: Strict Regulation vs. Environmental Education*

From this perspective, we adopt a comparative approach that enables us to compare the impact of strict regulations with that of an educational approach based on human capital and awareness-raising from an early age.

This analysis is based on work in the economics of education and sustainable development, which shows that investment in human capital produces long-term effects on individual and collective behavior, highlighting a bias in traditional regulatory approaches, which assume that a strict legal constraint guarantees that environmental objectives will be achieved. In reality, costs and incentives need to be rethought to incorporate a structuring educational lever.

This chapter, therefore, aims to fill the gaps in the literature in this area, by emphasizing the importance of a structured educational framework that guarantees an effective and lasting environmental ethic.

3 Development: Environmental Regulation and Its Limitations: Between Effectiveness and Constraints

Market failures in environmental matters remain one of the most pressing issues in contemporary economic debates. Public policies aim to correct these inefficiencies through environmental regulations Falque (1991), that take various forms : ecological taxes, emission quotas, binding standards, and tradable permit markets Carattini et al. (2017). These instruments seek to internalize negative externalities such as pollution and ecosystem degradation that are not accounted for by economic agents, leading to market imbalances and inefficient resource allocation.

While these instruments have demonstrated effectiveness in many instances, they also present structural limitations that warrant careful examination. These constraints are not merely conceptual but are also shaped by institutional quality, enforcement mechanisms, and the political will that conditions their implementation (Acemoglu et al., 2012). Analyzing both the successes and failures of environmental regulation is essential to understanding the conditions necessary for its effectiveness Aghion et al. (2014).

For instance, Sweden's carbon pricing, introduced in the 1990s, is often cited as a successful case of environmental taxation. Through a well-calibrated design and strong institutional support, Sweden significantly reduced its emissions while maintaining steady economic growth (Pareliussen and Purwin, 2023). Similarly, the U.S. Clean Air Act of 1970 led to substantial improvements in air quality through stringent standards and rigorous enforcement mechanisms (Greenstone, 2004). These examples illustrate that regulation can yield significant results when properly designed and implemented within a robust institutional framework. However, these successes should not overshadow the challenges associated with implementing such policies. This section examines the theoretical approaches to environmental regulation, their successes and limitations, and the debate between strict regulation and behavioral education as drivers of change.

3.1 *Presentation of Stylized Facts: Critical Observations on Regulation and Sustainability*

Before engaging in a detailed analysis of public policy instruments or sustainable intervention models, it is helpful to recall a set of stylized facts drawn from recent literature. These insights highlight the paradoxes and limitations of approaches based solely on regulation and emphasize the need for more integrated strategies based on education, ethical governance, and the reduction of transaction costs.

3.1.1 *Strict Regulation Does Not Systematically Enhance Environmental Ethics*

The work of Gneezy and Rustichini (2000) demonstrated that strict rules can paradoxically undermine individuals' intrinsic moral motivation by replacing internal commitment with external constraint. This critique now extends to regulatory tools such as Regulatory Impact Assessment (RIA).

Camille Lanssens (2024), in her in-depth study of RIA implementation in Belgium, shows that despite a sophisticated regulatory framework promoted by the OECD, institutional practices remain largely disconnected from the initial goals. RIA has failed to improve regulatory quality or foster a genuine culture of public ethics and responsibility in rulemaking.

Saltelli and Di Fiore (2023), in their edited volume *The Politics of Modelling*, reinforce this critique by illustrating how technical regulatory tools (models, indicators, simulations) often create an illusion of rationality and objectivity, while concealing political trade-offs and power asymmetries. This form of "governance by numbers," when imposed without collective ownership, has counterproductive ethical effects by reducing real-world complexity to disconnected metrics.

This gap between formal regulatory intensity and actual impact reflects a broader pattern: when regulation is perceived as bureaucratic or artificial, it can foster distrust or disengagement rather than strengthening collective ethics. In short, regulatory tools fail to induce lasting ethical change if they are neither understood nor embraced by stakeholders.

This underscores the need for participatory and educational approaches, where rules are co-constructed and understood as tools of social cohesion, rather than as technical constraints to be circumvented.

3.1.2 High Transaction Costs Encourage Short-term Opportunism

Williamson (1985) highlighted that transaction costs related to information, negotiation, monitoring, and contract enforcement can discourage long-term commitments and promote opportunistic strategies. Stéphane Callens (2023) revisits this notion from a historical and institutionalist perspective. Building on the foundational work of Coase and Williamson, he argues that transaction cost and property rights theory forms the backbone of governance structure analysis.

From this angle, any social organization including environmental policies can be interpreted as a form of contract. The "Homo Contractor" described by Williamson replaces the classical "Homo Economicus," shedding light on the strategic behavior of actors in imperfect institutional settings. When property rights are ill-defined or rules unenforceable, coordination and monitoring costs rise, paving the way for opportunistic behavior, especially in areas involving diffuse collective interests such as environmental protection Cialdini (2003).

Recent empirical research by Espinosa Díaz et al. (2023) on agri-environment-climate measures (AECMs) shows that high transaction costs especially those related to accessing information, understanding procedures, and administrative complexity are major barriers to farmers' participation. These costs deter voluntary adoption and reinforce short-term logics at the expense of long-term engagement. Furthermore, Slaev (2023) revisits Coasian thinking within the context of sustainability, stressing the importance of reducing transaction costs not only through organizational efficiency, but also via a clearer division of roles between governments, NGOs, and market actors. For Slaev, sustainability can only be achieved if institutions facilitate low-cost cooperative interactions aligned with shared long-term objectives.

Thus, the effective management of natural resources or climate policies depends on reducing transaction costs through robust institutions, contractual transparency, and long-term incentive alignment.

3.1.3 *Environmental Regulations Are Often Ineffective in Weak Institutional Contexts*

According to Acemoglu and Robinson (2012), the effectiveness of regulation depends heavily on the strength of institutions. In governance-challenged settings, even well-intentioned formal rules are often poorly implemented, distorted, or ill-suited. This is illustrated by the work of Ouattara and Sadia (2025) on the integration of artificial intelligence (AI) in educational human resource management in Côte d'Ivoire. While AI offers administrative optimization opportunities, it raises major ethical concerns in contexts where regulation is underdeveloped and governance is fragile: algorithmic biases, lack of transparency, insufficient data protection mechanisms, and the marginalization of human dimensions in decision-making.

The authors call for specific regulation based on regular audits, human oversight, and rigorous governance to avoid reinforcing existing inequalities. This reinforces the idea that, without strong institutions Acemoglu and Robinson (2008). and adapted control mechanisms, even the most promising technological or environmental innovations can lead to new forms of inefficiency or injustice.

3.1.4 *Environmental Education Fosters Intrinsic Motivation for Sustainable Behavior*

Research in behavioral economics (Deci & Ryan, 2000) has shown that educational processes help foster intrinsic motivation, which is more stable and less dependent on external incentives or sanctions. This is particularly true in the environmental domain, where Education for Sustainable Development (ESD) plays a pivotal role.

However, as Jean-Michel Lex and other environmental education practitioners in Belgium have noted, the term ESD is now marked by conceptual tensions. It is often confused or rivaled by other approaches such as

Environmental Education (EE), Citizenship Education (CE), or Global Citizenship Education (GCE). These distinctions reflect fundamentally different visions of educational engagement.

ESD is, in fact, an integrative approach at the intersection of environment, social justice, ethics, health, and democracy. Its effectiveness lies in its ability to break down disciplinary silos and encourage critical thinking about dominant economic systems thereby avoiding technocratic or greenwashed visions of "sustainable development."

Analysis shows that ESD transforms both individual and collective representations: it does not merely convey technical knowledge but cultivates values, postures, co-responsibility, and active citizenship. This reinforces the idea that only a critical, interdisciplinary, and systemic form of education can generate ecologically responsible behaviors rooted in durable ethics rather than external norms.

3.1.5 Education Reduces the Opportunity Costs of Ecological Transition

Education plays a key role in transforming individual preferences toward more sustainable choices by helping citizens better understand the long-term stakes of ecological transition. When people are made aware of the deferred benefits of sustainability such as climate stability, public health, or intergenerational equity, they are more willing to accept short-term trade-offs. Education helps redefine the balance between immediate self-interest and future collective well-being.

Several empirical studies confirm that environmental education boosts intrinsic motivation for pro-environmental behavior (Estrada-Araoz et al., 2023), reducing the perceived personal loss or sacrifice associated with green choices. Well-designed education thus lowers the cognitive and social costs of behavioral change and makes sustainable decisions more accessible both individually and collectively Coase, (2013).

Recent reviews in behavioral psychology also emphasize that educational interventions based on global interdependence, active learning, and civic participation are the most effective in fostering durable long-term behavioral change (Rau et al., 2022).

3.1.6 Social Norms Shaped Through Education Drive Behavioral Spillovers

As Elinor Ostrom (1990) observed, social norms shared within a community can exert more effective and lasting influence than top-down formal rules, particularly in the management of common-pool resources. Environmental education plays a central role in building these collective norms, cultivating shared values, ethical awareness, and voluntary long-term engagement.

This perspective is further explored by Roudil (2024), who highlights the existence of a "hidden curriculum" in school programs, shaped by the opaque influence of ideological groups. This phenomenon undermines the development of critical thinking suited to the systemic challenges of the Anthropocene. Roudil thus calls for a renewal of educational content to form citizens capable of grasping the complexity of economic theories and engaging meaningfully with contemporary environmental, social, and political challenges.

3.1.7 *Regulation-Based Public Policies Are Less Sustainable Without Mindset Shifts*

International institutions such as the OECD stress that environmental and social regulations alone are ineffective without complementary educational efforts fostering learning and accountability. This holds true in broader educational contexts. For instance, Bellita Bayard (2024), in her analysis of the Haitian education system, reveals a significant governance gap in the design and implementation of curricula, particularly those related to citizenship education.

This gap stemming from the fragility of a weak state creates a disconnect between political intent and educational reality. It illustrates how formal programs (regulation) alone are insufficient to transform mindsets: effective implementation requires strong leadership, solid governance, and adaptation to local contexts. Without these levers, education remains ineffective in shaping collective norms and civic skills just as environmental education cannot thrive without local institutional support.

3.1.8 *Educational Approaches Are Less Costly Long Term Than Repressive Regulatory Tools*

A meta-analysis by Bénabou and Tirole (2006) suggests that external incentives (fines, taxes) are effective in the short run but costly to sustain. In contrast, education-based strategies focused on learning and empowerment offer a better long-term return on investment.

This idea is supported by recent work on the efficiency index of public education spending, which shows that a significant portion of educational investment can be "self-financed" through improved student outcomes, increased future income, and related tax gains (France Stratégie, 2024). Targeted initiatives such as class size reductions, tutoring, or intensive teacher training yield lasting impacts on equity and system efficiency, and are more cost-effective than less focused policies such as grade repetition or general resource increases.

These findings underscore that the return on educational spending depends more on its effectiveness than its absolute level. In times of budget constraints and ecological transition, education remains a high-leverage strategy for social and environmental sustainability provided it is well-designed and context-specific.

3.2 The Pigouvian Approach: Corrective Taxation with Mixed Outcomes

The Pigouvian approach originates from Arthur Cecil Pigou's seminal work (*The Economics of Welfare*, 1920). According to this perspective, negative externalities create a gap between private and social costs. A tax proportional to the damage caused known as a Pigouvian tax helps align these costs and incentivizes economic agents to adopt more sustainable behaviors (Pigou, 2017). However, the effectiveness of this approach depends on the ability to accurately assess environmental costs. This assessment often relies on uncertain economic models. For example, estimates of the social cost of carbon vary significantly depending on underlying assumptions, which can lead to either ineffective under-taxation or excessive taxation that stifles innovation (Pindyck, 2013).

Another critical issue is the regressive nature of environmental taxation. In France, the carbon tax sparked strong opposition, particularly from the *Gilets Jaunes* movement, as it disproportionately affected low-income households that allocate a larger share of their budget to energy consumption (Baranzini et al., 2017). However, studies suggest that these effects can be mitigated through well-designed redistribution mechanisms (Sterner, 2012). Furthermore, the rigidity of environmental taxes poses an additional constraint. A tax set at a fixed level can quickly become obsolete in response to technological and economic developments, limiting its impact on innovation (Baumol & Oates, 1988).

3.3 The Coasian Approach: Property Rights and Negotiation

Ronald Coase (*The Problem of Social Cost*, 1960) proposed an alternative approach based on clearly defined property rights and negotiation among stakeholders. According to the Coase theorem, if property rights are well-defined and transaction costs are low, parties can reach an optimal solution without government intervention.

This approach inspired the creation of emissions trading markets, such as the European Union Emissions Trading System (*EU-ETS*). However, in practice, these markets have experienced failures, including significant fluctuations in carbon prices and manipulations by major polluters, thereby reducing their effectiveness (Ellerman et al., 2010). Moreover, transaction costs can be prohibitive when numerous actors are involved. For example, in local environmental disputes, affected communities often lack the necessary resources to negotiate effectively with polluting corporations (Medema, 2014).

3.4 The Stiglitzian Perspective: Information Asymmetry and State Intervention

Joseph Stiglitz (2000) highlights the importance of information asymmetries and advocates for proactive state intervention. Unlike Pigou and Coase, he argues that perfect information is rarely available and that firms tend to underestimate or conceal the true costs of their polluting activities.

This perspective justifies the implementation of strict standards and enhanced monitoring mechanisms. For instance, vehicle emissions regulations have led to significant reductions in air pollution, but they also entail high costs and technical challenges in enforcement (Stavins, 2007).

Nevertheless, strict regulation also has its drawbacks. Regulatory capture is a major concern: some corporations exert substantial lobbying power to shape regulations in their favor, thereby undermining their effectiveness (Stiglitz, 2017).

While regulation plays a crucial role, it is not always sufficient to drive lasting behavioral changes. Environmental education is often proposed as a more effective long-term alternative (Gifford & Nilsson, 2014). Establishing an ecological culture through awareness campaigns and behavioral incentives can promote sustainable practices without excessive coercion (Thaler & Sunstein, 2008). Nevertheless, the effectiveness of education depends on several factors, including the institutional context and individuals' ability to integrate this information into their daily decision-making. A combination of well-designed regulation and educational strategies appears to be the most effective approach for maximizing the impact of environmental policies.

3.5 *Synthesizing Approaches and Policy Debates in Environmental Governance*

The comparison table (Table 10.1) of approaches and policy implications highlights the key characteristics of three main strategies used to address environmental externalities. Each of these approaches is based on distinct principles and employs specific instruments to achieve its objectives. The Pigouvian approach, grounded in the imposition of environmental taxes, aims to internalize externalities by applying costs to polluting activities. This incentivizes economic actors to adopt more environmentally friendly behaviors. However, this approach faces significant limitations, such as the difficulty in accurately assessing environmental costs and the regressive effects on low-income households, which are particularly affected by taxes like carbon taxes. The Coasian approach, on the other hand, emphasizes negotiation between stakeholders when property rights are clearly defined. This approach is characterized by its flexibility and decentralization, allowing for tailored solutions. Nonetheless, it is hindered by high transaction costs, as well as information and power asymmetries between the parties,

Table 10.1 Comparative analysis of economic approaches to environmental regulation

Criteria	*Pigouvian approach*	*Coasian approach*	*Stiglitzian approach*
Fundamental principle	Corrective tax to internalize the externality	Negotiation between parties if rights are defined	Public regulation in the face of information asymmetries
Instruments used	Environmental taxes, subsidies	Property rights, permit markets	Standards, quotas, increased surveillance
Advantages	Simplicity, continuous incentive for innovation	Flexibility, decentralized solution	Consideration of information asymmetries
Limitations	Complex cost estimation, regressive effects	Transaction costs, bargaining inequalities	Risk of regulatory capture, bureaucratic costs

Source: Performed by the authors

which can lead to inequitable negotiations. Finally, the Stiglitzian approach advocates for strong public regulation to correct information asymmetries and prevent regulatory capture. This approach includes strict standards and quotas, as well as increased monitoring. Albeit it also suffers from limitations, notably high administrative costs and the risk of rigidity in the face of innovation, as overly strict standards may hinder the emergence of new technologies. Thus, this table illustrates not only the theoretical advantages of each approach but also the practical challenges they pose in the implementation of environmental policies.

3.6 *Contemporary Debates and Future Prospects*

The contemporary debate on environmental regulation focuses on the evaluation of available approaches, taking into account their effectiveness and the practical challenges they present. The optimization of environmental policies depends largely on the socio-economic context and political priorities. In this context, several dimensions need to be taken into account to determine which approach to adopt.

Many economists, such as Goulder and Parry (2008), advocate a mixed approach that combines environmental taxes and regulatory instruments to maximize the effectiveness of the fight against externalities. This combination seeks to leverage the strengths of each instrument while compensating for their respective limitations. While taxes offer strong economic incentives to reduce pollution, they often face challenges in accurately estimating external costs and can have regressive effects on low-income households (Baranzini et al., 2017). Conversely, command-and-control regulations help correct information asymmetries and ensure that environmental goals are met Chiroleu-Assouline (2007), but they can be rigid and costly, potentially hampering innovation.

Recent empirical work confirms the relevance of such hybrid approaches. Kohlscheen et al. (2024) demonstrate that combining carbon pricing with complementary measures enhances the reduction of CO2 emissions across diverse policy environments. Furthermore, public acceptance remains a critical factor in the success of these policies. Fabre et al. (2025) and Andre et al. (2024) show that support for environmental measures increases significantly when policies are perceived as fair and aligned with social norms, especially when redistributive mechanisms are included. As Sacko (2024) emphasizes, low-income households are disproportionately affected by carbon taxes, which reinforces the need for compensatory instruments, such as direct transfers or targeted tax relief.

Moreover, trust in scientific institutions and environmental awareness also play a role in policy acceptability. A recent study by Vecina et al. (2024) reveals that eco-anxiety and trust in science positively influence individuals' willingness to engage in pro-environmental behavior, suggesting that psychological and informational factors should not be overlooked in policy design. In parallel, Aghion, Cette, and Cohen (2014) underline that stable and predictable carbon pricing stimulates clean innovation by creating favorable investment conditions. However, excessively rigid standards may hinder technological breakthroughs, calling for flexible policy frameworks that both guide behavior and encourage long-term innovation.

4 A Sustainable Alternative: Education as a Driver of Environmental Ethics

In the current context of the global environmental crisis, traditional regulatory approaches through coercive policies and strict legislation often face limitations, both economically and socially. While these measures can produce immediate effects, they do not necessarily lead to profound mental shifts or sustained awareness of ecological issues. It is in this context that education, particularly education for sustainable development (ESD), emerges as an alternative solution, not only more economically viable in the long term but also more capable of inducing lasting behavioral changes.

This section explores why education represents a more sustainable alternative to strict regulations, examines the mechanisms through which it transforms individual and collective behaviors, and compares the costs and benefits of both approaches in terms of long-term sustainability. Through an analysis of existing theories and empirical evidence, we will show that investment in education and awareness can generate significant and lasting benefits for both the environment and society.

4.1 Why Prioritize Education over Coercion?

4.1.1 Education and the Internalization of Environmental Externalities

One of the main arguments in favor of environmental education is its ability to internalize the negative externalities associated with polluting behaviors. In an unregulated market, economic agents do not account for the environmental impact of their actions, leading to overconsumption of resources and ecosystem degradation (Pigou, 1920).

The traditional response to this market failure involves imposing taxes or setting emission quotas to correct these externalities. However, while these measures may be effective in the short term, they do not address the root causes of the problem: the lack of awareness among individuals and businesses regarding environmental issues. Environmental education helps modify preferences and behaviors upstream, thereby reducing the need for costly state interventions and fostering a more effective internalization of negative externalities.

4.1.2 Education and Human Capital Theory: A Profitable Investment for the Green Economy

The human capital approach (Becker, 1964) provides insight into the contribution of environmental education to sustainable development. According to this theory, investment in training and knowledge enhances individuals' productivity and fosters innovation. Applied to the environmental domain, higher levels of education lead to faster adoption of clean technologies and a smoother transition to a green economy.

Empirical studies confirm that countries that have heavily invested in environmental education experience more sustainable growth and greater resilience to ecological shocks (Hanushek & Woessmann, 2012). By training workers capable of designing and implementing environmentally friendly technological solutions, education reduces dependency on polluting industries and accelerates the development of innovative sectors such as renewable energy and the circular economy.

4.2 *Mechanisms of Individual and Collective Transformation: Role of Intertemporal Preferences and Education in Environmental Decision-Making*

A major challenge of the ecological transition lies in the fact that the benefits of sustainable actions are often delayed over time, while their costs are immediate. According to intertemporal preference theory (Frederick, Loewenstein & O'Donoghue, 2002), individuals tend to undervalue future benefits in favor of immediate gains, which hinders the adoption of sustainable behaviors Geller (2002).

Environmental education plays a crucial role in modifying perceptions of long-term costs and benefits. By enhancing understanding of climate issues and emphasizing the deferred impacts of present actions, it helps reduce discounting bias and encourages more responsible decisions. This mechanism is particularly relevant in developing countries, where economic constraints often lead individuals to prioritize immediate gains at the expense of natural resource preservation.

Another economic justification for education lies in its multiplier effect on collective behaviors. Inspired by Lucas's (1988) work on human capital, this mechanism suggests that the knowledge and behaviors acquired by an individual diffuse within their social and professional circles, generating positive externalities at the societal level.

Several empirical studies confirm that environmental education strengthens intrinsic motivation to adopt pro-environmental behaviors (Estrada-Araoz et al., 2023) by reducing the perception of personal loss or sacrifice associated with ecological practices. Well-designed education decreases the perceived cognitive and social costs of change, making sustainable decisions more accessible both at the individual and community levels.

Moreover, recent reviews in behavioral psychology emphasize that educational interventions based on global interdependence, active learning, and civic participation are the most effective in promoting long-term sustainable behaviors (Rau et al., 2022).

In the environmental domain, studies indicate that individuals aware of ecological issues influence their peers, colleagues, and communities, thereby fostering the collective adoption of sustainable practices (Ostrom, 1990). In contrast, strict regulation does not produce this ripple effect, as it relies on constraint rather than voluntary adherence to sustainability principles.

4.3 Cost-Benefit Comparison: Education vs. Strict Regulation

4.3.1 The Hidden Costs of Coercive Policies: Market Distortions and Carbon Leakage

While strict environmental regulations can effectively limit certain forms of pollution, they entail significant economic costs. First, they create market distortions by increasing production costs for businesses, potentially reducing their competitiveness, especially in countries where regulations are more lenient (Porter & van der Linde, 1995).

Another major issue is carbon leakage: when overly stringent environmental regulations are imposed in one country, certain industries relocate production to regions with laxer standards, thereby limiting the overall impact of the policy and sometimes worsening global pollution levels (Babiker, 2005).

4.3.2 Education as an Investment with Increasing Returns

Conversely, environmental education functions as an investment with increasing returns. According to endogenous growth models (Romer, 1990), the accumulation of knowledge fosters innovation and generates productivity gains that far outweigh initial costs. Applied to the ecological transition, this principle implies that every dollar invested in environmental education yields long-term benefits through the adoption of clean technologies, reduced energy costs, and improved economic resilience to climate crises.

Furthermore, environmental education helps lower the costs of compliance with new ecological standards for businesses. By training a skilled and environmentally conscious workforce, it facilitates the adoption of sustainable practices without requiring costly coercive measures for both the state and companies.

4.3.3 Comparing the Environmental Education and the Strict Environmental Regulations

Environmental education and strict regulation represent two distinct approaches to addressing environmental issues, each with its advantages and limitations, both in terms of costs and long-term benefits (Table 10.2).

Regarding implementation costs, environmental education has the advantage of relatively low initial investment. Resources are primarily allocated to developing educational programs, training teachers, and implementing awareness campaigns. However, these investments provide

Table 10.2 Comparison of environmental education and strict regulation approaches

Criteria	*Environmental education*	*Strict regulation*
Implementation costs	Low initial investment costs, but long-term benefits	High administrative and adaptation costs
Long-term benefits	Behavioral transformation, support for sustainable policies	Immediate effects, but limited in duration
Societal impact	Enhanced civic participation and collective responsibility	Imposed compliance, no change in mentality
Flexibility and adaptability	Allows for progressive evolution and adjustments	Rigidity of imposed standards

Source: Performed by the authors

long-term benefits by sustainably changing individual and collective behaviors (Tilbury, 2011). In contrast, strict regulation results in high administrative costs due to the need for rigorous monitoring, industry compliance, and managing the impacts of regulation (Baumol & Oates, 1988). These costs can represent a significant burden in the short term, while their long-term results remain limited unless accompanied by a transformation in mentalities.

Regarding long-term benefits, environmental education fosters behavioral transformation over the long term, supporting the adoption of sustainable practices and civic engagement. Studies show that younger generations aware of environmental issues are more likely to adopt ecological behaviors throughout their lives (Shutaleva et al., 2021). This approach is also more flexible and adaptable, allowing for adjustments based on new scientific knowledge and environmental challenges. In contrast, while strict regulations provide immediate effects, they are often limited in duration. Without a profound change in social culture and attitudes toward ecological practices, these effects may be temporary, with individuals attempting to circumvent the regulations if they become too restrictive (Janmaimool & Khajohnmanee, 2020).

The societal impact of environmental education is also deeper, as it encourages active societal involvement in solving environmental problems. Education strengthens collective responsibility and encourages greater civic participation. Individuals aware of environmental issues are more likely to engage in community initiatives and support sustainable policies (Stern, 2000). In contrast, strict regulations impose compliance without necessarily encouraging a change in attitude or commitment to the environmental cause. When individuals comply with standards, it is not always out of conviction but to avoid sanctions.

Finally, in terms of flexibility and adaptability, education allows for progressive evolution, enabling individuals to adapt to new information and adjust their behaviors in line with evolving environmental knowledge (Jaoul-Grammare & Stenger, 2023). It provides a more dynamic and continuous approach, unlike strict regulations, which often lack flexibility in the face of technological innovations and the rapid evolution of environmental challenges. Regulation can quickly become obsolete, limiting the possibilities for innovation in certain sectors, thus preventing the emergence of more sustainable alternative solutions.

While strict regulation may offer immediate results, environmental education constitutes a more sustainable and flexible solution for transforming behaviors over the long term and enabling more effective management of environmental challenges.

5 Toward an Integrated Model for Sustainable Environmental Ethics

5.1 A Potential Complementarity Between Education and Regulation?

To promote sustainable environmental ethics, an integrated model combining education and regulation emerges as a potential solution. The complementarity between these two approaches lies in the idea that education creates a solid foundation for sustainable behaviors, while regulation provides a structural framework to limit harmful actions to the environment. Education for sustainable development, in particular, has the power to transform mindsets and lead to more environmentally friendly practices over the long term (Sterling & Orr, 2001; Tilbury, 2011). According to Gough et al. (2000), an educated population in sustainable development is more likely to support environmental policies and actively contribute to the responsible management of natural resources.

A hybrid approach combining economic incentives and strict regulations may prove particularly effective in promoting environmentally friendly behaviors. Economic incentives, such as environmental taxes or subsidies for green technologies, encourage individuals and businesses to adopt sustainable practices by making ecological choices more financially attractive (Goulder & Parry, 2008). At the same time, strict regulations are necessary to prevent immediate and irreversible environmental damage by imposing clear limits on polluting activities (Sterner & Coria, 2013). For instance, studies show that the introduction of carbon emission taxes, coupled with subsidies for renewable energy, can accelerate the transition to a green economy by making clean energy sources more competitive compared to fossil fuels (Nordhaus, 2019).

This combination of incentives and regulations creates a framework where economic actors are motivated to innovate and invest in sustainable solutions while ensuring that minimum environmental protection standards are met (Acemoglu et al., 2012). Moreover, strict regulation alone can lead to economic inefficiencies and provoke social resistance, while economic incentives offer flexibility that allows businesses and consumers to gradually adapt their behaviors (Jaffe et al., 2005). The experience of emissions trading markets, such as the one implemented in the European Union, illustrates this dynamic by associating emission quotas with the possibility of trading these quotas on a market, which maximizes economic efficiency while achieving environmental goals (Ellerman et al., 2010).

Furthermore, education plays a crucial role in enhancing the effectiveness of these hybrid policies. A well-informed population is more likely to understand environmental issues and adopt behaviors aligned with sustainability goals (Baum & Gross, 2017). Educational programs focused on sustainable development raise citizens' awareness of the consequences of their actions and encourage them to support and comply with environmental policies (Tilbury, 2011). A study by Kollmuss and Agyeman (2002) shows that environmental education significantly impacts ecological decision-making and individual responsibility, particularly when integrated from an early age in the education system.

Finally, flexibility in implementing these policies allows them to be adapted to local contexts and technological developments. By considering cultural and socio-economic specificities, the measures adopted gain legitimacy and effectiveness, fostering broader societal support (Jordan et al., 2005). This adaptability is essential for addressing the complex and evolving environmental challenges we face today Assembly, U. G. (1972). Moreover, the work of Ostrom (2010) highlights the importance of participatory governance, where local actors are involved in natural resource management and in defining regulations, thus improving their acceptability and long-term effectiveness.

It is therefore imperative to establish a hybrid framework that goes beyond merely juxtaposing these two strategies but encourages their interaction. Education motivates individuals to adopt environmentally friendly behaviors, while regulation imposes limits and incentives to ensure that these behaviors are indeed adopted at the societal level.

5.2 *Implications for Policymakers and Policy Recommendations*

For environmental policies to be truly effective and sustainable, they must be based on governance that is both inclusive and accountable. Education for sustainable development must be at the heart of any political strategy, as it is the foundation for mobilizing citizens and businesses toward change. According to the Brundtland Report (1987), sustainable development cannot be achieved without the active participation of all stakeholders, including citizens who understand environmental issues. Education for sustainable development thus helps create a society where sustainability is integrated into daily values and behaviors (Fien & Tilbury, 2002).

From a political perspective, policymakers must encourage an approach to environmental governance that relies on collective accountability and active citizen participation (Aatany and Tarbalouti, 2024). Governance based on participatory principles helps overcome the mistrust often observed toward strict public policies (Kooiman et al., 2005). The participatory framework proposed by the hybrid model gives citizens an active role in the development and implementation of policies, thus strengthening the effectiveness and sustainability of reforms.

Recent studies (Abdarssoul, 2024; Rakotoarijaona, 2016) have shown that active participation of local communities in the management of natural resources enhances the sustainability of environmental policies. By integrating education on environmental management into this process, governments can foster a long-term transformation of social behaviors, which is much more sustainable than relying solely on restrictive regulations.

Policy recommendations to promote sustainable change must therefore include educational strategies that raise awareness of environmental issues from an early age, as well as regulatory frameworks that are sufficiently flexible to adapt to new knowledge and technological developments. Such a hybrid approach can maximize the benefits of both education and regulation, combining the effects of citizen awareness with the binding mechanisms of legal norms.

6 Conclusion

This chapter has explored the complex intersection of environmental ethics, policy design, and social transformation, offering a nuanced analysis of regulatory and educational models as essential drivers of systemic change in environmental governance. In the context of the ongoing global environmental crisis, it has become increasingly apparent that traditional regulatory approaches, while critical in certain circumstances, exhibit inherent limitations. These limitations include high administrative burdens, rigidity, and, most importantly, their failure to induce the deeper cognitive shifts necessary for long-lasting behavioral transformation. While regulations may successfully curb specific polluting behaviors, they are not sufficient to foster a profound societal consciousness or to initiate the long-term, sustainable change required to address the root causes of ecological degradation.

In contrast, our analysis highlights the immense potential of environmental education, particularly education for sustainable development (ESD), as a transformative, long-term solution. Education, by instilling a robust environmental ethic and raising awareness of the multifaceted ecological challenges, offers a uniquely sustainable pathway for behavioral change. The value of education lies not only in its ability to shift individual and collective behaviors, but also in its role as a critical lever for enhancing the social acceptability of environmental policies. Through its participatory and preventative nature, education fosters a more flexible, adaptive evolution of values and actions that align with the urgent demands of environmental sustainability. It empowers individuals to make informed, ethically grounded decisions, thereby reinforcing collective responsibility for sustainable development.

However, the central insight of this chapter is the argument that the effectiveness of environmental governance cannot be solely dependent on the binary choice between regulation and education. Instead, the integration of both approaches is essential to maximize their complementary effects. A hybrid framework combining the structural rigor of regulations with the adaptive and inclusive nature of education represents the most effective model for achieving sustainable governance. By involving citizens actively in the decision-making process and equipping them with the tools to understand and act on environmental issues, this dual approach ensures that policies are not only more effective in achieving environmental goals but also more legitimate in the eyes of society.

Our contribution fills a critical gap in the existing economic literature by offering a conceptual and practical framework that marries regulatory and educational strategies, highlighting the economic and social advantages of an integrated governance model. While much of the literature focuses on regulatory instruments as the primary tool for environmental governance, our analysis brings to the fore the under-explored potential of education as a sustainable economic lever. We argue that education is not merely a supplementary tool but a foundational component of long-term environmental strategies that can reduce the need for more costly and rigid regulations. The economic benefits of this approach, as we have demonstrated, include reduced administrative costs, greater market flexibility, and the fostering of innovation, which are essential for adapting to future environmental challenges.

For policymakers, the implications of this work are profound. It is evident that future governance strategies must prioritize the active participation of civil society, with education as a core pillar of sustainability policies. Emerging studies reinforce the idea that participatory governance, where citizens are engaged in shaping policies, not only enhances the effectiveness of these policies but also catalyzes social and environmental innovation. Therefore, investing in education is not simply an ethical or moral imperative, but an economically sound strategy for fostering a proactive, informed society that is equipped to drive long-term sustainability.

This conclusion calls for a transformative approach to environmental governance, one that redefines the role of education in shaping sustainable futures. We advocate for a model where education and regulation are not seen as competing forces but as complementary elements that together create a more effective system of environmental governance. By focusing on both the cognitive and behavioral dimensions of sustainability, this hybrid model represents an optimal path forward. It integrates the benefits of both policy tools while ensuring that economic efficiency and social justice are not mutually exclusive.

References:

Aatany, M., & Tarbalouti, E. (2024). Untangling the link between political regime and economic growth: A multidimensional analysis through a recent systematic literature review. *International Journal of Strategic Management and Economic Studies (IJSMES)*, *3*(2), 598–607. https://doi.org/10.5281/zenodo.10877949

Abdarssoul, N. (2024). Méthodes d'Évaluation des Impacts Environnementaux et leur efficacité dans les projets de développement: études de cas comparatives. *Journal of Business and Technologies, 1*(3), 3). https://jobt.org/index.php/publications/article/view/17

Acemoglu, D., & Robinson, J. (2008). *The role of institutions in growth and development* (Vol. 10). World Bank. https://doi.org/10.1596/978-0-8213-8100-7

Acemoglu, D., & Robinson, J. (2012). Institutions, political economy and growth. Nobel prize 2012 presentations.

Acemoglu, D., Aghion, P., Bursztyn, L., & Hemous, D. (2012). The environment and directed technical change. *American economic review, 102*(1), 131–166.

Aghion, P., Cette, G., & Cohen, É. (2014). Chapitre 2. Penser autrement. *Hors collection*, 51–93. https://doi.org/10.3917/oj.aghio.2014.01

Aghion, P., Hemous, D., & Kharroubi, E. (2014). *The effect of environmental regulations on innovation. Journal of Environmental Economics and Management.* https://doi.org/10.1016/j.jmoneco.2013.12.003

Agoulon, A., Butet, A., Hoch, T., Perez, G., Plantard, O., Verheyden, H., & Vourc'h, G. (2016). *3. Dynamique des populations de tiques et liaison avec les facteurs environnementaux. Tiques et maladies à tiques: biologie, écologie évolutive, épidémiologie* (pp. 85–112). *IRD Editions.* https://doi.org/10.4000/books.irdeditions.9027

Alaamri, N., Tarbalouti, E., & El, A. I. D. I. (2024). Repercussions on innovation in education for sustainable development: A theoretical reflection. *International Journal of Research and Innovation in Social Science, 8*(3s), 2311–2325. https://doi.org/10.47772/IJRISS.2024.803169S

Andre, P., Boneva, T., Chopra, F., & Falk, A. (2024). Misperceived social norms and willingness to act against climate change. *Review of Economics and Statistics*, 1–46.

Assembly, U. G. (1972). *Resolution 2997 (XXVII): Institutional and Financial Arrangements for International Environmental Cooperation'.* A/RES/27/2997 15 Dec. https://doi.org/10.1017/S0020782900056679

Babiker, M. H. (2005). Climate change policy, market structure, and carbon leakage. *Journal of international Economics, 65*(2), 421–445.

Baranzini, A., Van den Bergh, J. C., Carattini, S., Howarth, R. B., Padilla, E., & Roca, J. (2017). Carbon pricing inclimate policy: seven reasons, complementary instruments, and political economy considerations. *Wiley Interdisciplinary Reviews: Climate Change, 8*(4), e462.

Baum, C. M., & Gross, C. (2017). Sustainability policy as if people mattered: Developing a framework for environmentally significant behavioral change. *Journal of Bioeconomics, 19*, 53–95. https://doi.org/10.1007/s10818-016-9238-3

Baumol, W. J., & Oates, W. E. (1988). *The theory of environmental policy*. Cambridge University Press.

Bayard, B. (2024). *Le déficit de gouvernance du système éducatif haïtien: Une obstruction au projet de mise en oeuvre du curriculum de l'éducation à la citoyenneté du nouveau secondaire*. Le Lys Bleu Éditions.

Becker, G. S. (1964). Human capita. New York: National Bureau of Economic Research.

Bénabou, R., & Tirole, J. (2006). Incentives and prosocial behavior. *American economic review, 96*(5), 1652–1678.

Brundtland, G. H. (1987). Our common future Call for action. *Environmental Conservation, 14*(4), 291–294. https://doi.org/10.1017/S0376892900016805

Callens, S. (2023). Chapitre 6. Droits de propriété, coûts de transaction. *Hors collection*, 99–115.

Carattini, S., Baranzini, A., Thalmann, P., Varone, F., & Vöhringer, F. (2017). Green taxes in a post-Paris world: Are millions of nays inevitable? *Environmental and Resource Economics, 68*, 97–128. https://doi.org/10.1007/s10640-017-0133-8

Chiroleu-Assouline, M. (2007). Efficacité comparée des instruments de régulation environnementale.

Cialdini, R. B. (2003). *Crafting normative messages to protect the environment*. Current Directions in Psychological Science. https://doi.org/10.1111/1467-8721.01242

Coase, R. H. (2013). The problem of social cost. *The Journal of Law and Economics, 56*(4), 837–877. https://doi.org/10.1086/674872

Dijksterhuis, A., Smith, P. K., Van Baaren, R. B., & Wigboldus, D. H. (2005). The unconscious consumer: Effects of environment on consumer behavior. *Journal of Consumer Psychology, 15*(3), 193–202. https://doi.org/10.1207/s15327663jcp1503_3

Ellerman, A. D., Convery, F. J., & De Perthuis, C. (2010). *Pricing carbon: The European Union emissions trading scheme*. Cambridge University Press. https://doi.org/10.1017/CBO9780511760803

Espinosa Diaz, S. (2023). Analyzing transaction costs in agri-environment-climate measures.

Espinosa Diaz, S., Riccioli, F., Di Iacovo, F., & Moruzzo, R. (2023). Transaction costs in agri-environment-climate measures: A review of the literature. *Sustainability, 15*(9), 7454.

Estrada-Araoz, E. G., Gallegos Ramos, N. A., Paredes Valverde, Y., Quispe Herrera, R., & Mori Bazán, J. (2023). Examining the relationship between environmental education and pro-environmental behavior in regular basic education students: A cross-sectional study. *Social Sciences, 12*(5), 307. https://doi.org/10.3390/socsci12050307

Fabre, A., Douenne, T., & Mattauch, L. (2025). Majority support for global redistributive and climate policies. *Nature Human Behaviour*, 1–12.

Facchini, F., & Falque, M. (2022). Droit de propriété et protection de l'environnement. *Journal des libertés.*

Falque, M. (1991). La puissance publique: garante ou destructrice de l'environnement? *Journal des économistes et des études humaines, 2*(1), 103–122.

Fien, J., & Tilbury, D. (2002). The global challenge of sustainability. *Education and Sustainability: Responding to the Global Challenge, 1.*

France Stratégie. (2024). Guide de l'évaluation socioéconomique des investissements publics. Action publique. Recherche et Pratiques, (2), 60–60.

Frederick, S., Loewenstein, G., & O'donoghue, T. (2002). Time discounting and time preference: A critical review. *Journal of economic literature, 40*(2), 351–401.

Geller, E. S. (2002). The challenge of educating for sustainable behavior. *Journal of Social Issues.* https://doi.org/10.1111/0022-4537.00176

Gifford, R., & Nilsson, A. (2014). Personal and social factors that influence pro-environmental concern and behaviour: A review. *International Journal of Psychology, 49*(3), 141–157. https://doi.org/10.1002/ijop.12034

Girault, Y., & Sauvé, L. (2008). L'éducation scientifique, l'éducation à l'environnement et l'éducation pour le développement durable Croisements, enjeux et mouvances. *aster.*

Gneezy, U., & Rustichini, A. (2000). Pay enough or don't pay at all. *The Quarterly journal of economics, 115*(3), 791–810.

Gough, S., Scott, W., & Stables, A. (2000). Beyond O'Riordan: Balancing anthropocentrism and ecocentrism. *International Research in Geographical and Environmental Education, 9*(1), 36–47. https://doi.org/10.1080/10382040008667642

Goulder, L. H., & Parry, I. W. (2008). Instrument choice in environmental policy. *Review of Environmental Economics and Policy.* https://doi.org/10.1093/reep/ren005

Greenstone, M. (2004). Did the Clean Air Act cause the remarkable decline in sulfur dioxide concentrations? *Journal of Environmental Economics and Management, 47*(3), 585–611. https://doi.org/10.1016/j.jeem.2003.12.001

Guertin, M. A., Poirier-Ghys, G., Reed, M. G., Dupras, J., & Rajotte Sauriol, G. (2019).L'éducation relative à l'environnement dans la Réserve de biosphère du mont Saint-Hilaire: observations et tensions. *Éducation relative à l'environnement. Regards-Recherches-Réflexions, 15*(1). doi:https://doi.org/10.4000/ere.3810

Hanushek, E. A., & Woessmann, L. (2012). Do better schools lead to more growth? Cognitive skills, economic outcomes, and causation. *Journal of economic growth, 17*(4), 267–321.

Heath, M. K., Gleason, B., Mehta, R., & Hall, T. (2024). More than knowing: Toward collective, critical, and ecological approaches in educational technology research. *Educational Technology Research and Development, 72*(5), 2519–2541. https://doi.org/10.1007/s11423-024-10223-9

Jaffe, A. B., Newell, R. G., & Stavins, R. N. (2005). A tale of two market failures: Technology and environmental policy. *Ecological economics, 54*(2-3), 164–174

Janmaimool, P., & Khajohnmanee, S. (2020). Enhancing university students' global citizenship, public mindedness, and moral quotient for promoting sense of environmental responsibility and pro-environmental behaviours. *Environment, Development and Sustainability, 22*(2), 957–970. https://doi.org/10.1007/s10668-018-0228-6

Jaoul-Grammare, M., & Stenger, A. (2023). L'éducation, un déterminant essentiel des comportements pro environnementaux.

Jordan, A., Wurzel, R. K., & Zito, A. (2005). The rise of 'new'policy instruments in comparative perspective: Has governance eclipsed government? *Political Studies, 53*(3), 477–496. https://doi.org/10.1111/j.1467-9248.2005.00540.x

Kohlscheen, E., Moessner, R., & Takats, E. (2024). Effects of carbon pricing and other climate policies on CO_2 emissions. arXiv preprint arXiv:2402.03800.

Kollmuss, A., & Agyeman, J. (2002). Mind the gap: Why do people act environmentally and what are the barriers to pro-environmental behavior? *Environmental Education Research, 8*(3), 239–260. https://doi.org/10.1080/13504620220145401

Kooiman, J., Jentoft, S., Bavinck, M., & Pullin, R. (2005). *Fish for life: Interactive governance for fisheries* (p. 432). Amsterdam University Press. https://doi.org/10.5117/9789053566862

Lanssens, C. (2024). L'analyse d'impact de la réglementation (AIR) en Belgique, 1997-2023. *Courrier hebdomadaire du CRISP*, 25922593(7), 1–124

Liu, C., Bardaka, E., & Paschalidis, E. (2023). Sustainable transport choices in public transit access: Travel behavior differences between university students and other young adults. *International Journal of Sustainable Transportation, 17*(6), 679–695. https://doi.org/10.1080/15568318.2022.2084656

Lucas Jr, R. E. (1988). On the mechanics of economic development. *Journal of monetary economics, 22*(1), 3–42.

Medema, S. G. (2014). Economics and institutions. *Revue économique, 65*(2), 243–261. https://doi.org/10.3917/reco.652.0243

Nordhaus, W. (2019). Climate change: The ultimate challenge for economics. *American Economic Review, 109*(6), 1991–2014. https://doi.org/10.1257/aer.109.6.1991

Ostrom, E. (1990). Governing the commons: The evolution of institutions for collective action. Cambridge universitypress.

Ostrom, E. (2010). Beyond markets and states: Polycentric governance of complex economic systems. *American Economic Review, 100*(3), 641–672. https://doi.org/10.1257/aer.100.3.641

Ouattara, M. O., & Sadia, M. A. (2025, May). Enjeux Éthiques De L'utilisation De L'ia Dans La Gestion Des Ressources Humaines Educative En Côte D'ivoire. In *7ème Colloque International Du Raiffet Du 5 Au 8 Mai 2025 À L'ipnetp D'abidjan En Côte D'ivoire.*
Pareliussen, J., & Purwin, A. (2023). Climate policies and Sweden's green industrial revolution. Documents de travail du Département des Affaires économiques de l'OCDE.
Pigou, A. C. (1920). Co-operative societies and income tax. *The Economic Journal, 30*(118), 156–162.
Pigou, A. (2017). *The economics of welfare.* Routledge. https://doi.org/10.4324/9781351304368
Pindyck, R. S. (2013). Climate change policy: What do the models tell us? *Journal of Economic Literature, 51*(3), 860–872. https://doi.org/10.1257/jel.51.3.860
Porter, M. E., & Linde, C. V. D. (1995). Toward a new conception of the environment-competitiveness relationship. *Journal of economic perspectives, 9*(4), 97–118
Rakotoarijaona, N. H. (2016). Gestion communautaire des ressources naturelles et vulnérabilité à la pauvreté de la population locale: cas de Didy et de Vohitsara. In *XXXIIèmes Journées du développement ATM 2016: Catastrophes, vulnérabilités et résiliences dans les pays en développement.*
Rau, H., Nicolai, S., & Stoll-Kleemann, S. (2022). A systematic review to assess the evidence-based effectiveness, content, and success factors of behavior change interventions for enhancing pro-environmental behavior in individuals. *Frontiers in Psychology, 13,* Article 901927. https://doi.org/10.3389/fpsyg.2022.901927
Riordan, M. H., & Williamson, O. E. (1985). Asset specificity and economic organization. *International journal of industrial organization, 3*(4), 365–378.
Romer, P. M. (1990). Endogenous technological change. *Journal of political Economy, 98*(5, Part 2), S71–S102
Roudil, A. (2024). *Vers un curriculum possible et pertinent d'une éducation à une économie durable au temps de l'anthropocène.* Doctoral dissertation, Université de montpellier.
Ryan, R. M., & Deci, E. L. (2000). Self-determination theory and the facilitation of intrinsic motivation, social development, and well-being. *American psychologist, 55*(1), 68.
Sacko, N. D. A. (2024). Les défis/obstacles rencontrés par les gouvernements dans l'application de la taxe carbone: l'approche canadienne.
Saltelli, A., & Di Fiore, M. (Éds.). (2023). *The politics of modelling: Numbers between science and policy.* Oxford University Press.
Schultz, P. W. (2011). Conservation means behavior. *Conservation Biology, 25*(6), 1080–1083. https://doi.org/10.1111/j.1523-1739.2011.01766.x

Shutaleva, A., Martyushev, N., Nikonova, Z., Savchenko, I., Abramova, S., Lubimova, V., & Novgorodtseva, A. (2021). Environmental behavior of youth and sustainable development. *Sustainability, 14*(1), 250.

Slaev, A. D. (2023). The transaction costs of sustainability: Coase's Proviso and the roles of environmentalists and the government. *Sustainability, 15*(4), 3096.

Stavins, R. (2007). Environmental economics.

Sterling, S., & Orr, D. (2001). *Sustainable education: Re-visioning learning and change* (Vol. 6). Green Books for the Schumacher Society. ISBN-10: 1870098994.

Stern, P. C. (2000). New environmental theories: Toward a coherent theory of environmentally significant behavior. *Journal of Social Issues, 56*(3), 407–424. https://doi.org/10.1111/0022-4537.00175

Sterner, T. (2012). Distributional effects of taxing transport fuel. *Energy Policy, 41*, 75–83. https://doi.org/10.1016/j.enpol.2010.03.012

Sterner, T., & Coria, J. (2013). *Policy instruments for environmental and natural resource management*. Routledge. https://doi.org/10.4324/9781315780894

Stiglitz, J. E. (2000). The contributions of the economics of information to twentieth century economics. *The quarterly journal of economics, 115*(4), 1441–1478.

Stiglitz, J. E. (2006). Un autre monde: contre le fanatisme du marché. ISBN-13: 978-2-213-62748-9.

Stiglitz, J. E. (2017). Industrial policy, learning, and development. *The Practice of Industrial Policy: Government–Business Coordination in Africa and East Asia*, 23–39. https://doi.org/10.1093/acprof:oso/9780198796954.003.0002

Stiglitz, J. E., & Stiglitz, J. E. (2000). Economics of the public sector. ISBN-10: 0393966518.

Stiglitz, J. E., & Weiss, A. (1981). Credit rationing in markets with imperfect information. *The American Economic Review, 71*(3), 393–410. https://www.jstor.org/stable/1802787

Takeuchi, H., & Tanaka, H. (2020). Water reuse and recycling in Japan History, current situation, and future perspectives. *Water Cycle, 1*, 1–12. https://doi.org/10.1016/j.watcyc.2020.05.001

Thaler, R., & Sunstein, C. (2008). Nudge: Improving decisions about health, wealth and happiness. In *Amsterdam Law Forum; HeinOnline: Online* (p. 89). ISBN: 978-0-141-04001-1.

Tilbury, D. (2011). Higher education for sustainability: A global overview of commitment and progress. *Higher Education in the World, 4*(1), 18–28. ISBN: 978-0-230-53555-8.

United Nations. (1992). Rio declaration on environment and development (A/CONF. 151/26 (Vol. I)).

Vecina, M. L., Alonso-Ferres, M., López-García, L., & Díaz-Silveira, C. (2024). Eco-anxiety and trust in science in Spain: Two paths to connect climate change perceptions and general willingness for environmental behavior. *Sustainability, 16*(8), 3187.

CHAPTER 11

Generation Z's Attitudes on Global Climate Justice: A Case Study

Christine Faith Murillo, Yvonnie Antolin, Lance Guiang, Charles Rj Arroyo, Kathleen Mariel Parreñas, Phil Rodolf Quisel, Chardrey Tan, Mearlyn Vargas, and Kenneth Lisbo

1 Introduction

The twenty-first-century climate crisis imposes economic, social, and environmental burdens on poorer, hotter, and lower-lying nation-states (Tol, 2018). The global climate justice framework addresses these impacts on vulnerable populations (Kumar, 2024), recognizing this contemporary crisis as an environmental, social, and ethical issue exacerbating systemic inequalities (Heyen, 2022). Grounded in Thomas Pogge's global justice

C. F. Murillo • Y. Antolin • L. Guiang • C. R. Arroyo • K. M. Parreñas • P. R. Quisel • C. Tan • M. Vargas • K. Lisbo (✉)
Department of Political Science, Mindanao State University – General Santos, General Santos, Philippines
e-mail: christinefaith.murillo@msugensan.edu.ph; yvonnie.antolin@msugensan.edu.ph; lance.guiang@msugensan.edu.ph; charlesrj.arroyo@msugensan.edu.ph; kathleenmariel.parrenas@msugensan.edu.ph; philrodolf.quisel@msugensan.edu.ph; chardrey.tan@msugensan.edu.ph; mearylyn.vargas@msugensan.edu.ph; kenneth.lisbo@msugensan.edu.ph

K. Hirech (ed.), *Environmental Education and Sustainable Development*, https://doi.org/10.1007/978-3-032-05760-0_11

framework, this chapter supports his critique of the existing global order, which perpetuates systemic injustices, fueling environmental degradation and social inequality (Pogge, 2008). Also, he argues that wealthy nations are more responsible for addressing these imbalances since their actions disproportionately contribute to environmental destruction (Pogge, 2010).

In the Global South, the Philippines is highly susceptible to the effects of climate change, and pursuing climate justice is one of its priorities (Lazarito et al., 2024). It experiences intensified tropical typhoons, rising sea levels, storm surges, and extreme heat (Rogayan, 2024), threatening biodiversity, food security, public health, and the economy (Reynon, 2021). In 2013, Typhoon Haiyan (Yolanda) caused the largest storm surge in the nation's recorded history, causing significant destruction in Visayas (Soria et al., 2016). Supplementing this, the 2021 Climate Risk Index reported that the Philippines ranked 17th as the country most affected by extreme weather events.

Correspondingly, the Philippines is also vulnerable to slow-onset climate changes, specifically increased temperatures and prolonged droughts that affect food security and exacerbate societal problems (Berse, 2024). The extreme heat conditions affected areas of agriculture, causing crop damage and education, affecting millions of students (World Weather Attribution, 2024). In General Santos City, Southern Philippines, the disruption caused by the heatwaves of 2023 and 2024 prompted the local government and Mindanao State University—General Santos (MSU-GenSan) to suspend face-to-face classes and shift to online modalities (Rebollido, 2024). The City's Department of Education reported that at least 2557 students complained of severe headaches, and 291 experienced nosebleeds due to intense heat (Rebollido, 2024). This phenomenon highlights the need for climate adaptation and increased awareness of climate activism among young people. Studies suggest that access to climate change education increases students' participation in climate activism. However, universities need to implement an environmental education curriculum to increase the pro-environmental attitudes of the younger generation (Almansa-Martínez et al., 2024).

For the purposes of this chapter, Generation Z is a significant demographic group born roughly between 1997 and 2012 (Beresford Research, 2023). Literature pointed out that they perceive climate change as a threat causing anxiety and necessitating a sense of urgency toward environmental advocacy (Wijaya & Kokchang, 2023). Unlike the older generations, their political engagements and activism are influenced mainly by digital platforms. This provides them with spaces for mobilization, awareness, and pressure on politicians on a global scale (Surmacz et al., 2024). Filipino Generation Z also engaged in climate activism in the Philippines, exemplified by youth-led organizations like the Youth Advocates for Climate Action Philippines, which was established in 2019 to create an alliance of youth organizations to combat the detrimental effects of climate change. They campaign for collective efforts to facilitate systemic change (Youth Advocates for Climate Action, 2024).

These movements indicate Filipino Generation Z engaged with climate issues through direct action and youth-led organizations. These efforts are essential in examining how they behave and navigate their generational role in combating the detrimental effects of climate change. For the Generation Z students enrolled at MSU-GenSan, who experienced the adverse impacts of climate change, this concern implies an imperative discussion and a call for action. Advancing this assertion, this chapter investigated the attitudes of Generation Z students at MSU-GenSan toward global climate justice and their perceived need to engage in climate activism actively within Thomas Pogge's framework. This chapter explored the following areas: their knowledge and awareness of global climate justice and environmental policies and the perceived need to get involved in climate activism, such as education, funding, and networking opportunities.

This chapter contributes to the existing research on climate change studies in the Philippines by looking at the attitudes of Generation Z students who experienced firsthand the repercussions of climate-induced disasters such as typhoons and heatwaves. Furthermore, this chapter is consistent with the Sustainable Development Goal (SDG 13) which emphasizes urgent action to counter climate change and its impacts. This chapter aims to improve education and awareness of climate activism, specifically to integrate it into the curriculum and acquire funding and networking opportunities for the Global South's younger generations.

2 Global Justice on Climate Change

This chapter was framed from Thomas Pogge's Global Justice theory and offers an analytical framework to understand the attitudes of Generation Z students toward climate justice at MSU-GenSan, Southern Philippines. Pogge (2008), in his seminal work *World Poverty and Human Rights: Cosmopolitan Responsibilities and Reforms*, argued that the international economic order and political institutions mainly served the interests of wealthy countries in the Global North. He proposed that this system perpetuates systemic inequalities that disadvantage poorer countries in the Global South. This analysis aligns with current climate change debates, as it shows how wealthy nations have contributed significantly to greenhouse gas emissions while poorer nations, which contribute less, suffer more from rising temperatures (United Nations Office of Disaster Risk Reductions, 2022). In light of this, Generation Z is vital in demanding global justice especially in the Global South, for example the Philippines, and must adopt measures, mainly through higher education institutions, to integrate climate justice and activism into curriculum and operations. This call responds to the continuing global injustice that leaves the Global South vulnerable to climate change effects.

Pogge (2008) stressed the need to rethink international policies and governance to ensure fair responsibility-sharing and address global problems. He challenges conventional systems managing climate change, urging systemic reform in the global order. The call to transform global institutions that govern economic and environmental relations gives Generation Z a normative basis to question intergenerational equity and demand global accountability. These theoretical underpinnings resonate with Generation Z, a cohort known for its drive for social justice, sustainability, and grassroots movements (Confetto et al., 2023).

Continuing this discussion, Pogge's framework indirectly responds to climate change, which requires shared responsibility among nations. From these insights, individual action is also key to achieving climate justice. Therefore, education must equip the youth with the knowledge and skills for climate action. Studies recommend fully integrating climate education into curricula so students can grasp environmental policies and justice frameworks (Bartlett et al., 2022). This knowledge gap limits their involvement in climate activism. Supporting this, funding student-led initiatives empowers youth to sustain engagement in local climate projects. Finally, networking opportunities can foster collaboration between student organizations, NGOs, and other stakeholders in collective climate advocacy.

3 Methodology

This study utilized a quantitative research framework that analyzed the Generation Z students' attitudes toward global climate justice at MSU-GenSan. A survey questionnaire was employed to understand the

Table 11.1 Likert scale interpretation

Likert-scale description	*Likert scale*	*Likert-scale Interval*	*Interpretation*
Strongly Agree	5	4.21–5.00	Very high
Agree	4	3.42–4.20	High
Neutral	3	2.61–3.40	Moderate
Disagree	2	1.81–2.60	Low
Strongly Disagree	1	1.00–1.80	Very low

respondents' awareness of climate change and their perception of education, funding, and networking opportunities for climate activism. The survey included student participants from different university colleges to acquire a measurable data set. To assess the attitudes of Generation Z students toward global climate justice, this study employed a structured survey questionnaire utilizing a five-point Likert scale. This scale allowed respondents to express their level of agreement with various statements. The Likert scale ranges from "Strongly Disagree" (1) to "Strongly Agree" (5), with corresponding interval scores to determine the degree of agreement or perception. Table 11.1 presents the specific intervals and their corresponding interpretations, which guided the analysis of the collected data (Table 11.1).

Moreover, the survey was administered through Google Forms to maximize participation and ensure inclusivity among the university students. Descriptive statistics, including mean and standard deviation, were utilized to analyze the central tendencies and variability of the participants' responses. The survey questionnaire was also adapted and modified from the existing literature on climate justice and aligned to the local context of Southern Philippines, validated by two political science professors, one philosophy professor, one psychology professor, and one climate activist.

Furthermore, the sampling distribution was outlined utilizing proportionate stratified random sampling to ensure fair representation from each college based on the 2024–2025 data. The total number of enrolled students is 9676 based on the Office of the University Registrar's data. Using Slovin's formula with a 5% margin of error, the computed sample size was 384 students.

This methodological approach ensured that more prominent colleges contributed more respondents while smaller colleges remained adequately represented. The calculated sample sizes provided a balanced distribution, maintaining the reliability and generalizability of the study's findings while accurately reflecting MSU-GenSan's student demographics. Ethical approval was sought from university authorities, consent forms were obtained from all the respondents, and their responses were treated with utmost confidentiality. The participants' anonymity was maintained through unique identifiers, and the questionnaires were securely stored using a two-way process for security authentication. To supplement the ethical and data management protocol, the study adhered to the Philippine laws through compliance with the Data Privacy Act of 2012 of the Republic of the Philippines (R.A. No. 10173). This chapter also ensured that the findings were used solely for academic purposes and that the participants were informed that they had the right to withdraw from their participation in the study at any stage without prejudice.

4 Results and Discussion

This chapter examined the perceptions and attitudes of Generation Z students at MSU-GenSan toward climate justice. The responses provide insight into their knowledge of global climate policies, perceived need for educational initiatives, financial support, and collaborative opportunities to strengthen climate action. Table 11.2 summarizes the mean and standard deviation for each indicator under the four categories, revealing the trends and general sentiments of the respondents regarding their role in climate activism.

4.1 Climate-Conscious or Clueless? Generation Z's Awareness About Climate Justice

Awareness about global climate justice contributes to Generation Z's participation in environmental advocacy, policy discourse, and sustainable practices. As the generation most affected by climate change, it necessitates their understanding of climate justice and willingness to engage and support demand for systemic reforms for the global order. A significant remark from the data is the strong agreement that "climate change affects everyone" (Mean = 4.74, SD = 0.76). This proves the study of Ariestya et al. (2022), as both studies claimed that Generation Z shows a strong climate change awareness. This implies that Generation Z students at MSU-GenSan widely acknowledge the universal impact of climate change.

Table 11.2 Generation Z's attitudes toward climate justice at MSU-GenSan

Category	*Indicators*	*Mean*	*Standard deviation*	*Significance level*
Awareness	Climate change has a universal impact on all individuals	4.74	0.76	Strongly Agree
	Awareness of global rules on climate change is essential	3.96	0.80	Agree
	Understanding climate justice and international environmental policies is crucial	3.9	0.84	Agree
	There is a lack of sufficient awareness about global climate justice	4.13	0.94	Agree
	Finding reliable information about global climate justice is challenging	3.42	0.99	Agree
	Current plans to combat the detrimental effects of climate change are insufficient	3.89	1.05	Agree
	Increased discussions on global climate justice and international environmental policies are necessary	4.58	0.68	Strongly Agree
	Global regulations play a crucial role in addressing climate change	4.59	0.64	Strongly Agree
	Academic discussions should incorporate global climate justice more inclusively	4.41	0.72	Strongly Agree
	Higher education institutions should implement comprehensive programs and discussions on global climate justice	4.36	0.76	Strongly Agree
Education	Climate change education enhances Generation Z's knowledge and engagement in activism	4.22	0.83	Strongly Agree
	Access to climate and environmental resources encourages participation in activism	4.13	0.84	Agree
	Integrating climate education across disciplines boosts student engagement in climate activism	4.21	0.81	Strongly Agree
	More climate-related events and seminars in the university would motivate Generation Z to engage in activism	4.16	0.82	Agree
	Mentorship from experienced climate activists increases Generation Z's likelihood of participating in activism	4.14	0.82	Agree

(*continued*)

Table 11.2 (continued)

Category	*Indicators*	*Mean*	*Standard deviation*	*Significance level*
Funding	Financial support is essential for Generation Z students to engage in climate activism	4.11	0.84	Agree
	Government funding for student-led climate activism projects would enhance its effectiveness	4.20	0.77	Agree
	Allocating funds for climate activism is important to sustaining student-led initiatives	4.02	0.81	Agree
	Generation Z students would actively participate in climate justice campaigns with adequate financial support	4.13	0.77	Agree
	More students would engage in climate activism if additional funding opportunities for climate-related projects were available	4.04	0.80	Agree
Networking	Networking opportunities are important for Generation Z students to strengthen climate justice movements	4.12	0.77	Agree
	Climate justice networking events would foster interdisciplinary collaboration among students	4.12	0.77	Agree
	Building connections with climate activists and organizations is important for creating impactful initiatives	4.20	0.74	Agree
	A platform for discussion and idea-sharing with climate activists would support student engagement	4.20	0.74	Agree
	Strong professional and social networks would increase Generation Z's participation in climate activism	4.22	0.78	Strongly Agree

However, participants only have moderate familiarity with the "global rules on climate change" (Mean = 3.96, SD = 0.80) and "climate justice and international environmental policies." Persada et al. (2019) also observed that although technology and social media were the catalysts of Generation Z's awareness, the available information with different interpretations and understanding of issues like climate change can be a challenge to them. This implies the need for more organized climate education and integrating climate governance lessons within the university curricula.

The respondents' agreement with "insufficient awareness about global climate justice" (Mean = 4.13, SD = 0.84) reinforces the knowledge gap identified by this study. It implies that many Generation Z students still perceive that their awareness of climate change is inadequate. Substantiating this finding, respondents believe that "finding reliable information about global climate justice is challenging" (Mean = 3.42, SD = 0.94). This reinforced the studies of Sloan Morgan (2023) that the younger generation faced obstacles in contributing to justice-focused climate action. Both studies also suggest that there is a need for empowering mechanisms to increase engagement in climate justice.

Moreover, participants strongly agree that "there should be more discussions on global climate justice and international environmental policies" (Mean = 4.58, SD = 0.68). This suggests a high demand for significant engagement and dialogue within the academic community. They also strongly agreed that "Global regulations play a crucial role in addressing climate change" (Mean = 4.59, SD = 0.64). This corroborates the study of Shohel and Sall (2024), as these studies put forward the integrations of academic discussion on climate change as a form of social justice that keeps global sustainability and development. Also, these findings imply that higher education institutions like MSU-GenSan must create opportunities for students to be exposed to policy discussions, workshops, research, and community engagements in climate justice.

In light of this analysis, participants also demand "greater inclusion of global climate justice in academic discussions" (Mean = 4.41, SD = 0.72), which exemplifies the need to integrate climate justice education into the curriculum. This aligns with the idea that Generation Z becomes integrated into public spaces as leaders and stakeholders require exposure to government frameworks (Tan et al., 2023). Generation Z students also strongly agree that "Higher education institutions should offer comprehensive programs and discussions on global climate justice" (Mean = 4.36, SD = 0.83), illustrating higher education institutions' proactive expectations to promote climate education. Lastly, the Generation Z students were highly aware of global climate justice, based explicitly on this section's overall mean and standard deviation (M = 4.20, SD = 0.77). This indicates that this young cohort, in context, recognizes the detrimental effects of climate change and the implications of environmental problems.

4.2 *Climate Justice in the Classroom: Emancipating Generation Z Students in Climate Education*

Climate activism among the younger generation, especially Generation Z, has gained momentum as these cohorts recognize their role in addressing environmental concerns. According to Neas et al. (2022), one of the key limitations and overemphasis of the current debate on youth climate activism is that it is overly focused on mass mobilizations. One of the dominant discussions of the literature was connected to the actions of young climate activist Greta Thunberg, who influenced the global Fridays to Future movement climate activism. This movement encouraged young people to organize school strikes and walkouts for climate action. However, these situations are only one element of the role of Generation Z in climate action. This section will discuss how access to knowledge, resources, interdisciplinary approaches for formal and informal curricula, experiential learning, and proper mentorship influences students' engagement.

An important observation from the findings is the strong agreement that "Climate change education enhances Generation Z's knowledge and engagement in activism" (Mean = 4.22, SD = 0.83) and "Access to climate and environmental resources encourages participation in activism" (Mean = 4.13, SD =0.84). This suggests that Generation Z students recognize the role of formal education in honing their understanding. This can be strengthened if the higher institution has enough resources to enable climate justice engagement. Furthermore, Pogge (2008) remarked that institutional injustices prevent communities from accessing essential resources like education that can facilitate collective action. Generation Z students strongly agreed that "Integrating climate education across disciplines boosts student engagement in climate activism" (Mean = 4.21, SD = 0.81). The Generation Z students also agreed that "More climate-related events and seminars would motivate them to engage in activism" (Mean = 4.16, SD = 0.82) and "Mentorship from experienced climate activists increases their likelihood of participating in activism" (Mean = 4.14, SD =0.82). This highlights the necessity of continuous experiential learning and engagement of younger cohorts to develop a sense of informed climate citizenship values. Karpudewan and Khan (2017) believe experiential-based climate education will improve students' knowledge and motivation to protect the environment. This implies that the Generation Z students at MSU-GenSan perceived that activities would provide platforms for students to engage with climate and environmental experts, policymakers, and fellow activists and create a culture of collective responsibility.

Lastly, the overall perception of the perceived need for education for climate activism indicates a mean score and standard deviation of (M = 4.17, SD = 0.82). This implies that these respondents perceived knowledge as important in facilitating change. This section demonstrates that higher education institutions must integrate climate education across disciplines, access resources, and prioritize experiential learning through educational events and mentorship programs. Pogge supports the idea that systemic changes at the institutional level are necessary to facilitate structural inequalities that prevent appropriate climate activism.

4.3 *Investing in Climate Action Initiatives: The Role of Funding Support in Empowering the Generation Z Students*

Funding was also crucial in advancing climate action initiatives by empowering higher education institutions to implement appropriate and contextualized action plans. This section delves into how Generation Z students at MSU-GenSan, Southern Philippines, recognized the importance of funding in student-led climate activism. This analysis posits that financial support can address structural inequalities and empower the younger generation to be integrated into climate action.

The Generation Z students agreed that "Financial support is essential for Generation Z students to engage in climate activism" (Mean = 4.11, SD = 0.84). This strengthens Pogge's argument (2008) that structural inequalities within the present global order limit the capacity of underdeveloped nations to contribute to global justice efforts. Furthermore, Generation Z students supported the idea that "Government funding for student-led climate activism projects would enhance their effectiveness" (Mean = 4.20, SD = 0.77). They received the highest mean score for this section. It indicates the respondents' perception of the consensus on the importance of state support for climate action initiatives.

The respondents recognized that "Allocating funds for climate activism is important to sustaining student-led initiatives" (Mean = 4.02, SD = 0.81). This implies the importance of financial sustainability in guaranteeing the long-term impact of climate activism in higher education institutions. This position indicates that the absence of consistent funding might put student-led programs at risk, making them short-lived. Additionally, the respondents agreed that "Generation Z students would actively participate in climate justice campaigns with adequate financial support" (Mean = 4.13, SD =0.77). It accentuates the crucial roles of funding availability and student involvement in climate justice initiatives. In context, Generation Z has an overall mean score and standard deviation (M = 4.10, SD = 0.84) on the perceived necessity for financial support in climate activism.

4.4 *From Collaboration to Action: How Networking Strengthens Generation Z's Climate Justice Efforts*

Networking through youth-led organizations can strengthen the political interest of the younger generations to engage in climate justice initiatives. This section highlights that networking provides a platform for discussion and idea-sharing, which enhances engagement by creating spaces for students to be integrated into climate collective action across different generations.

This section affirms that networking is one of the key components that will strengthen climate activism (Mean = 4.12, SD = 0.77). This resonates with the perspective of Pogge's (2008) notion of global justice, which highlighted the collective responsibility of individuals and nations to deal with global inequalities, including climate change. Also, Li et al. (2023) postulated that network communication can significantly increase the environmental awareness and engagement of the younger generation in ecological conservation.

Furthermore, respondents agreed that "Climate justice networking events would foster interdisciplinary collaboration among students" (Mean = 4.12, SD = 0.77). This validates the finding of Bruin et al. (2019) that interdisciplinary collaboration helped increase communication about climate change mitigation and impacts. They also agreed that "Building connections with climate activists and organizations is important for creating impactful initiatives" (Mean = 4.20, SD = 0.72). This suggests that higher education institutions must focus on facilitating external collaborations between external sectors and stakeholders to strengthen the effectiveness of student-led climate initiatives.

In addition, the Generation Z students agreed that "A platform for discussion and idea-sharing with climate activists would support student engagement" (Mean = 4.20, SD = 0.74). This harmonized with Carrick et al. (2022) that through deliberation and participatory democracy, decision-making would create spaces for dialogue, inclusivity, process orientation, and empowerment in engagement. Their perception indicates that through these spaces, higher education institutions strengthen the university's role in climate advocacy and could motivate greater student participation in climate justice initiatives.

Generation Z students also strongly adhere to the statement that "Strong professional and social networks would increase Generation Z participation in climate activism" (Mean = 4.22, SD = 0.78). This implies that social networks between younger generations and other generational cohorts are essential. Specifically, they provide them access to resources, emotional support, and strategic guidance. Enriching this, Generation Z in context has an overall mean score and standard deviation (M = 4.17, SD = 0.82) on the perceived importance of networking opportunities in climate activism. This suggests an agreement on the necessity for collaboration, mentoring, and external partnerships between NGOs, business sectors, and other organizations for linkages and cross-sector alliances.

4.5 *Composite Mean*

The composite mean was calculated for the four key dimensions explored in this study to provide a comprehensive summary of Generation Z students' attitudes toward climate justice at MSU-GenSan. The composite mean indicates the respondents' overall perception, offering insights into the most significant areas (Table 11.3).

Table 11.3 Composite mean of Generation Z's perceptions of climate justice

Category	*Mean scores*	*Interpretation*
Awareness	4.20	High
Education	4.17	High
Funding	4.10	High
Networking	4.17	High
Composite mean	**4.16**	**High**

The results from the table indicate that all four dimensions received a high composite mean score, suggesting that Generation Z students at MSU-GenSan demonstrate a strong awareness of climate justice and a willingness to engage in climate activism. The highest-rated category, awareness (M = 4.20), confirms that students recognize the urgency and significance of climate-related issues. However, despite their awareness, gaps remain in their understanding of global climate policies, signaling the need for more structured educational initiatives. Both education (M = 4.17) and networking (M = 4.17) were rated similarly, highlighting the role of academic institutions and social connections in empowering students to participate in climate action. The findings suggest that students perceive climate education as essential for strengthening their activism, while networking allows them to collaborate with external organizations, mentors, and policymakers.

While slightly lower than the others, the funding category (M = 4.10) remains a crucial concern. Financial constraints may hinder students from actively participating in climate initiatives, as many require resources, event organization, and campaign materials. This finding reinforces the need for government support, university grants, and sponsorships to sustain student-led climate action.

Overall, the composite mean of 4.16 confirms that while Generation Z students at MSU-GenSan are highly engaged in climate justice, they require institutional support regarding education, funding, and networking to sustain their efforts. These findings emphasize the importance of structural interventions in higher education institutions to empower students as active agents in climate activism.

5 Conclusion

This chapter's analysis highlights Generation Z students' attitude toward climate activism at MSU-GenSan, Southern Philippines, through the four identified dimensions: awareness, education, funding, and networking. These components were framed through the present literature and debates on climate justice and the framework postulated by Pogge's Global Justice. The researchers analyzed these components to understand their need to engage in climate justice movements. Awareness is the primary foundation for Generation Z in context, as experiential learning on climate-related issues influences their behaviour and motivation to participate in the climate action movements and initiatives. Education strengthens this foundation by calling on higher education institutions like MSU-GenSan to hone students' scientific knowledge and critical thinking skills through climate pedagogy. Also, funding solidifies the call that activism moves beyond discourse but requires resources and financial support. Lastly, networking guarantees experiential learning from experts, external organizations like NGOs, and business sectors that intensify and sustain their climate movement initiatives through intergenerational collaboration and partnerships.

The study put forward the argument of the global justice framework of Thomas Pogge that, in the context of climate change, it is inherently connected with systemic inequalities; this moral obligation requires institutional support, financial resources, and networking. Moving forward, higher education institutions like MSU-GenSan should strengthen their role as agents and collaborators of student-led movements by:

Strengthening climate awareness programs through university-sponsored campaigns and advocacy:

- Integrate climate education for both informal and formal curricula to hone students' appropriate competencies and skills for evidence and contextualized-based advocacy;
- Prioritize long-term financial support mechanisms, which include grants, crowdfunding, and public–private collaborations. This amplifies the sustainability of financial aspects for student-led climate programs.
- Create linkages and networking opportunities through intergenerational collaborations with NGOs, business sectors, and other external organizations or groups to mobilize a stronger and more resilient climate movement across generations.

Acknowledging the significance of the attitudes of Generation Z students at MSU GenSan toward global climate justice can help apprehend the adverse effects of climate change and attain global climate justice. Also, future research can explore the role of digital climate activism among Generation Z in climate advocacy. Through these areas, higher education institutions will equip the present cohort that dominates the tertiary institution with the necessary knowledge, skill sets, resources, and spaces to deal with systemic environmental problems. Furthermore, this chapter is consistent with the Sustainable Development Goal (SDG 13) which emphasizes urgent action to counter climate change and its impacts. This chapter aimed to improve education and awareness of climate activism, specifically integrating it into the curriculum and acquiring funding and networking opportunities for the Global South's younger generations.

References

Almansa-Martínez, A., López-Gómez, S., & Castillo-Esparcia, A. (2024). Climate change literacy and commitment in Spanish university students. *Journal of Communication Management.* https://doi.org/10.1108/jcom-07-2022-0081

Ariestya, A., Paramitha, G., & Elmada, M. A. G. (2022). Climate change awareness of Gen Z: The influence of frame and jargon on online news. *Jurnal Studi Komunikasi, 6*(3), 753–770. https://doi.org/10.25139/jsk.v6i3.5287

Bartlett, M., Larson, J., & Lee, S. (2022). Environmental justice pedagogies and self-efficacy for climate action. *Sustainability.* https://doi.org/10.3390/su142215086

Beresford Research. (2023). Age range by generation—Beresford research. https://www.beresfordresearch.com/age-range-by-generation/.

Berse, K. B. (2024). *Climate emergency in the Philippines: Impacts and imperatives for urgent policy action.* Springer Nature. https://link.springer.com/book/10.1007/978-981-99-7804-5

Bruin, W., Bruin, W., & Morgan, M. (2019). Reflections on an interdisciplinary collaboration to inform public understanding of climate change, mitigation, and impacts. *Proceedings of the National Academy of Sciences, 116*, 7676–7683. https://doi.org/10.1073/pnas.1803726115

Carrick, J., Bell, D., Fitzsimmons, C., Gray, T., & Stewart, G. (2022). Principles and practical criteria for effective participatory environmental planning and decision-making. *Journal of Environmental Planning and Management, 66*, 2854–2877. https://doi.org/10.1080/09640568.2022.2086857

Confetto, M., Covucci, C., Addeo, F., & Normando, M. (2023). Sustainability advocacy antecedents: How social media content influences sustainable behaviours among Generation Z. *Journal of Consumer Marketing.* https://doi.org/10.1108/jcm-11-2021-5038

Heyen, D. (2022). Social justice in the context of climate policy: Systematizing the variety of inequality dimensions, social impacts, and justice principles. *Climate Policy, 23*, 539–554. https://doi.org/10.1080/14693062.2022.2142499

Karpudewan, M., & Khan, N. (2017). Experiential-based climate change education: Fostering students' knowledge and motivation towards the environment. *International Research in Geographical and Environmental Education, 26*, 207–222. https://doi.org/10.1080/10382046.2017.1330037

Kumar, Y. (2024). Climate Justice in India: Addressing the Impact of Climate Change on Vulnerable Populations. *Idealistic Journal of Advanced Research in Progressive Spectrums* (IJARPS) eISSN–2583-6986, 3(11), 148–157. https://journal.ijarps.org/index.php/IJARPS/article/view/474/441

Lazarito, A. G., Legaspi, L. L., & Loiweha, P. (2024). Pathways to sustainability: A neo-Gramscian exploration of just transition initiatives in Thailand and the Philippines. *Graduate Review of Political Science and Public Administration Journal, 3*(1), 1–20. https://so07.tci-thaijo.org/index.php/GRPSPAJ/article/view/4253

Li, C., Li, C., Zhang, X., & Ge, Y. (2023). Influence of network communication on college students' awareness of environmental protection. *Environmental Engineering and Management Journal.* https://doi.org/10.30638/eemj.2023.171

Neas, S., Ward, A., & Bowman, B. (2022). Young people's climate activism: A review of the literature. *Frontiers in Political Science, 4*, 940876. https://doi.org/10.3389/fpos.2022.940876

Persada, S. F., Miraja, B. A., & Nadlifatin, R. (2019). Understanding the generation z behavior on D-learning: A Unified Theory of Acceptance and Use of Technology (UTAUT) approach. *International Journal of Emerging Technologies in Learning, 14*(5), 20–33. https://doi.org/10.3991/ijet.v14i05.9993

Pogge, T. (2008). *World poverty and human rights* (2nd ed.). Polity Press. https://www.wiley.com/en-us/World+Poverty+and+Human+Rights%2C+2nd+Edition-p-9780745641447

Pogge, T. (2010). Keynote address: Poverty, climate change, and overpopulation. https://digitalcommons.law.uga.edu/cgi/viewcontent.cgi?article=1132&context=gjicl.

Rebollido, R. (2024). Heat-induced ailments affect thousands of General Santos students. *Rappler.* https://www.rappler.com/philippines/mindanao/heat-induced-ailments-affect-thousands-general-santos-students-april-2024/.

Reynon, J. (2021). Climate change in the Philippines. *Journal of Latin American Sciences and Culture.* https://doi.org/10.52428/27888991.v3i4.142

Rogayan, D. (2024). Addressing the escalating heat index in the Philippines: A call to action. *Journal of Public Health.* https://doi.org/10.1093/pubmed/fdae073

Shohel, M. M. C., & Sall, W. (2024). Climate change education for sustainable development and social justice: Students as the co-creators of an undergraduate module. In *Practitioner research in college-based education* (pp. 203–232). IGI Global. https://www.igi-global.com/chapter/climate-change-education-for-sustainable-development-and-social-justice/356054

Sloan Morgan, O., Melchior, F., Thomas, K., & McNab-Coombs, L. (2023). Youth and climate justice: Representations of young people in action for sustainable futures. *The Geographical Journal.* https://doi.org/10.1111/geoj.12547

Soria, J., Switzer, A., Villanoy, C., Fritz, H., Bilgera, P., Cabrera, O., Siringan, F., Maria, Y., Ramos, R., & Fernandez, I. (2016). Repeat storm surge disasters of Typhoon Haiyan and its 1897 predecessor in the Philippines. *Bulletin of the American Meteorological Society, 97*, 31–48. https://doi.org/10.1175/BAMS-D-14-00245.1

Surmacz, T., Wierzbiński, B., Kuźniar, W., & Witek, L. (2024). Towards sustainable consumption: Generation Z's views on ownership and access in the sharing economy. *Energies.* https://doi.org/10.3390/en17143377

Tan, E., Wanganoo, L., & Mathur, M. (2023). Generation Z, sustainability orientation and higher education implications: An ecopedagogical conceptual framework. *Journal of Applied Learning and Teaching, 6*(1), 314–323. https://doi.org/10.37074/jalt.2023.6.1.ss2

Tol, R. (2018). The economic impacts of climate change. *Review of Environmental Economics and Policy, 12*, 4–25. https://doi.org/10.1093/reep/rex027

United Nations Office for Disaster Risk Reductions. (2022). Impacts of super typhoons and climate change. *PreventionWeb.* https://www.preventionweb.net/news/impacts-super-typhoons-and-climate-change

Wijaya, D. I., & Kokchang, P. (2023). Factors Influencing Generation Z's Pro-Environmental Behavior towards Indonesia's Energy Transition. *Sustainability, 15*(18), 13485. https://doi.org/10.3390/su151813485

World Weather Attribution. (2024). Climate change made the deadly heatwaves that hit millions of highly vulnerable people across Asia more frequent and extreme. https://www.worldweatherattribution.org/climate-change-made-the-deadly-heatwaves-that-hit-millions-of-highly-vulnerable-people-across-asia-more-frequent-and-extreme/

Youth Advocates for Climate Action. (2024). Youth advocates for climate action Philippines. https://yacap.org/about/

CHAPTER 12

Blue Education: Shaping the Future of Environmental Education

Giambattista Bufalino

1 Introduction

The ocean has long held a profound place in human history not only as a physical expanse but also as a source of cultural, economic, and scientific knowledge. From early trade routes and exploration to philosophical and spiritual symbolism, the sea has served as a metaphor for mystery, power, and the boundless nature of the world (Abulafia, 2019). Despite its centrality in shaping civilizations and ecosystems alike, the ocean remains underrepresented in educational discourse, especially in the field of environmental education.

In recent decades, educational systems have increasingly focused on environmental sustainability, but primarily from a terrestrial perspective. "Green education" centered on land-based issues such as deforestation, soil degradation, and terrestrial biodiversity has been successfully embedded in school curricula, policy frameworks, and public awareness campaigns (Orr, 1992). However, this emphasis has often come at the expense

G. Bufalino (✉)
Department of Educational Sciences, University of Catania, Catania, Italy
e-mail: gbufalino@unict.it

K. Hirech (ed.), *Environmental Education and Sustainable Development*, https://doi.org/10.1007/978-3-032-05760-0_12

of marine and freshwater ecosystems, leading to a significant gap in how sustainability is taught and understood. Although oceans regulate climate, support biodiversity, and sustain millions of livelihoods, they are frequently sidelined in environmental education initiatives (Dillon, 2014).

This chapter introduces Blue Education, an emerging framework that aims to rebalance environmental education by emphasizing the vital importance of aquatic ecosystems. Defined as an interdisciplinary and ecocentric approach, Blue Education promotes ecological literacy that integrates knowledge of oceans, rivers, and lakes with cultural, ethical, and philosophical dimensions of sustainability. While related concepts such as Ocean Literacy and marine environmental education focus primarily on knowledge and awareness, Blue Education expands the scope to include emotional, cultural, and historical connections with water environments (Mokos et al., 2022; Regazzoni, 2022).

The main objective of this chapter is to critically explore Blue Education as a transformative and necessary shift in sustainability education. It examines how the dominance of land-centric approaches often rooted in colonial and anthropocentric worldviews has contributed to the marginalization of aquatic systems in both theory and practice. Drawing on posthumanist theory, traditional ecological knowledge (TEK), and the evolving field of the Blue Humanities, the chapter argues for a more inclusive, equitable, and holistic environmental pedagogy.

To understand the urgency and relevance of this shift, we begin by examining how conventional environmental education has historically excluded marine systems, and how this exclusion reflects broader epistemological and cultural biases. Blue Education, as this chapter will demonstrate, seeks to correct this imbalance by fostering deeper ecological connections and promoting sustainable action that includes, rather than ignores, the life below water.

2 Decolonizing Environmental Education

Historically, environmental education has been dominated by a land-centric perspective, reinforcing an anthropocentric and colonial bias that prioritizes terrestrial ecosystems over marine and freshwater environments (Whyte, 2018). The exclusion of marine studies from mainstream environmental curricula has resulted in a critical gap in understanding the interconnectivity of terrestrial and aquatic ecosystems. The United Nations Sustainable Development Goal (SDG) 14, "Life Below Water," underscores the necessity of protecting oceans, seas, and marine resources for sustainable development (United Nations, 2015). However, despite this recognition, ocean conservation remains underrepresented in formal education systems (Dillon, 2014).

Blue Education challenges this paradigm by advocating for an inclusive and integrated environmental education model that recognizes the interdependence of all ecosystems. The shift from a "green" to a "blue" framework seeks to dismantle the hierarchical perception that privileges land-based conservation efforts while neglecting aquatic environments. Scholars have pointed out that this oversight stems from colonial-era perspectives that viewed the ocean primarily as a resource for exploitation rather than a complex ecological system requiring stewardship (Mokos et al., 2022). To counteract this "ocean blindness," Blue Education calls for systemic changes in educational policies, curricular reforms, and the incorporation of interdisciplinary approaches to environmental literacy.

One of the most significant aspects of decolonizing environmental education is integrating Indigenous Traditional Ecological Knowledge (TEK). Many educational institutions have begun incorporating TEK into their curricula, particularly in marine science and environmental sustainability programs. For example, in New Zealand, the integration of Māori knowledge systems into marine education has enhanced students' understanding of ecological stewardship by emphasizing sustainable fishing practices and marine conservation rooted in centuries-old traditions. Similarly, in Canada, Indigenous-led environmental initiatives have been incorporated into school programs, allowing students to engage in hands-on learning about traditional water management and conservation techniques. Indigenous communities have long practiced sustainable marine stewardship, balancing ecological health with resource use. For example, Pacific Islander navigation and fishing practices reflect a profound understanding of ocean currents, biodiversity, and seasonal changes, offering valuable insights for modern conservation strategies (Zidny et al., 2020). Integrating TEK into contemporary educational frameworks not only diversifies knowledge systems but also fosters a more ethical and culturally sensitive approach to ocean conservation

3 The Cultural Significance of the Sea

The sea has long influenced human civilization, serving as a medium for exploration, trade, and cultural exchange (Abulafia, 2019). From the Phoenician and Greek maritime traditions to the great voyages of exploration in the Age of Discovery, the sea has shaped human migration patterns, economic systems, and knowledge production. Ancient Greek literature, such as Homer's *Odyssey*, portrays the ocean as a powerful force shaping human destiny, reinforcing its symbolic and mythological significance (Ianni, 1996). Similarly, in religious traditions, the sea represents both chaos and divine power, as exemplified in the parting of the Red Sea in the Bible and Varuna's dominion in Hindu cosmology (Coomaraswamy, 1994).

In addition to its spiritual and mythological dimensions, the sea has played a crucial role in shaping artistic, literary, and philosophical thought. From Shakespeare's "The Tempest" to Melville's *Moby-Dick*, literary works reflect the sea's dual role as a source of inspiration and existential challenge (Melville, 2002; Mack, 2013). This cultural legacy underscores the importance of integrating marine studies into environmental education to foster a deeper appreciation for the historical and symbolic significance of the ocean.

Folklore and oral traditions also reflect humanity's deep connection with the sea. The Sicilian legend of Colapesce, in which a young boy sacrifices himself to support Sicily's crumbling foundations, exemplifies the reverence and mythical associations surrounding the sea (Pitrè, 2016). Recognizing these narratives within Blue Education fosters a cultural and emotional connection to marine environments, deepening student engagement by making marine conservation personally and culturally relevant. By integrating folklore, mythology, and historical narratives into the curriculum, educators can create more immersive learning experiences that resonate with students. This approach not only enhances retention but also fosters a sense of identity and responsibility toward marine stewardship, as students see their own cultural heritage reflected in environmental education, reinforcing the need for its preservation (Guarracino, 2007). Beyond its cultural and mythological significance, it is essential to explore how educational movements have sought to reintroduce the ocean into environmental discourse.

4 The Evolution of Ocean Literacy

The Ocean Literacy (OL) movement, which originated in the United States around 2000, emerged as a concerted effort by oceanographers and educators to address a critical gap in primary and secondary education: the absence of a systematic inclusion of ocean sciences in school curricula (Schoedinger et al., 2010). Rooted in the foundational concept of literacy, OL initially focused on knowledge acquisition and the integration of marine and oceanic studies into the broader educational framework. A pivotal milestone occurred in 2002 with the Oceans for Life conference, organized by the College of Exploration and the National Geographic Society, which catalyzed discourse on embedding ocean science topics in education (Santoro et al., 2017). In 2004, representatives from ocean science, education, and policymaking sectors reached a consensus on the definition of OL, culminating in the publication of "Ocean Literacy: The Essential Principles of Ocean Science K-12 for Marine Educators" (National Oceanic and Atmospheric Administration, NOAA, 2013). This landmark document delineated key principles, fundamental concepts, and age-specific learning objectives, articulating the knowledge an ocean-literate individual in the United States should possess by the conclusion of secondary education.

While originally conceived as a knowledge-centric construct, OL has evolved into a multifaceted, interdisciplinary framework that transcends traditional educational paradigms. This progression reflects a profound philosophical shift, emphasizing not merely the accumulation of information but the cultivation of a holistic understanding of the ocean (Mokos et al., 2022). OL now incorporates a diverse array of perspectives, including scientific, historical, geographical, ethical, cultural, and gender-focused dimensions, fostering a more nuanced appreciation of the ocean's intrinsic value. At its core, OL embodies an epistemic virtue, cultivating respect, wonder, and stewardship. It challenges the anthropocentric paradigms that have historically dominated environmental discourse, advocating instead for a biocentric and ecological worldview that acknowledges the ocean as an indispensable component of Earth's life-support system (Regazzoni, 2022). This reconceptualization necessitates sustainable practices that honor the intricate interdependencies inherent within marine ecosystems, reframing humanity's relationship with the natural world (Huntington, 2020).

The definition of an ocean-literate individual captures this shift, positing that such a person "understands the fundamental concepts about the functioning of the ocean, can communicate about the ocean in a meaningful way, and is able to make informed and responsible decisions regarding the ocean and its resources" (NOAA, 2013, p. 5). The ultimate objective of OL initiatives is to instigate behavioral change, fostering sustainable actions among citizens to address and mitigate the challenges facing marine environments. One of the movement's most pressing challenges is overcoming societal "ocean blindness," a consequence of superficial or inadequate understanding of marine systems (Koulouri et al., 2022). To counteract this, innovative educational tools and expanded access to marine data are essential, enabling educators and communicators to deepen public engagement with ocean sciences and sustainable marine development (Costa & Calderina, 2018).

Over the years, the OL movement has evolved into a global initiative, underpinned by international collaboration and dissemination efforts. In 2011, Portugal became one of the first nations to adapt the US OL framework through the Conhecer o Oceano initiative, which successfully integrated ocean content into national science education programs (Copejans et al., 2012). The European trajectory of OL gained momentum in 2012 with the inaugural European Conference on Ocean Literacy in Bruges, which convened high-level policymakers, marine education experts, and international stakeholders (Mokos et al., 2022). This conference also marked the establishment of the European Marine Science Educators Association, a pivotal organization dedicated to fostering international collaboration, equipping educators, and advancing marine education across Europe. Such initiatives have been instrumental in elevating Ocean Literacy within European educational and public discourse (Fauville et al., 2019).

Further advancing OL, the United Nations hosted a conference from June 5 to 9, 2017, in support of Sustainable Development Goal 14 "Life Below Water," which aims to conserve and sustainably use oceans, seas, and marine resources (United Nations, 2015). This conference prominently promoted the concept of OL, emphasizing its importance in achieving sustainable ocean management. UNESCO has also played a significant role in advancing OL through the initiative Ocean Literacy for All, which aims to establish a global partnership recognizing the ocean's crucial role in maintaining the planet's health (Santoro et al., 2017). In effect, oceans play a fundamental role in maintaining planetary health by regulating climate patterns, producing over 50% of the world's oxygen, and serving as carbon sinks that absorb approximately 30% of anthropogenic carbon dioxide emissions (Claudet et al., 2020). Furthermore, oceans support millions of livelihoods through industries such as fishing, tourism, and maritime trade, making their preservation essential for economic and social stability (Fauville et al., 2019).

The OL movement continues to expand globally, encompassing diverse cultural and geographic contexts. Rather than providing a universal, prescriptive framework, it prioritizes adaptability, offering guidance and examples that can be tailored to the specific needs of different regions. This flexibility ensures that OL remains relevant and effective in engaging diverse audiences, ultimately fostering a collective commitment to marine stewardship and sustainability (Zidny et al., 2020; McKinley et al., 2023). While Ocean Literacy has laid a strong foundation for marine awareness in education, Blue Education builds upon and expands these principles, integrating philosophical, ethical, and cultural dimensions.

5 From Ocean Literacy to Blue Education

Building upon the foundational work of Ocean Literacy (OL), Blue Education represents an evolution in environmental understanding, one that broadens its scope by integrating humanistic, ethical, and ontological dimensions into our relationship with aquatic environments (Regazzoni, 2022). While OL has primarily focused on increasing scientific awareness and encouraging stewardship through factual knowledge, Blue Education seeks to cultivate a more integrated, value-based, and experiential connection to marine systems.

This paradigm shift aligns closely with principles of deep ecology, which prioritize the intrinsic worth of all living systems, regardless of their utility to humans (Devall & Sessions, 2007). By challenging anthropocentric perspectives, Blue Education urges a reconsideration of what counts as knowledge and whose voices are included in shaping sustainable futures.

This approach also resonates with posthumanist environmental thought, which critiques the hierarchical separation between humans and the more-than-human world. It reframes ethical considerations by recognizing agency in non-human entities, such as oceans, rivers, and marine life, emphasizing their roles not merely as resources but as co-inhabitants of shared ecosystems (MacCormack, 2012).

Further conceptual richness comes from the Blue Humanities, an emerging interdisciplinary field that explores the symbolic, aesthetic, and affective dimensions of human–water relationships through literature, philosophy, visual arts, and cultural history (Mentz, 2024). This perspective brings emotional and narrative depth to marine education, encouraging learners to engage with water not only as a scientific subject but also as a source of meaning, memory, and identity.

New theoretical developments such as the Hydrocene, a proposed epoch that centers water as a decisive force in shaping ecological and social realities, invite educational systems to rethink time, agency, and environmental change through the lens of water (Bailey-Charteris, 2024). Such frameworks underscore the urgency of incorporating hydrological thinking into pedagogical strategies that acknowledge fluidity, interconnection, and mutual dependence.

Moreover, Blue Education complements fields like ecopedagogy and place-based education, which encourage learners to develop a deep understanding of their local environments and ecosystems. It adds a crucial aquatic dimension to these practices, extending their reach from forests and fields to coasts, rivers, wetlands, and coral reefs. This convergence fosters a pedagogy of immersion both literal and metaphorical where emotional engagement and critical reflection coexist.

By integrating these varied perspectives, Blue Education cultivates a more responsive and adaptive form of environmental learning, one capable of addressing complex ecological challenges while nurturing a profound sense of planetary belonging and care (Sandrini, 2021).

6 Practical Pathways for Blue Education

The effectiveness of Blue Education lies in its ability to bridge theory and practice, transforming philosophical concepts into tangible learning experiences. This approach integrates various educational strategies, including formal instruction, experiential learning, technological advancements, and community-driven initiatives, ensuring its applicability across diverse educational settings. By adopting these methodologies, Blue Education fosters ecological literacy, environmental responsibility, and a sustainable relationship with aquatic ecosystems.

6.1 Formal Education

Blue Education has the potential to significantly reshape curricula by incorporating marine science, cultural perspectives, and ethical considerations. One notable example is the European Blue Schools Initiative (Gallez et al., 2024), a project endorsed by the European Commission and UNESCO's Ocean Literacy framework. This initiative encourages schools to adopt marine-focused curricula by implementing hands-on projects that engage students with real-world environmental challenges. Participating schools undertake interdisciplinary activities that promote critical thinking, creativity, and collaboration, addressing key issues such as marine pollution, biodiversity loss, and climate change. For instance, students might engage in local water quality assessments, develop sustainable fishing strategies, or create campaigns to reduce plastic consumption. Schools that successfully implement these principles receive recognition as "Blue Schools," underscoring their commitment to marine conservation and environmental stewardship.

6.2 Informal Education

Beyond the classroom, informal educational experiences expand the reach of Blue Education, offering immersive and interactive opportunities to engage with marine environments. Museums, aquariums, and citizen science programs provide platforms for experiential learning, allowing participants to explore marine biodiversity and conservation issues firsthand. Institutions such as the Oceanário de Lisboa and the Monterey Bay Aquarium facilitate interactive exhibits and educational workshops designed to deepen public understanding of ocean ecosystems. Additionally, virtual reality experiences and environmental documentaries, such as *Chasing Coral* (n.d.), offer audiences a visual and emotional connection to the challenges facing marine environments, inspiring greater advocacy and action. Citizen science initiatives also play a critical role, enabling individuals to contribute to marine conservation efforts by collecting data on pollution levels and biodiversity, fostering both scientific engagement and environmental responsibility.

6.3 Experiential Learning

Experiential learning is a cornerstone of Blue Education, as it fosters direct interaction with aquatic environments, promoting a sense of responsibility and interconnectedness. Activities such as coastal clean-ups, wetland restoration, and marine biodiversity monitoring provide participants with hands-on conservation experience. Outdoor adventure programs, including kayaking and snorkeling excursions, allow learners to observe marine life in its natural habitat, fostering a deeper appreciation for ocean ecosystems. Moreover, technological innovation has expanded the reach of marine education, making Ocean literacy more accessible. Interactive educational games teach principles of sustainable resource management, while open-access platforms such as the Ocean Biogeographic Information System (Vandepitte et al., 2011) offer real-time marine data for students and researchers to analyze oceanic systems and contribute to conservation efforts. The success of Blue Education is further reinforced through strategic partnerships at both the global and local levels. Programs such as the United Nations Ocean Decade of Marine Science for Sustainable Development facilitate interdisciplinary collaboration to address marine challenges and advance sustainability efforts. Non-governmental organizations, marine research institutions, and private sector stakeholders play a crucial role in providing funding, expertise, and resources for educational initiatives, ensuring the widespread implementation of Ocean Literacy programs (United Nations, 2015).

6.4 Barriers and Opportunities

Despite its many advantages, the widespread adoption of Blue Education faces notable challenges. Limited financial resources, lack of trained educators, and systemic educational constraints often impede its integration, particularly in underserved regions. To address these barriers, governments and academic institutions must invest in educator training programs, curriculum development, and infrastructure enhancements to support marine education. Additionally, fostering a cultural shift in societal attitudes toward aquatic ecosystems is essential. Historically, oceans and freshwater bodies have been perceived primarily as economic resources; however, Blue Education advocates for a more holistic perspective that recognizes their ecological and cultural significance. Public awareness campaigns and early education initiatives are critical for fostering a collective sense of responsibility toward marine conservation and sustainability (Regazzoni, 2022).

7 Conclusion

Blue Education represents a necessary and transformative evolution in environmental education. By shifting the focus from a predominantly terrestrial perspective to one that embraces the full complexity of aquatic ecosystems, it addresses critical gaps in sustainability discourse and practice. This chapter has outlined how Blue Education integrates ecological, cultural, and ethical dimensions, drawing on frameworks such as ocean literacy, Indigenous knowledge systems, and the Blue Humanities to reframe the relationship between people and the sea. The arguments presented underscore the importance of fostering a holistic and inclusive approach to environmental learning, one that values the symbolic, practical, and existential significance of marine and freshwater environments. From curricular reform and experiential learning to community-based engagement, the pathways explored highlight the potential of Blue Education to inspire deeper ecological awareness and action. In a time of growing environmental uncertainty, the cultivation of ocean-literate and water-conscious citizens is not a luxury but a necessity. Blue Education offers both a conceptual shift and a practical roadmap to reimagine how we educate for a sustainable future, one that truly reflects the interconnectedness of all life on Earth.

References

Abulafia, D. (2019). *The boundless sea: A human history of the oceans.* Allen Lane.

Bailey-Charteris, B. (2024). *The Hydrocene: Eco-aesthetics in the age of water.* Taylor & Francis Group.

Chasing Coral. (n.d.). Chasing Coral documentary. https://chasingcoral.com/

Claudet, J., et al. (2020). A roadmap for using the UN decade of ocean science for sustainable development in support of science, policy, and action. *One Earth, 2*(1), 34–42. https://doi.org/10.1016/j.oneear.2019.10.012

Coomaraswamy, A. K. (1994). *Induismo e buddismo.* Rusconi.

Copejans, E., Rouch, F., & Fauvielle, G. (2012). The European marine science educators association (EMSEA): Towards a more ocean literate Europe. *The Journal of Marine Education, 28*(2), 43–46.

Costa, S., & Calderina, R. (2018). Bibliometric analysis of ocean literacy: An underrated term in the scientific literature. *Marine Policy, 87*, 149–157.

Devall, B., & Sessions, G. (2007). *Deep ecology: Living as if nature mattered.* Gibbs Smith.

Dillon, J. (2014). Environmental education. In N. G. Lederman & S. K. Abell (Eds.), *Handbook of research on science education* (pp. 497–514). Routledge.

Fauville, G., Strang, C., Cannady, M. A., & Chen, Y. F. (2019). Development of the international ocean literacy survey: Measuring knowledge across the world. *Environmental Education Research, 25*(2), 238–263.

Gallez, E., Mujica, C. P. F., Gadeyne, S., Canters, F., & Baró, F. (2024). Nature-based school environments for all children? comparing exposure to school-related green and blue infrastructure in four European cities. *Ecological Indicators, 166*, 112374.

Guarracino, S. (2007). *Mediterraneo: Immagini, storie e teorie da Omero a Braudel.* Mondadori.

Huntington, H. P. (2020). Using traditional ecological knowledge in science: Methods and applications. *Ecological Applications, 10*(5), 1270–1274.

Ianni, P. (1996). *Il mare degli antichi.* Dedalo.

Koulouri, P., Mogias, A., & Gerovasileiou, V. (2022). Ocean literacy across the Mediterranean Sea region in the era of 2030 agenda and the decade of ocean science for sustainable development (2021–2030). *Mediterranean Marine Science, 23*(2), 266–269.

MacCormack, P. (2012). *Posthuman ethics.* Ashgate.

Mack, J. (2013). *The sea: A cultural history.* Reaktion Books.

McKinley, E., Burdon, D., & Shellock, R. J. (2023). The evolution of ocean literacy: A new framework for the United Nations Ocean Decade and beyond. *Marine Pollution Bulletin, 186*, 114467.

Melville, H. (2002). *Moby-Dick* (H. Parker & H. Hayford, Eds.). W. W. Norton and Company.

Mentz, S. (2024). *An introduction to the blue humanities.* Routledge.

Mokos, M., et al. (2022). Navigating ocean literacy in Europe: 10 years of history and future perspectives. *Mediterranean Marine Science, 23*(2), 277–288.

National Oceanic and Atmospheric Administration (NOAA). (2013). *Ocean literacy: The essential principles and fundamental concepts of ocean sciences for learners of all ages.* https://oceanliteracy.unesco.org/wp-content/uploads/2020/09/OceanLiteracyGuide_V3_2020-8x11-1.pdf

Orr, D. (1992). *Environmental literacy: Education as if the earth mattered.* E. F. Schumacher Lectures.

Pitrè, G. (2016). *Cola Pesce e altre fiabe e leggende popolari siciliane.* Donzelli.

Regazzoni, S. (2022). *Oceano. Filosofia del pianeta.* Ponte delle Grazie.

Sandrini, S. (2021). Professioni pedagogiche. A sostegno di una transizione verde e umana. *Form@re, 21*(2), 6–21.

Santoro, F., Selvaggia, S., Scowcroft, G., Fauville, G., & Tuddenham, P. (Eds.). (2017). *Ocean literacy for all: A toolkit.* UNESCO.

Schoedinger, S., Tran, L. U., & Whitley, L. (2010). From the principles to the scope and sequence: A brief history of the ocean literacy campaign. *Current: The Journal of Marine Education, 3*, 3–7.

United Nations. (2015). *Transforming our world: The 2030 agenda for sustainable development.* https://sdgs.un.org/goals/goal14

Vandepitte, L., Hernandez, F., Claus, S., Vanhoorne, B., De Hauwere, N., Deneudt, K., et al. (2011). Analysing the content of the European Ocean Biogeographic Information System (EurOBIS): available data, limitations, prospects and a look at the future. *Hydrobiologia, 667*, 1–14.

Whyte, K. (2018). Settler colonialism, ecology, and environmental injustice. *Environment and Society, 9*(1), 125–144.

Zidny, R., Sjöström, J., & Eilks, I. (2020). A multi-perspective reflection on how indigenous knowledge and related ideas can improve science education for sustainability. *Science & Education, 29*(1), 145–185.

PART III

Policies, Community Engagement and Global Perspectives

CHAPTER 13

Promoting Community Engagement and Environmental Awareness Through Educational Initiatives

Raed Atef and Bremananth Ramachandran

1 Introduction

Universities serve as incubators for future professionals, leaders, and citizens who are equipped with the knowledge and values needed to advance sustainable development (Awashra, 2012). In Palestine, universities bear a unique responsibility to foster environmental awareness, shaped by the country's socio-political instability, environmental degradation, and occupation-linked restrictions. While there has been a formal integration of environmental education into university curricula, non-formal education avenues remain significantly under-utilized. This chapter critically examines how Palestinian universities can redress this imbalance by leveraging non-formal learning approaches that are rooted in local realities and global sustainability goals.

R. Atef (✉)
Independent Researcher, Ramallah, Palestine

B. Ramachandran
College of Business Administration, A'Sharqiyah University, Ibra, Oman

K. Hirech (ed.), *Environmental Education and Sustainable Development*, https://doi.org/10.1007/978-3-032-05760-0_13

Despite well-meaning efforts, environmental education at Palestinian universities faces persistent structural and ideological barriers. These include financial limitations, political instability, weak institutional commitment, and a lack of meaningful partnerships with surrounding communities (Leal Filho et al., 2021). Most initiatives continue to privilege formal instructional models, sidelining informal education mechanisms such as extracurricular programming, community engagement, and the strategic use of digital and traditional media (Kelly et al., 2022). Although select institutions have engaged in community projects and collaborations with non-governmental organizations (NGOs), these efforts remain fragmented and unanchored by a coherent theoretical or strategic framework (Awashreh, 2020).

To address these limitations, this chapter adopts a transformative learning perspective (Mezirow, 1997) to understand how informal environmental education can reshape perceptions, values, and behaviors. Transformative learning theory provides a lens for analyzing how students critically examine assumptions, engage in reflective discourse, and arrive at new ecological consciousness through experiential learning. This is complemented by decolonial and Indigenous environmental frameworks, which critique the dominance of Eurocentric environmental paradigms and highlight the potential of context-sensitive, culturally embedded, and resistance-informed educational practices particularly vital in the Palestinian context marked by occupation, environmental dispossession, and restricted sovereignty.

Further, this chapter draws on posthumanist perspectives to question anthropocentric assumptions embedded in environmental education and explore how digital technologies, including AI-based classification tools, can be ethically integrated into educational practices without reinforcing technocratic or extractive models.

In doing so, the chapter argues for an interdisciplinary, justice-oriented, and decolonial model of informal education that empowers both students and communities. Palestinian universities can advance this model by developing sustained extracurricular activities, place-based learning, media initiatives, and strategic partnerships with civil society and government bodies (Awashreh, 2018).

The objectives of this chapter are to:

- Foster critical recognition of global and local environmental injustices
- Promote community engagement in sustainable practices

- Prepare students to influence environmental policy and decision-making.
- Advance informal education through workshops, multimedia strategies, and community-based collaborations (Awashreh, 2025).

The chapter addresses two central research questions:

1. How do Palestinian universities encourage environmental knowledge and civic engagement through informal learning?
2. How can informal environmental education be strategically integrated to enhance community-based sustainability?

The chapter is organized as follows: Section 2 outlines the research methodology; Section 3 presents the findings; and Sect. 4 offers a discussion and conclusion, including implications and recommendations for practice and policy.

Existing literature increasingly calls for integrating community-centered and interdisciplinary approaches to environmental education, particularly in regions affected by socio-political conflict (Gough, 2016; Khasnabis et al., 2010). Informal learning practices such as public campaigns, youth-led projects, and grassroots workshops have been instrumental in contexts like Lebanon and Jordan, where partnerships between universities and government institutions have bolstered environmental awareness and public participation (Sadovska et al., 2024). Drawing on these experiences, this chapter argues that Palestinian universities must move beyond institutional boundaries to embed sustainability into everyday practices. While challenges such as political instability, resource scarcity, and institutional fragmentation persist (Hasan & Buheji, 2024), an integrated, theoretically informed approach to informal education can serve as a transformative vehicle for environmental justice and sustainable development.

2 Methods

2.1 *Research Design*

This study adopts a convergent mixed-methods design that integrates both qualitative and quantitative approaches to provide a holistic understanding of environmental education initiatives in Palestinian universities.

The qualitative component involves the thematic analysis of secondary sources, while the quantitative component utilizes machine learning techniques to classify institutional sustainability impacts based on predefined indicators.

This integrated approach allows for triangulation and enhances the validity and depth of the findings, ensuring both interpretive insight and empirical robustness (Renjith et al., 2021). The design aligns with transformative learning theory and participatory environmental education frameworks that emphasize evidence-based and community-driven practices.

2.2 Data Collection

The qualitative part is primarily founded on secondary data, which is obtained from scholarly and non-scholarly sources, i.e., peer-reviewed journals, reports of NGOs, and university research.

- **Qualitative Component**

The qualitative data are derived from a structured **document analysis** of peer-reviewed academic articles, policy reports, NGO documentation, and institutional publications. These sources were selected based on relevance to non-formal environmental education in Palestine and regional comparators. Inclusion criteria ensured that documents addressed one or more of the six sustainability indicators used in the quantitative component.

- **Quantitative Component**

For the quantitative analysis, data were systematically compiled for 178 Palestinian institutions, including universities and colleges. Each institution was scored on six environmental sustainability indicators:

- Community Engagement
- Developing Partnerships
- Empowering Women
- Raising Awareness on Carbon Emissions
- Youth Environmental Stewardship
- Obstacles

- Each institution's performance was categorized into one of three impact levels: high, medium, or low. To achieve this classification, we applied a supervised machine learning algorithm, specifically, a pattern recognition model using random forest classification, integrated with deep learning components. The dataset consists of 6 × 178 input features corresponding to the six indicators and 3 × 178 output targets representing the three impact categories.
- The dataset was randomly split into three sets:
  - Training set: 124 samples
  - Validation set: 27 samples
  - Testing set: 27 samples
- This division supports rigorous performance evaluation and reduces overfitting (Fig. 13.1).

2.3 Data Analysis

The data analysis involved a multi-step machine learning pipeline, blending traditional classification with neural network optimization. The

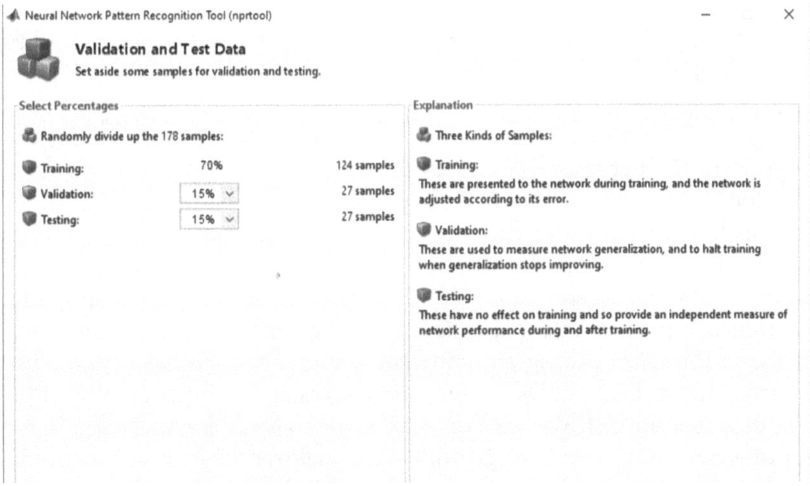

Fig. 13.1 Validation and test data for training, validate, and testing the neural network

analysis aimed to identify institutional patterns in environmental initiatives and classify their overall impact.

2.3.1 Deep Learning Implementation

We employed a supervised learning model built on a random forest classifier and enhanced with a deep learning neural network architecture. The model included:

- Input layer: six neurons corresponding to the six sustainability indicators
- Hidden layer: 20 neurons with sigmoid activation functions and optimized weights and biases
- Output layer: three neurons representing high-, medium-, and low-impact categories

The model followed a scaled conjugate gradient backpropagation protocol, ensuring optimization of training and preventing overfitting.

- Training phase: The model adjusted weights based on error minimization from 124 input samples.
- Validation phase: Generalization was monitored using 27 samples to halt training when performance plateaued.
- Testing phase: Performance was independently evaluated with a final set of 27 samples to confirm accuracy.

Figures 13.1 to 13.3 illustrate the training, network architecture, and classification process used in this study.

Input 124 samples have been presented to the deep learning model during the training, and the model is adjusted according to its errors. During the validation, it is utilized to measure the deep learning generalization, and to halt the training when generalization stops improving. Testing is a process of independent measurement of deep learning network performance during and after the training process have been completed. Figure 13.2 shows learning network size has been employed for this proposed model. We have used 20 neurons for hidden layers. It has a set of weights (W) and bias (B) with sigmoidal transformation functions.

The input layer has six neurons for Community Engagement, Developing Partnerships, Empowering Women, Raising Awareness on Carbon Emissions, Youth Environmental Stewardship, and Obstacles.

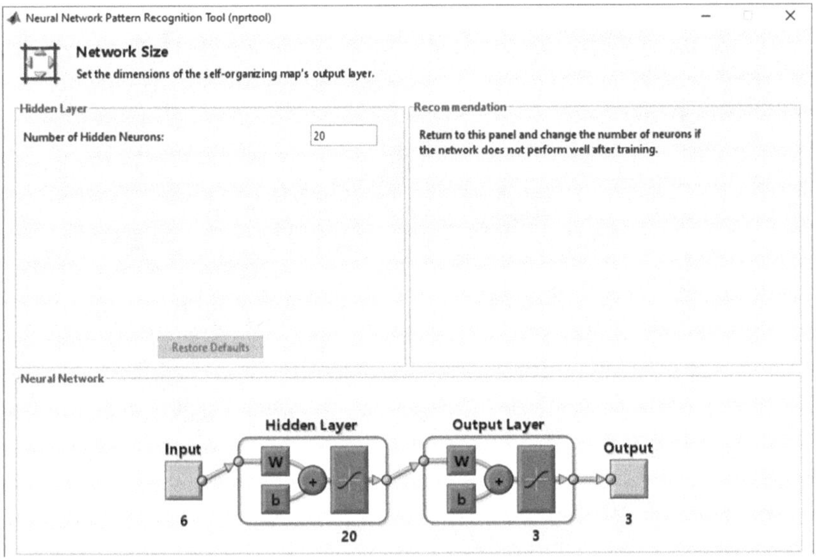

Fig. 13.2 Neural network model for educational initiatives among Palestinian universities and colleges

The three neurons have employed output layers such as high, medium, and low impacts. Figure 13.3 depicts the process of the train network to classify the inputs according to the targets. Fusion with random forest classifiers has been performed with scaled conjugate gradient backpropagation. The results of the proposed deep learning model are discussed in Sect. 4.

3 Results Analysis Using Deep Learning

In this proposed research model, we have employed a neural network pattern recognition tool. This tool is primarily used to classify the input data set according to the targets. We have collected 178 samples. One hundred and twenty-four, 27, and 27 samples have been employed for training, validation, and testing the network. The following sections discuss the parameters and their result implications.

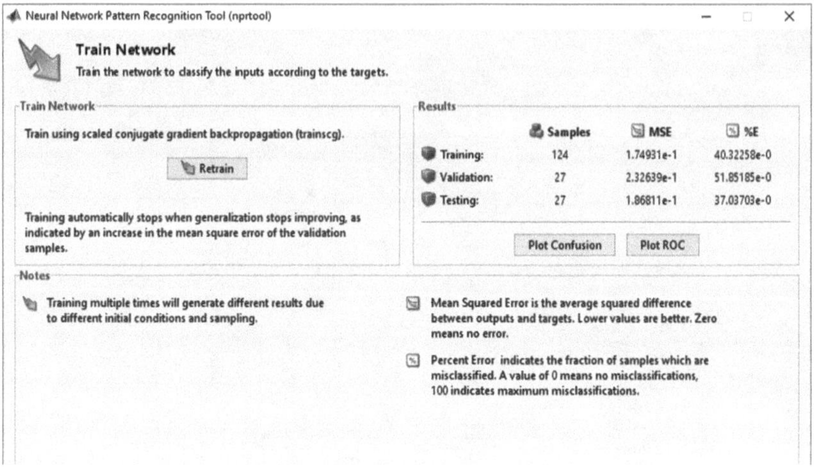

Fig. 13.3 Train network for deep learning process

3.1 Community Engagement

Events organized for the community on university campuses by Palestinian institutes focused on sustainability in 2024 demonstrated how informal environmental education can effectively initiate community participation and engagement in an effort to increase awareness toward sustainability. Multiple community parts activities would be offered and would occupy diverse learning modes like hand-on experience, expert lecture, and activity-based learning in order to reach higher levels of knowledge and participation in environmental activity (Bethlehem University, 2024). Nowadays, environmental sustainability is of high concern in universities. Bethlehem University introduced new academic courses that include the Bachelor's Program in Technology for Environmental Sustainability. The novelty of this course consists in providing the students with theoretical knowledge and practical skills to solve the most vital challenges of environmental concern with the use of technological means (PIBS a, 2024). Table 13.1 shows and summarizes the key findings on environmental sustainability initiatives in Palestine.

Table 13.1 Key findings on environmental sustainability initiatives

Section	*Key findings*
Community Engagement	Informal environmental education at Bethlehem University effectively enhanced sustainability awareness through hands-on activities and expert-led discussions
Developing Partnerships	Collaborative efforts between universities and NGOs strengthened environmental and climate justice initiatives, with an emphasis on applied research and community-based solutions
Empowering Women	Training programs provided women with knowledge and skills in sustainable agriculture, environmental program management, and leadership, leading to the establishment of a women's environmental center
Raising Awareness on Carbon Emissions	Workshops engaged experts and the public in discussions on carbon emissions, mitigation strategies, and interdisciplinary collaboration
Youth Environmental Stewardship	Interactive field trips and experiential learning activities helped foster environmental awareness and responsibility among young participants
Obstacles	The Israeli occupation has caused environmental degradation, water scarcity, and pollution, with significant impacts on agriculture, biodiversity, and Palestinian livelihoods

Sources: Developed by authors

Table 13.2 Environmental awareness through educational initiatives at Palestinian universities and colleges

Evaluation stages	*MSE*	*PE*
Training (124)	1.74931e-1	40.32258e-0
Validation (27)	2.33639e-1	51.85185e-0
Testing (27)	1.8681e-1	37.03703e-0

3.2 *Research Model Evaluation*

This model yields two types which are utilized for evaluation of the model. The first is the mean square error (MSE). It has indicated the average squared difference between output and target. Percentage error (PE) is another error method that has provided a fraction of samples which are misclassified among the samples in the network. Table 13.2 depicts the evaluation errors of the proposed model.

To evaluate the network model by testing the larger number of samples, we have utilized the evaluate network AI model. It is an option to test the proposed model on more data, and then decide if network performance is good enough. The overall error rates have been obtained by the network 1.85487e-1 and 41.57303e-0 for MSE and PE, respectively. Figure 13.4 shows the educational sustainability data set for the parameters such as Community Engagement, Developing Partnerships, Empowering Women, Raising Awareness on Carbon Emissions, Youth Environmental Stewardship, and Obstacles. It reveals educational sustainability with these factors among 178 samples.

Figure 13.5 depicts validation performance for the proposed models with best performance at 0.23264 in Epoch 7. This model has been trained in 13 epochs measured with MSE.

The proposed AI-deep learning model for environmental sustainability achieved a training performance with a gradient of 0.011248 and six validation checks at epoch 13, and its performance was further evaluated using confusion matrices for the training, validation, and testing datasets. It provides the trade-off between output and target classes. This matrix representation reveals the classes such as high, medium, and low factors of Promoting Community Engagement and Environmental Awareness through Educational Initiatives at Palestinian Universities and Colleges. It classifies as true positive, false positive, true negative, and false negative

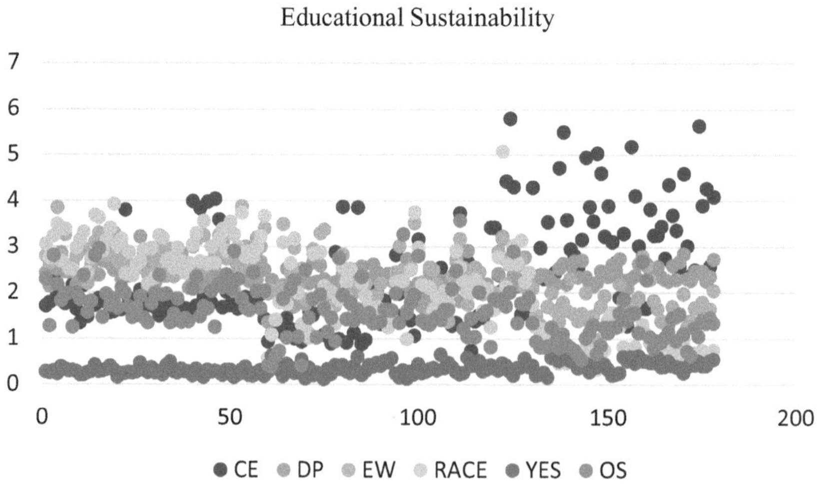

Fig. 13.4 Parameters of educational sustainability

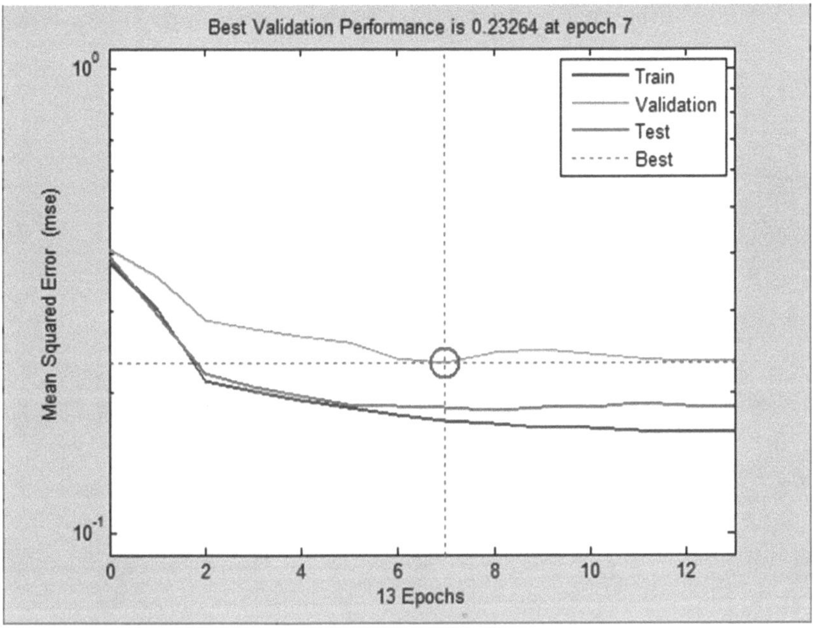

Fig. 13.5 Validation performance for the proposed models

cases among three target classes. Among 124 samples of the training set, 43 have high impacts on education sustainability, 30 are medium impact, and two are low impact. The overall high impact education sustainability is 58.4%. Figure 13.6 depicts the ROC of the proposed AI-deep learning model between true positive and false positives of training, validation, testing, and overall. It reveals that the proposed model can be used for any kind of environmental awareness evaluation because the results are above the area of curve with equal-error rate.

3.3 Developing Partnership

This can be seen within the context of partnerships that become very important for the spirit of collaboration between universities and NGOs toward the advancement of environmental and climate justice in Palestine. In 2023, a commitment was made through the signing of the MOU, in jointly addressing those challenges through research, education, and

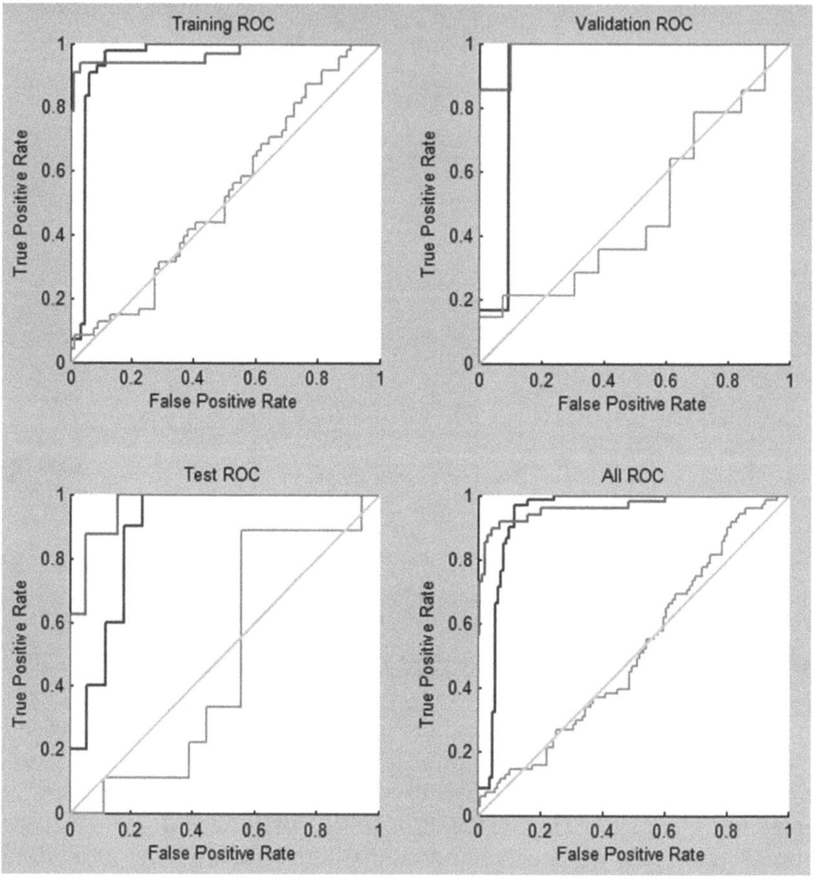

Fig. 13.6 ROC of the proposed AI-deep learning model with true positive and false positives

community-based initiatives. From now on, the partnership intends to offer sustainable solutions that are likely to meet the aims of the local communities and that of the general environmental landscape, through reinforcing academia and grassroots organizations (PIBS b, 2024).

The second essential element of partnership is the emphasis on applied research with practical solutions to on-the-ground environmental problems. The integrated approach at the center ensures that climate adaptation strategies are evidence-based and community-led, connecting academic knowledge with hands-on fieldwork. With support coming both internationally and locally, it accesses a broad network to make the initiative larger, more effective, and even longer-lasting (Ardoin et al., 2020).

From the broader perspective, the mission can introduce clearer and more affective environmental sustainability into education, research, and community engagement. Partnerships with these universities can facilitate the development of additional projects and activities aimed at creating a greener future, while NGOs remain committed to ensuring that these initiatives result in concrete and meaningful interventions at the grassroots level. This coming together empowers the course of environmental and climate justice and builds the grounds for a sustainable and just future in Palestine even further (Cairney et al., 2023). This indeed is a leap forward in the struggle for environmental and climate justice. The partnership therefore provides a model through which universities and NGOs may collaborate for sustainable environmental impact to enable such communities to adapt better to the challenge of climate change (Middlesex University, 2024).

3.4 *Empowering Women in Environmental Leadership*

More significantly, a set of collaboration events and workshops has been conducted with local organizations, municipalities, and NGOs for training women to perform leading environmental activities. Meetings showcased that women are highly active as environmentalists and leaders, showing the high potential contribution toward sustainability by seeking improvements in knowledge and enhancing better environmental protection. Much is pending in terms of women's engagement in this sphere, which is yet to be enhanced (UNEP, n.d.)

Sustainable agriculture topped the events where skills on useful methods in farming that are considered eco-friendly were developed, including resource conservation and climate-smart agricultural practices.. Over and above the theoretical, women allowed themselves to be trained appropriately so as to gain deeper insights into sustainable land management, organic farming, and traditional tools in agriculture (Balabantaray, 2023). This exposure, apart from enhancing their technical skills, also encouraged the conservation and revitalization of indigenous knowledge in agriculture that has been passed down through generations.

It also introduced the participants to environmental program management, thereby giving the necessary tools for the design, implementation, and management of community-based initiatives in sustainability. Building on leadership skills, the training sought to enhance a connectedness network of women environmental advocates able to catalyze positive change within their respective communities. Discussions also outlined the wider challenges facing women in accessing resources and decision-making opportunities in environmental governance and included strategies for overcoming these (Çakmakçı et al., 2023). In addition, one outcome of the initiative was the establishment of a women's environmental center,—a site of consolidated learning, potential skill development, and community address for further progress on environmental activities. This center does not allow women to apply their sustainable projects but helps them to engage in joint action in the solving of urgent environmental problems. Mentorship, capacity-building programs, and continued consultation with local authorities and civil society are some of the ways in which the center is expected to play a major role in improving women's involvement in environmental decision-making (Seyfang et al., 2014).

Equipped with knowledge, practical skills, and leadership, this project strives for long-term social justice and environmental sustainability. Women's increased involvement in environmental sustainability initiatives not only enhances the environmental quality of their immediate environment but also strengthens the overall cause of environmental justice in Palestine. Through sustained partnership and support, such projects pave the way toward a more inclusive and sustainable future, where women are respected as valuable stakeholders in environmental policy and practice decision-making (Balabantaray, 2023).

3.5 *Raising Awareness of Carbon Emissions*

Universities and non-governmental organizations raise awareness regarding anthropogenic carbon dioxide emissions and challenges on mitigation. One recent workshop brought biodiversity advocates, environmental professionals, and climate enthusiasts who explored the most recent state of carbon emissions and their environmental effects, as well as strategies toward reduction (Leal Filho et al., 2023).

NGOS raise issues of major present sources of emissions: industrial processes, transportation, and energy production. The necessity of urgent mitigation was underlined in the session, discussing environmental, economic, political, and social challenges using scientific data and real-world examples. The workshop underlined the scientific debate and the interdisciplinary collaboration among scientists, policy analysts, business, and civil society. Open discussions allowed the exchange of ideas among participants on options such as clean energy technologies, carbon capture systems, and policies that will reduce dependence on fossil fuels (Dwivedi et al., 2022). The online format of the event guaranteed that it would reach every part of the globe, proving again how strong virtual platforms are for environmental education and global participation. Melding scientific insights with interactive discussions and virtual engagement, the workshop clearly stated the ongoing need for dialogue and common action on climate change. Climate change and promoting a sustainable future (Mohammed et al., 2024).

3.6 Instilling Environmental Stewardship in Youth

Outdoor activities included the participation of members, particularly the young, in practical training with a view to achieving practical knowledge regarding various botanical species, habitat protection, and other environmental concerns. Such learning will help to enhance their relationship with nature and eventually create an attachment to the environment, which is a vital ingredient in long-term stewardship. The interactive field trip made the participants anything but passive onlookers, able to apply their knowledge actively in real time: the identification of plant species, discussing their role in the ecosystem, and recognition of changes in the environment all of that enhances appreciation for biodiversity and the interdependence of ecosystems. By taking them outdoors, this activity reaffirmed the need to use the environment responsibly and individual efforts toward conservation. The successful experiment demonstrated that non-formal education could be utilized to increase environmental awareness. Less abstract than courses might be in a school room, the firsthand experience of learning from nature simplified such courses to be easier and more memorable to apply. The project was also to demonstrate how environmental education could actively involve different groups in society young people, family, and women translating theoretical knowledge into a sense of responsibility and practice.

Beyond the field trip alone, such initiatives have shown how much further the impact of experiential learning in effecting behavior change and sustainable community practices has reached. These findings therefore reinforce further investment in these kinds of programs in the empowerment of communities and the strengthening of environmental awareness for the betterment of global sustainability outcomes.

3.7 *Obstacles*

Israeli occupation has led to widespread environmental degradation in Palestine, which is widely known as "environmental apartheid" (Buheji & Al-Muhannadi, 2023). One of the critical issues is water scarcity since Israel gets access to over 80% of the water resources of the West Bank and Palestinian villages have only 19 gallons per capita per day, which is far below WHO's minimum. Restrictions on well-drilling and rainwater harvesting promote desertification, while Israeli settlements discharge untreated sewage, contaminating land and water (Bismuth, 2016).

Agriculture, an important sector, has been greatly impacted. Over 247,000 acres of land have been taken out of agriculture since 1967, and 2.5 million trees uprooted, including one million olive trees. Settler terror also disrupts agriculture, reducing food security and loss of biodiversity (Al Botmeh & Saadeh, 2024). Twice the amount of per capita solid waste produced by Palestinians is emitted by the settlements, including toxic industrial waste and 52 million cubic yards of untreated sewage dumped annually on Palestinian soil, causing dire health issues (Randles, 2024).

The separation barrier has encouraged deforestation, loss of biodiversity, and fragmentation of ecosystems and severed farmers from access to land. Approximately 30% of the West Bank is dominated by military zones, keeping Palestinians out of cultivation and grazing (United Nations, 2004). In Gaza, over 96% of water is unsafe for consumption due to pollution, worsened by Israeli bombardments on power stations and sewage facilities, leading to mass disease and misery in wetlands (Raouf & Ruizi, 2023).

In spite of such environmental harms, Israel "greenwashes" its reputation by claiming sustainability improvements while draining Palestinian resources. Its natural gas economy benefits from Palestinian waters, and its military-industrial complex is a significant environmental polluter (Salazar et al., 2022). The occupation's effects reach globally and regionally, requiring policy changes and global intervention for justice and sustainability (Mason, 2013).

Scientific discourse continues to address these challenges. Interdisciplinary collaboration in climate change, carbon sequestration, and clean technology has been emphasized by a recent workshop (Hampton & Whitmarsh, 2023). Environmental education programs such as those of the Cremisan Salesian Monastery promote environmental awareness and conservation activism (Hassan, 2025). These environmental injustices are addressed through policy reform, science, and awareness to provide a sustainable future (Muneeb et al., 2022).

4 Discussion and Conclusion

Palestinian universities bear the role of being a pioneer to bring out environmental consciousness and sustainability through normalizing sustainability in curricula and civil engagement. Within challenging contexts such as political unrest, resources limitations, and red tape constraints, universities are positioned to drive environmental change through integrated research, practicum-oriented teaching, and planned partnerships (Monna et al., 2022).

Universities must link learning about the environment to real action so that sustainability principles extend beyond the field of academic argument. Field experiences, workshops, and local activities facilitate experiential learning, and long-term accountability for the environment (Passarelli & Kolb, 2011). Israeli occupation exacerbates environmental degradation, limiting the availability of water, land conservation, and opportunities for study-based field experiences (Institute for Middle East, 2022). University–NGO–government partnerships enhance the sustainability education of marginalized groups, such as women, through conducting activities like sustainable agriculture workshops (Abu Hanieh et al., 2015). Incorporation of environmental justice into the curriculum can also further empower students to shape policy reform and sustainable development (Chu & Karr, 2017).

Cyberspace accommodates global cooperation that overcomes logistical barriers to environmental education. Institutions of higher learning must initiate green campus initiatives, waste prevention, and water conservation programs as models of societal conduct. Government and institutional patronage scaling is critical for tackling economic and political challenges (Blewitt, 2010). Future research will need to examine emerging strategies, such as virtual reality and e-learning, to enhance environmental education. Comparative studies with universities in similar regions will provide best-practice lessons. Despite challenges, Palestinian universities remain at the core of advancing ecological sustainability through interdisciplinary collaboration, community outreach, and policy influence. Securing partnerships and finance will be critical in advancing environmental education toward long-term regional and international impact.

References

Abu Hanieh, A. M., Abdelall, S., Krajnik, P., & Hasan, A. (2015). Industry-academia partnership for sustainable development in Palestine. *Procedia CIRP, 26*, 109–114. https://doi.org/10.1016/j.procir.2014.07.184

Al Botmeh, S., & Saadeh, I. (2024). Palestinian agriculture, food security, and incomes in the context of genocide. In *International Scientific Symposium: Priorities for Palestine's Economy in the Midst of War*, 38. https://mas.ps/cached_uploads/download/2024/12/02/paper-2-1733151676.pdf

Ardoin, N. M., Bowers, A. W., & Gaillard, E. (2020). Environmental education outcomes for conservation: A systematic review. *Biological Conservation, 241*, 108224. https://doi.org/10.1016/j.biocon.2019.108224

Awashra, R. A. (2012). *Governance of NGOs (with reference to national goals) in the Occupied Palestinian Territory: A case study of the West Bank.* Flinders University of South Australia, School of Social and Policy Studies.

Awashreh, R. (2018). Palestinian NGOs: External governance, stakeholders, and accountability. *Journal of Governance and Public Policy, 5*(2). https://doi.org/10.18196/jgpp.v5i2.4306

Awashreh, R. (2020). Palestinian perspectives on foreign aid. *International Journal of Research—GRANTHAALAYAH, 8*(6), 236–251. https://doi.org/10.29121/granthaalayah.v8.i6.2020.539

Awashreh, R. (2025). Navigating leadership dynamics: Insights and strategies across domains. In M. Rahman, G. Koolmon, S. Shaikh, & A. Munna (Eds.), *Optimizing leadership and governance in higher education* (pp. 203–232). IGI Global Scientific Publishing. https://doi.org/10.4018/979-8-3693-3534-5.ch010

Balabantaray, S. R. (2023). Women's leadership and sustainable environmental initiatives: A macroscopic investigation from the ecofeminism framework. *International Journal of Multidisciplinary Research and Growth Evaluation, 4*(4), 1039–1046.

Bethlehem University. (2024). Community engagement for environmental awareness through PIBS. https://www.bethlehem.edu/2024/10/14/community-engagement-for-environmental-awareness-through-pibs/

Bismuth, C. (2016). Water resources, cooperation and power asymmetries in the water management of the lower Jordan Valley: The situation today and the path that has led there. In *Society-water-technology: A critical appraisal of major water engineering projects* (pp. 189–204). Springer. https://doi.org/10.1007/978-3-319-18971-0_13

Blewitt, J. (2010). Higher education for a sustainable world. *Education + Training, 52*(6/7), 477–488. https://doi.org/10.1108/00400911011068432

Buheji, M., & Al-Muhannadi, K. (2023). Mitigating risks of environmental impacts on Gaza—Review of precautions & solutions post (2023 war). *International Journal of Advanced Research in Engineering and Technology (IJARET), 14*(7), 15–47. https://doi.org/10.17605/OSF.IO/2JMCP

Cairney, P., Timonina, I., & Stephan, H. (2023). How can policy and policymaking foster climate justice? A qualitative systematic review. *Open Research Europe, 3*, 51. https://doi.org/10.12688/openreseurope.15719.2

Çakmakçı, R., Salık, M. A., & Çakmakçı, S. (2023). Assessment and principles of environmentally sustainable food and agriculture systems. *Agriculture, 13*(5), 1073. https://doi.org/10.3390/agriculture13051073

Chu, E. W., & Karr, J. R. (2017). Environmental impact: Concept, consequences, measurement. In *Reference module in life sciences* (pp. 1–6). Elsevier. https://doi.org/10.1016/B978-0-12-809633-8.02380-3

Dwivedi, Y. K., Hughes, L., Kar, A. K., Baabdullah, A. M., Grover, P., Abbas, R., Andreini, D., et al. (2022). Climate change and COP26: Are digital technologies and information management part of the problem or the solution? An editorial reflection and call to action. *International Journal of Information Management, 63*, 102456. https://doi.org/10.1016/j.ijinfomgt.2021.102456

Gough, B. (2016). Reflexivity in qualitative psychological research. Journal of Positive Psychology. pp. 1–2. ISSN 1743-9760 DOI: https://doi.org/10.1080/17439760.2016.1262615

Hampton, S., & Whitmarsh, L. (2023). Choices for climate action: A review of the multiple roles individuals play. *One Earth, 6*(9), 1157–1172. https://doi.org/10.1016/j.oneear.2023.08.006

Hasan, A., & Buheji, M. (2024). Education resilience under the occupation: Case of Palestine. International Journal of Youth. *Education, 8*(1), 33–45. https://doi.org/10.18576/ijye/080103

Hassan, N. E. (2025). The role of parents in fostering environmental stewardship in children. *International Journal of Trendy Research in Engineering and Technology, 9*(1), 56–63. https://doi.org/10.54473/IJTRET.2024.9107

Institute for Middle East Understanding. (2022). *Fact sheet: Israel's environmental apartheid in Palestine*. https://imeu.org/article/environmental-apartheid-in-palestine

Kelly, R., Elsler, L. G., Polejack, A., van der Linden, S., Tönnesson, K., Schoedinger, S. E., Santoro, F., Pecl, G. T., Palmgren, M., Mariani, P., Glithero, D., Evans, K., Cvitanovic, C., Cook, J., Bartram, J., & Wisz, M. S. (2022). Empowering young people with climate and ocean science: Five strategies for adults to consider. *One Earth, 5*(8), 861–874. https://doi.org/10.1016/j.oneear.2022.07.007

Khasnabis, C., Motsch, K. H., Achu, K., Al Jubah, K., Brodtkorb, S., Chervin, P., ... Lander, T. (2010). Community-based rehabilitation: CBR guidelines. https://www.ncbi.nlm.nih.gov/books/NBK310920/

Leal Filho, W., Sima, M., Sharifi, A., et al. (2021). Handling climate change education at universities: An overview. *Environmental Sciences Europe, 33*(109). https://doi.org/10.1186/s12302-021-00552-5

Leal Filho, W., Aina, Y. A., Dinis, M. A. P., Purcell, W., & Nagy, G. J. (2023). Climate change: Why higher education matters? Science of the Total Environment, 892, 164819. https://doi.org/10.1016/j.scitotenv.2023.164819

Mason, M. (2013). Climate change, securitisation and the Israeli–Palestinian conflict. *The Geographical Journal, 179*(4), 298–308. https://doi.org/10.1111/geoj.12034

Mezirow, J. (1997). Transformative learning: Theory to practice. New Directions for Adult and Continuing Education, 74, 5–12. https://doi.org/10.1002/ace.7401

Middlesex University. (2024). *A call to action for sustainability at Middlesex University on World Environment Day*. Middlesex University. https://www.mdx.ac.ae/news/a-call-to-action-for-sustainability-at-middlesex-university-on-world-environment-day.

Mohammed, S. Y., Aljanabi, M., & Gadekallu, T. R. (2024). Navigating the nexus: A systematic review of the symbiotic relationship between the metaverse and gaming. *International Journal of Cognitive Computing in Engineering, 5*, 88–103. https://doi.org/10.1016/j.ijcce.2024.02.001

Monna, S., Barlet, A., Haj Hussein, M., Bruneau, D., Juaidi, A., & Baba, M. (2022). Sustainability integration in Palestinian universities: A focus on teaching and research at engineering faculties. *International Journal of Sustainability in Higher Education, 23*(7), 1709–1729. https://doi.org/10.1108/IJSHE-08-2021-0338

Muneeb, A., Sheikh, A., & Sheikh, A. A. (2022). The environmental fallout of the Israel-Palestine conflict: A deep dive into the ecological impact of ongoing hostilities. *Journal of Survey in Fisheries Sciences, 8*(3), 624–630. https://doi.org/10.53555/sfs.v8i3.2932

Passarelli, A. M., & Kolb, D. A. (2011). The learning way: Learning from experience as the path to lifelong learning and development. In *The Oxford handbook of lifelong learning* (pp. 70–90.) https://doi.org/10.1093/oxfordhb/9780195390483.013.0028

PIBS a, Palestinian Institute for Biodiversity and Sustainability (2024). Programs of Education. https://www.palestinenature.org/education/

PIBS b, Palestinian Institute for Biodiversity and Sustainability (2024). Sustainable human and natural communities. https://www.palestinenature.org/

Randles, S. (2024). Ripple effects: Exploring the environmental impact of Israeli settlements' wastewater discharge. NRC Palestine. https://www.nrc.no/globalassets/pdf/reports/ripple-effects/exploring-environmental-impact-israeli-settlements-wastewater-discharge.pdf

Raouf, M., & Ruizi, Y. U. (2023). Environmental impacts of the current war in Gaza. Gulf Research Centre. https://www.grc.net/single-commentary/120

Renjith, V., Yesodharan, R., Noronha, J. A., Ladd, E., & George, A. (2021). Qualitative methods in health care research. *International Journal of Preventive Medicine, 12*(1), 20. https://doi.org/10.4103/ijpvm.IJPVM_321_19

Sadovska, V., Rastorgueva, N., Migliorini, P., & Melin, M. (2024). Engagement of stakeholders in action-oriented education for sustainability: A study of motivations and benefits and development of a process model. *The Journal of Agricultural Education and Extension*, 1–23. https://doi.org/10.1080/1389224X.2024.2415607

Salazar, C., Jaime, M., Leiva, M., & González, N. (2022). From theory to action: Explaining the process of knowledge, attitudes, and practices regarding the use and disposal of plastic among school children. Journal of Environmental Psychology, 80, 101777. https://doi.org/10.1016/j.jenvp.2022.101777

Seyfang, G., Hielscher, S., Hargreaves, T., Martiskainen, M., & Smith, A. (2014). A grassroots sustainable energy niche? Reflections on community energy in the UK. *Environmental Innovation and Societal Transitions, 13*, 21–44. https://doi.org/10.1016/j.eist.2014.04.004

UNEP, United Nations Environment Programme. (n.d.). *Women and the environment.* https://wedo.org/wp-content/uploads/womenandtheenvironment.pdf

United Nations. (2004). *United Nations international meeting on the impact of the construction of the wall in the occupied Palestinian territory including in and around East Jerusalem.* United Nations Office at Geneva. https://www.un.org/unispal/document/auto-insert-205608/

CHAPTER 14

Environmental Education and Communication: Challenges and Opportunities

Abdelilah Allaoui and Abdelhakim Mansoureddine

1 Introduction

Communication is essentially the basis for making people aware of their environment and for stimulating people to achieve a sustainable lifestyle (Ardoin & Bowers, 2020; Van Steenberghe, 2011; Moser, 2010). At a moment when environmental crises are becoming ever more urgent (Davis et al., 2018), the mode of delivery can have serious repercussions for individuals and society as a whole. To be effective, communication has to do more than inform; it has to inspire, mobilize, and create actual action to protect the planet (Yoshimura-Rank, 2013).

This chapter explores fundamentals of communication in environmental education from a theoretical perspective; the analysis considers basic

A. Allaoui (✉) • A. Mansoureddine
Faculty of Letters and Human Sciences, Mohammed V University, Rabat, Morocco
e-mail: abdelilah_allaoui@um5.ac.ma

K. Hirech (ed.), *Environmental Education and Sustainable Development*, https://doi.org/10.1007/978-3-032-05760-0_14

elements such as the sender, receiver, message, and context in any communication (Girault, 2018). It examines what the place of that communicative persuasion and education might be in connecting knowledge with action. Moreover, comparing traditional and digital media, this study investigates the optimum modalities that should be adopted for the dissemination of nuanced environmental messages, while also providing critical information about the subsequent efforts directed toward sustainable behavior by activating awareness-raising campaigns (Matteau, 2018; Sauriol, 2016).

However, communication needs to adapt pedagogical approaches to include more participatory and experiential methods, thereby promoting deep learning and sustained engagement (Eriksson, 2022; Lafitte, 2014; Armstrong et al., 2018). The central issue addressed in this chapter is therefore how to structure and use communication to maximize its impact in environmental education, while effectively mobilizing public opinion toward concrete and responsible actions.

In order to tackle this problem, the chapter is formatted in several sections. There are two main components to this: one is the theorized communication principles and their relevance to using analytical capabilities in the educational field. It then examines tools and materials from traditional media to digital platforms in terms of reach and effectiveness. A section is then dedicated to innovative teaching practices and the importance of non-verbal communication (Rouchouse, 2017). Finally, the study explores the participatory dimension of communication and its crucial role in civic engagement.

This work aims to answer the following question: To what extent can communication promote effective environmental education? To achieve this, a quantitative approach was adopted, based on the design and administration of a questionnaire. The project was submitted to a sample of students, allowing them to collect data on their perceptions, behaviors, and the role of communication strategies in environmental awareness. This methodology provides a sound basis for analyzing the impact of communication on environmental education (Mitchell et al., 2019; Kollmuss & Agyeman, 2002).

2 Theoretical Foundations of Communication in Environmental Education

The fundamental elements of communication that require translation across any medium also provide the foundation of environmental education (transmitter, receiver, message, channel, and context) (Sukma et al.,

2020). These elements influence the dissemination of environmental messages, attitudes, and beliefs, as well as the degree to which they can affect behavioral change on both individual and collective levels. Moreover, educational and persuasive communication plays a crucial role in disseminating knowledge and raising awareness, thereby helping to strengthen citizen engagement in environmental preservation.

As a first step, Jérôme Lafitte analyzes environmental controversies, stressing that the involvement of adult learners must not be limited to the mere transmission of information. He shows that such involvement remains limited: simply having the right attitude to participate in the process is not sufficient, as many approaches focus primarily on behavioral change. According to him, educational practices should evolve by focusing on environmental issues as one of the key pillars for mobilizing knowledge and fostering an informed understanding of current challenges (Lafitte, 2014).

Yves Girault expands this approach by examining the evolution of environmental education research programs in francophone countries. Influenced by international organizations such as UNESCO, he noted that early research often focused on environmental education. He highlights the tensions between values and objectives, particularly regarding the dissemination of knowledge through social networks. These processes raise important issues regarding the construction and dissemination of information, as well as their impact on perceptions of socio-environmental problems (Girault, 2018).

These analyses show a contrast between traditional approaches to environmental education and current socio-cultural realities. They highlight the importance of critical thinking and the active engagement of educators, while stressing the need to adapt communication practices and sources of information to new contexts (Zandvliet, 2019).

2.1 *The Key Concepts of Communication*

In environmental education, understanding the basics of communication, the sender–receiver relationship, the message, the channels, and how context affects information interpretation helps create effective strategies to raise awareness of environmental issues and promote sustainable behavior. The field is evolving rapidly now, shaped by digital technologies and social networks. In this regard, Matteau investigated how effectively these platforms can encourage people toward pro-environmental behavior (Matteau, 2018).

He cites a paradoxical trend: mobile devices are central to our lives, which may at times distract people from their connection to nature. The author performs a deep analysis in this field and studies how social media can be one of the most powerful devices in environmental campaigns. His analysis locates six aims, such as identifying environmental communication and discussing the most effective means of engaging citizens (Mallick & Bajpai, 2019).

The analysis makes an interesting comparison between traditional media and the world of social networks, stressing the challenges and opportunities. Matteau highlights things like cognitive and contextual factors that impact human behavior. Its evidential basis shows psychological and social obstacles that tend to be significant barriers to change, and it proposes the commissioning of targeted communication to bridge these hurdles, making campaigns more effective.

Matteau provides valuable insight into the interaction between new technologies and environmental engagement by laying the groundwork for communication strategies that are both relevant and impactful, and that take into account the social and psychological dynamics shaping civic action.

2.2 *Persuasive and Educational Communication*

In the late 1970s, Jack Mezirow saw transformative learning as central to adult education, particularly in helping individuals adapt and thrive in a changing world. His research on adult women returning to higher education revealed that profound learning often begins with a "disorienting dilemma," an experience that challenges one's existing worldview. Drawing on constructivism and critical theory, Mezirow described transformative learning as a process by which individuals revise their frames of reference through critical reflection and rational discourse.

Mezirow's work has since been expanded by other scholars. For instance, Paulo Freire emphasized the role of social justice and consciousness-raising, particularly among marginalized groups. Patricia Cranton highlighted authenticity and self-awareness, while Edward Taylor brought attention to the role of emotions, intuition, and culture.

The concept is interesting because it explains how individuals change the way they see themselves and the world through critical reflection, dialogue, and experience. Unlike conventional learning, which often focuses on acquiring knowledge or mastering skills, transformative learning is concerned with fostering deep, meaningful change in a person's beliefs, values, and assumptions. Transformative learning is not a linear process but involves a series of stages. Mezirow originally identified ten phases:

1. Experiencing a disorienting dilemma
2. Self-examination with feelings of fear, anger, guilt, or shame
3. Critical assessment of assumptions
4. Recognizing that one's discontent and transformation are shared
5. Exploring options for new roles or actions
6. Planning a course of action
7. Acquiring new knowledge and skills
8. Trying out new roles
9. Building competence and self-confidence
10. Reintegration into one's life with a new perspective

While Mezirow emphasized rationality and individual reflection, critics argue that this view is too narrow. Emotions, cultural context, spirituality, and relational dynamics also play a vital role. Recent developments in transformative learning theory seek to integrate these dimensions, making the theory more holistic and inclusive.

The field is currently evolving rapidly, shaped by digital technologies and social networks. In this regard, Matteau (2018) investigated how effectively these platforms can promote pro-environmental behaviors among individuals. He cites a paradoxical trend: mobile devices, while central to our lives, may at times distract people from their connection to nature. Using techniques and strategies designed to influence attitudes and guide behavior, persuasive and educational communication is a key part of environmental education. Its main objective is to raise public awareness of ecological issues while encouraging sustainable actions (Eriksson, 2022).

The importance of such communication is particularly evident at a time when pro-environmental behavior is essential for sustainability. For instance, the article "The Effects of Messages on Environmental Behaviour" by Miki Yoshimura-Rank points out a critical issue: despite rising awareness about the impact of human activities on the environment, it does not necessarily lead to behavioral changes. This raises important implications for how to more effectively communicate the gap (Yoshimura-Rank, 2013).

It's clear from Yoshimura-Rank that facts are not enough. Psychological and social barriers often stop people from acting in ways that are good for the world. To get past these, we need to understand why people do the things they do. Yoshimura-Rank suggests using public service announcements that mix persuasive messages with good social norms to get people to take action.

On the same topic, Davis et al.'s piece in Environmental Communication and Science Communication, called "Dual Roles of Environmental Communication in Science Communication," looks into the links between science (SC) and environmental (EC) communication. The authors stress how important it is for people from different fields to work together to improve civic participation and support epistemic justice. Environmental Communication (EC) tends to adopt a more political stance, which may restrict responses due to fear that emotional or behavioral strategies could distort risk perception. In contrast, Science Communication (SC) usually avoids an activist approach. Please keep the revised version accordingly. The gaps between Human–Robot Interaction and data analytics may be bridged so that their strengths may mutually support data-driven decision-making (Davis et al. 2018).

Lastly, Ardoin and Bowers (2020) build a case for developing ecological awareness very early on. Through her work in Early Childhood Environmental Education (ECEE), she demonstrates that outdoor learning and school-based programs that integrate environmental themes can be integral to children's development of a sense of connectedness to nature. They also acknowledge that further studies are needed, especially in adapting these educational practices to the cultural and socio-economic realities of different communities. Taken together, these insights underscore that simply providing facts does not ensure understanding or incorporation; to be effective, environmental communication must go beyond information transmission. It has to include awareness, education, and science. It requires a broad, long-term vision, and it must start in childhood to foster informed ecological awareness that motivates not only individual but also collective efforts toward a more sustainable future.

3 Communication Approaches and Tools in Environmental Education

3.1 Traditional and Digital Media

Media, awareness initiatives, and instructional methodologies are vital elements of environmental education communication. The media significantly influences public opinion. An example of a contentious problem in this field is the discourse surrounding neonicotinoids, as demonstrated in Geneviève Rajotte Sauriol's essay (2016). Journalists have to describe environmental problems objectively as information keepers, while taking into account the role of businesses and NGOs. This reveals the complexity of communication ecosystems where goals may conflict.

Etienne van Steenberghe pointed out that education stayed low on the environmental agenda even with rising environmental consciousness. He advocated a multidisciplinary framework that integrates public health into an environmental paradigm and stressed the need to turn information into action. Integrated education is necessary to make awareness-raising campaigns informative and engaging (van Steenberghe, 2011).

Katherine Matteau (2018) promotes political engagement through social media but questions its sustainability. Even if they thrive in public engagement and dissemination, these platforms are energy-intensive, raising questions in the long run in terms of environmental impact. Matteau advocates for a critical examination of the effectiveness and eco-friendliness of digital media (Matteau, 2018).

Environmental communications primarily occur through traditional and digital media, as well as education. According to the literature, a comprehensive and critical approach is necessary for these projects. If we are going to empower meaningful, lasting environmental change, then we ought to be looking at a medium's information-carrying capacity and environmental impact (Boulianne et al., 2020).

3.2 Awareness and Education Campaigns

They are of critical importance in promoting sustainable behavior and the protection of the environment. This has been made possible through different strategies, including visual storytelling, participatory initiatives with the inclusion of prominent personalities (Mansoureddine and Allaoui, 2024). Visual communication through persuasive videos or artful photos not only grabs the public's attention but also builds an emotional relationship with environmental issues, making them more relatable to ordinary people. Striking images of environmental damage, like polluted oceans or deforested lands, paired with real, practical solutions, can truly move people toward sustainable action.

Every day, citizens have the power to become change makers through local efforts like neighborhood workshops or clean-up campaigns. These grassroots projects raise awareness and bring people together, creating a strong sense of community while encouraging eco-friendly habits.

Strong players are also celebrities and influencers. Through social media, for example, or speech in more conventional venues, their trust and extensive reach allow them to magnify environmental themes and expand them, thereby resonating with a larger audience (Yalcin et al., 2020).

Amal-Mohamed-AIT (2020) notes that when taken together, community action, engaging narrative, and well-known public personalities will have more impact. Like attempts to reduce single-use plastics, niche projects show how carefully mixed outreach strategies can produce significant, persistent behavioral change.

4 Communication and Pedagogical Practices in Environmental Education

4.1 *The Importance of Communication in the Transmission of Knowledge*

Seydi (2024) emphasizes that communication is especially critical for knowledge transfer in environmental education. The ability of educators to convey clear messages, tailored to the specific needs of their audience, is fundamental to the effectiveness of verbal and written communication. The need for an interactive and precisely defined strategy is underlined, especially in order to guarantee a good awareness of environmental issues. This is demonstrated by specific instances whereby the assimilation of environmental knowledge reflects the direct effect of well-organized communication (Seydi, 2024).

Examining French-language learning among Guinean preschoolers in a bilingual environment, Seydi notes various aspects of educational communication. Supported by modified pedagogies, the research indicates that early exposures help to actively acquire the French language. Keys are social and cultural relationships; nonetheless, educational paths and socioeconomic levels differ.

Still, this study comes with certain limits. On the one hand, it highlights specific preschool facilities in Guinea, therefore restricting the relevance of results to other environments or sectors. Conversely, the study emphasizes linguistic and sociolinguistic aspects, therefore excluding the effect of emotional or motivating elements on learning.

Additionally, a methodology that relies on observation and questionnaires can introduce bias in educator and parental perceptions. Finally, in a multilingual environment, the lack of continuous research shows a difficulty in defining the change of language over time.

In essence, even if Seydi's study is valuable, it emphasizes the necessity of more research encompassing psychological aspects and the influence of long-term inclusive education policies to create a more comprehensive knowledge of learning in multilingual environments.

4.2 *Non-Verbal Communication and Experiential Learning*

Non-verbal communication, particularly through gestures, plays an essential role in experiential learning, as highlighted by Julie Rouchouse in her study, which emphasizes the importance of teaching gestures in the French as a Foreign Language classroom, showing how body movements, particularly of the arms and hands, facilitate access to meaning, especially for allophone students. She shows, through her own experiences in primary schools, that sign language, speech, and gestures can improve oral education and break down language barriers (Rouchouse, 2017).

The article by Rouchouse goes into great detail about gesture as a pedagogical tool for teachers. These gestures augment verbal explanations and provide an excellent substitute where speech cannot be used, even though much of the teacher's correction or encouragement involves spoken language. Her research, carried out in schools in Saint-Étienne, shows that using gestures during activities such as nursery rhymes or vocabulary learning helps to better retain information. However, it invites further study of individual differences in the perception of gestures and their application in various cultural and linguistic contexts (Rouchouse, 2017).

In a complementary approach, Mitchell et al. highlight the crucial role of movement and emotion in second language learning. Their study shows that integrating theatrical techniques into language workshops boosts learners' confidence while optimizing their oral production. By working together with the body and voice, teachers can refine students' prosodic awareness and enrich their emotional expression (Mitchell et al., 2019).

Both articles share a general idea: the link between body and speech is a powerful tool to make language learning more dynamic and, in the end, meaningful. In these theatrical or educational exercises, the incorporation of movement and gestures changes the nature of learning, since the information becomes more attainable and meaningful in a movement context. These strategies have a bright future in enhancing instruction and empowering language learners.

5 Case Study

Ben M'Sik High School, located in the delegation of Ben M'Sik Prefecture in Casablanca, stands out for its central role in the education of nearly 987 students, divided into 27 classes during the current school year 2024–2025. A majority of these students are girls (58.8%), while boys represent 41.2%. Students are mainly divided between the first year of the baccalaureate (54.4%) and the second year (44.7%). This demographic and academic diversity illustrates the richness of this institution, providing a relevant context for analyzing educational approaches and initiatives related to topics such as the environment and sustainable development.

5.1 *Methodology*

We choose the quantitative approach because it offers objectivity, precision, and the ability to generalize findings. The approach offers several key advantages that make it especially valuable in research requiring precision, objectivity, and generalizability. One of its primary strengths lies in its ability to produce measurable and statistically analyzable data, which enables researchers to identify patterns, test hypotheses, and determine relationships between variables with a high degree of accuracy.

In the educational field, the quantitative approach offers significant advantages in the field of education, particularly when the goal is to assess outcomes, measure performance, or inform policy. By relying on numerical data and statistical analysis, it enables educators and researchers to identify patterns, track student achievement, and evaluate the effectiveness of teaching methods or educational programs with a high level of precision.

To this end, a structured questionnaire was designed and administered to a sample of students at Ben M'Sik High School, located in the Ben M'Sik prefecture of Casablanca, during the 2024–2025 school year. The questionnaire developed for this study consisted primarily of closed-ended, multiple-choice questions, divided into several thematic sections: level of knowledge of environmental issues, interest in certain topics (such as climate change, renewable energy, etc.), preferred communication tools, participation in educational or ecological activities, and perception of the impact of courses on their daily behaviors. Evaluation scales were integrated to measure the degree of engagement or the importance given to certain actions.

It was distributed remotely, via students' WhatsApp groups, ensuring broad coverage within the target school. Students were invited to respond online using a Google Form, which facilitated structured and anonymous data collection. The responses were then automatically centralized in a spreadsheet for statistical analysis. This digital distribution method allowed for rapid, flexible, and secure participation, while preserving respondent confidentiality. Concerning the sample, a group of high school students was selected (first and second year of the baccalaureate), with a total of approximately 987 students. The detailed results are based on the responses obtained via the questionnaire.

Regarding the analysis, the results are often presented as percentages, which provide a clear and concise overview of the data and are commonly supported by graphs or charts to visually represent patterns, trends, or comparisons across groups. This visual presentation not only enhances understanding but also facilitates communication of findings to educators, policymakers, and stakeholders.

5.2 *Findings*

5.2.1 *Knowledge and Perception of Environmental Issues*

The results point to a general perception of a moderate level of knowledge among high school students, with 65.8% of them at the "medium" level. This majority reflects a certain familiarity with environmental issues, but also a significant margin for further understanding. The fact that about a quarter of students (26.3%) rate themselves at a good level is encouraging, as it suggests that a significant segment of the group feels competent in these subjects. However, the low representation of "Very good" (0.9%) and "Low" or "Very low" (7% in total) levels illustrates both the lack of acknowledged expertise and the scarcity of extreme gaps. These data suggest that high school students have a sufficient basis to build more

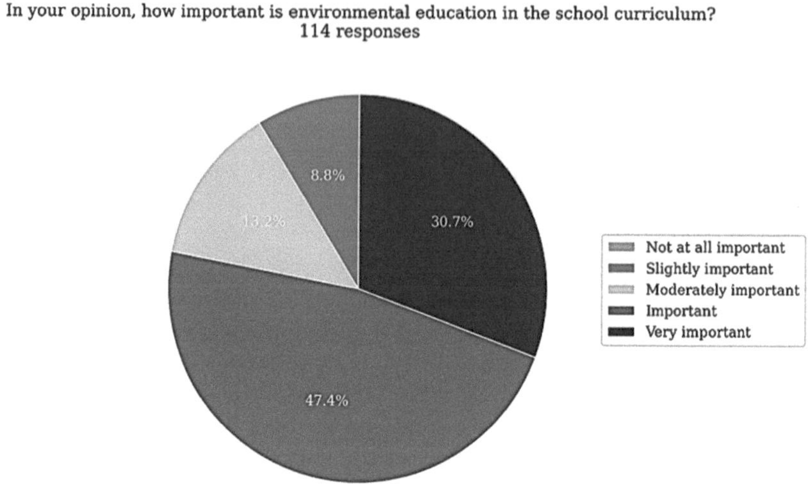

Fig. 14.1 Importance of environmental education

specialized skills through targeted educational initiatives and programs to deepen their engagement on environmental issues.

These results illustrated in Fig. 14.1 reflect a strong consensus among high school students on the importance of environmental education in the school curriculum, with almost 80% of respondents calling it "Important" or "Very Important." This reflects a growing awareness of environmental issues and a clear desire to place these issues at the heart of their educational journey. The small proportion of pupils who consider it to be "Medium important" (13.2%) or "Not important" (8.8%) points out that only a few minority voices perceive this education as a lower priority. The complete absence of responses indicating that it would be "not at all important" reinforces the idea of widespread support for its integration. These data underscore the need for education decision-makers to strengthen and structure environmental education to meet students' expectations while equipping them to meet today's ecological challenges.

The results in Fig. 14.2 show that the students interviewed pay particular attention to the most current and global environmental issues. Climate change is of most interest, with 44.7% of respondents citing it as a priority, followed closely by air, water, and soil pollution (43.9%) and renewable energy (43%). These choices reflect an awareness of global issues that have

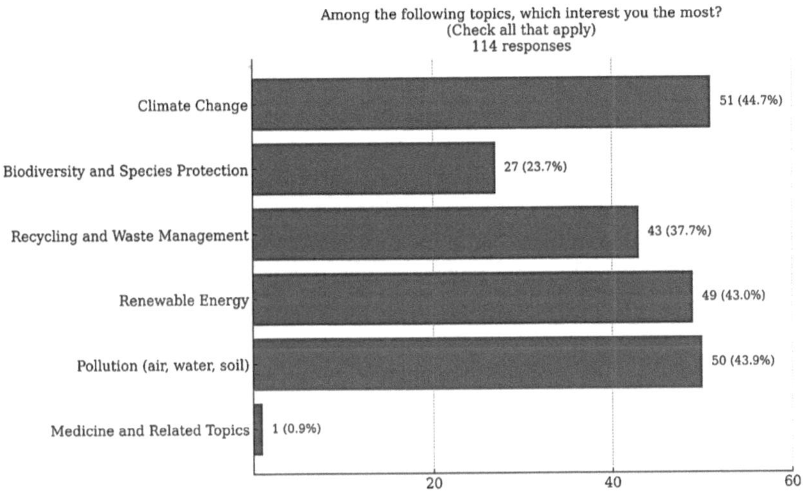

Fig. 14.2 Interesting topics for students. (Source: Authors 2025)

a direct impact on their daily lives. Recycling and waste management also account for a significant proportion (37.7%), reflecting a growing sensitivity to concrete and local actions to preserve the environment. On the other hand, biodiversity and species protection (23.7%) seem to be less involved, as do specific subjects such as environmental medicine (0.9%). These data indicate that students prioritize global and practical themes, pointing to ways to adapt environmental education programs to better suit their inter ests and broaden their curiosity toward other important issues.

5.2.2 *Communication Media and Tools*

The results clearly show that young people prefer dynamic and interactive ways to learn about the environment. Videos like documentaries and feature stories came out on top with 57.9% of responses, showing just how powerful visuals and emotions can be. Close behind are workshops and field trips (50.9%), which reveal a strong appetite for hands-on and immersive experiences. Social media and interactive tools like educational games also scored highly (36.8% each), showing just how vital digital platforms are in keeping students interested and involved.

In contrast, more traditional formats like online articles and newspapers came in much lower, with only 8.8% of students favoring them. This probably shows how differently pupils interact with material today compared to conventional forms. In other words, these techniques fail to effectively engage students.

What, therefore, should be the lesson? If we wish environmental education to really connect, we must give top priority to visual, interactive, and immersive forms that fit young people's learning style.

Social media is not being completely used, even if it offers possibilities. While many students consider it as increasing awareness, others think its impact is limited. This underlines a clear possibility: if we are to really change things, we have to rethink how we use these media. Thus, it is imperative to provide material that is not just new, relevant, and related to students' beliefs and interests, but also educational.

5.2.3 *Experiential Learning and Teaching Methods*

The responses show that a majority of students have never participated in experiential environmental activities at school, although a significant proportion have already benefited from such initiatives. This highlights the potential for expanding these immersive pedagogical approaches, which could enrich environmental education and foster more concrete student engagement with environmental issues. The data suggest that experiential learning activities play a key role in raising young people's awareness. Their impact is generally perceived as positive, with nearly 50% of respondents acknowledging their effectiveness, making them valuable educational tools.

However, to meet the expectations of all students and maximize the impact of such activities, efforts should be made to enhance their accessibility, increase their frequency, enrich their content, and ensure structured educational follow-up. Student enthusiasm for hands-on and immersive activities is particularly strong. Nature outings are the most preferred format, with 77% of students favoring them, revealing a strong desire to connect directly with the environment. Practical workshops such as gardening or recycling also appeal to 51.3% of respondents, underlining the attractiveness of interactive, manual activities.

Collaborative projects such as awareness campaigns or public space clean-up operations are supported by 44.2% of students, showing a clear interest in collective and concrete forms of engagement. By contrast, simulations and role-playing (22.1%) and conferences or debates (14.2%) attract less interest, likely due to their more theoretical and less dynamic nature. Overall, these results demonstrate the importance of participatory, field-based approaches to engage students and strengthen their learning in environmental education.

5.2.4 *Participatory Communication and Personal Engagement*

The responses indicate that the majority of students consider active participation in environmental education courses to be important or very important. This highlights the interest of students in interactive pedagogical approaches, such as group discussions and projects, which promote greater involvement and a deeper understanding of environmental issues.

The responses revealed that the majority of students interviewed had never participated in a school project that had a direct impact on the environment, such as recycling initiatives or neighborhood clean-ups. However, a sizeable proportion of students indicated that they had taken part in this type of activity. This highlights an untapped potential within schools to incorporate more concrete projects, which enable students to feel directly involved in protecting their environment. These initiatives could not only strengthen their understanding of ecological issues but also encourage responsible and sustainable behavior daily.

The majority of high school students (51.8%) say they are ready to take active action to protect the environment in their daily lives, reflecting a keen awareness and confidence in their ability to make a difference. Others, almost a third (28.1%), feel somewhat less confident, but remain open to more investment if given the right tools and encouragement.

However, 15.8% of students feel they can only contribute "a little," while a small minority of 4.4% feel they cannot "have any" impact. These figures reveal that some students feel constrained, probably by a lack of resources or information. This underlines the importance of highlighting simple and accessible actions they can take, while offering them tailored support to build their confidence and encourage them to take concrete actions that, taken together, can make a real difference.

The results illustrated in Fig. 14.3 reveal a strong interest among students in the creation of eco-clubs in high schools, which was supported by 75.4% of respondents. These clubs seem to be seen as an ideal way to engage together, exchange ideas, and take concrete action for the environment. The group projects, cited by 48.2%, also confirm the importance of collective work in mobilizing and involving students in concrete initiatives.

Social media campaigns, chosen by 37.7%, show that these platforms play a key role in capturing attention and raising awareness among young people. In contrast, lectures or testimonies by professionals (22.8%) are of

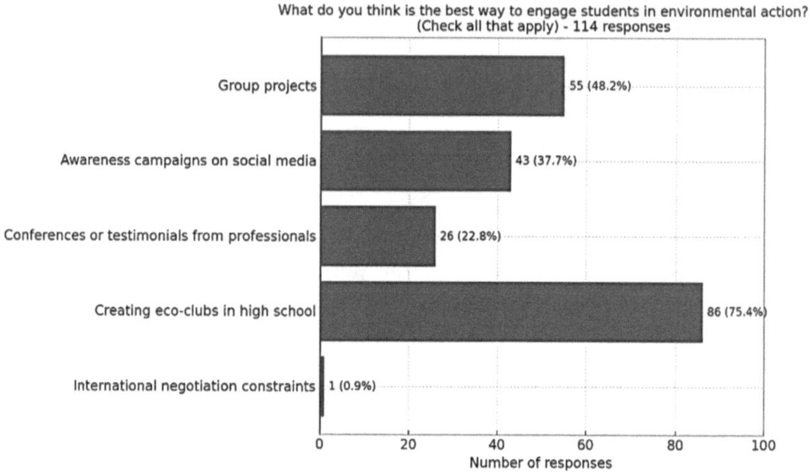

Fig. 14.3 Mobilizing pupils for environmental actions. (Source: Authors 2025)

less interest, perhaps because they seem less engaging or concrete for students. These results highlight the desire of young people to participate in local and collaborative actions, which enable them to feel genuinely involved in change.

5.2.5 *Assessment of the Impact of Communication Practices*

The results (Fig. 14.4) show that environmental education is beginning to influence students' habits, although progress remains to be made. Nearly four in 10 students (39.3%) say they have significantly changed their behavior, taking actions such as recycling or reducing plastic. A significant proportion (47.3%) said they had changed their practices "a little," reflecting ongoing awareness that could be amplified by more concrete and engaging initiatives.

However, almost 10% said they had not changed their habits, and 3.6% were not influenced at all. This shows that while environmental education courses are paying off, more needs to be done to encourage greater engagement. More immersive initiatives or initiatives tailored to the realities of students could help turn this awareness into lasting and profound actions.

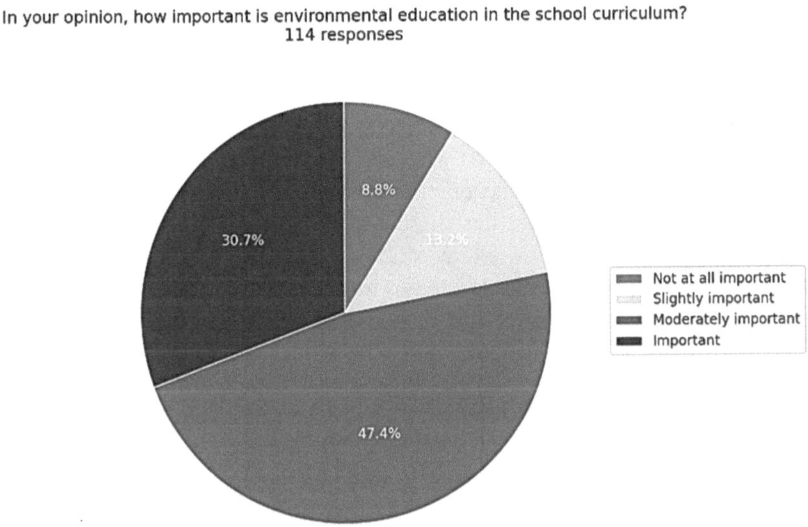

Fig. 14.4 Mobilizing pupils for environmental actions. (Source: Authors 2025)

The responses show that students adopt certain eco-responsible behaviors, but that commitment varies according to the actions. The majority (76.1%) save water and electricity, a simple and accessible effort that seems to be a priority for many. Waste sorting is the second highest (54.9%), reflecting a significant awareness of waste management.

However, practices such as reducing plastic use (28.3%) or using public transport and cycling (19.5%) are less widespread, which can be explained by practical constraints or a lack of targeted awareness. Finally, only 6.2% of students mention other initiatives, suggesting that more diversified campaigns and tailored solutions could encourage wider adoption of green behaviors daily.

The results show that environmental education courses have an overall positive impact on pupils, with almost 65% acknowledging a strong or very strong influence. This reflects the effectiveness of these lessons in raising awareness and encouraging sustainable practices. However, moderate or weak responses (34%) point to a need to adapt content to better reach certain audiences or strengthen concrete actions.

5.3 *Discussion*

The results clearly indicate how students see environmental education as well as their actual motivation for involvement. Inspired by John Stacey Adams's equity theory, the findings help to clarify how students evaluate their educational efforts against what they believe to be their return.

Most students believe that environmental education affects their daily choices, which is evidence that these courses alter actual behavior rather than only theoretical ones. Many students describe small but significant improvements, like turning off the tap to save water or developing a practice of garbage sorting. In their eyes, when the effort feels matched by impact, the message sticks, and that's when true environmental awareness starts to take root. But not everything clicks for everyone. Some students feel that the recognition or feedback for their efforts isn't quite there, and that may be why deeper or more complex habits like ditching single-use plastic or cycling instead of driving haven't yet taken hold for them.

Regarding learning styles, students express themselves loud and clear: they want to be active participants rather than merely passive listeners. Field outings (77%) and eco-clubs (75.4%), places where kids can get involved, see effects right away, and feel that their efforts count, are the venues where they gravitate. These encounters equip them with momentum and direction. More conventional forms like conferences and lectures, meanwhile, attract fewer. For what reason? Probably that students cannot instantly perceive how their involvement or presence results in anything specific.

Still, not everyone has the opportunity to participate, even with all this passion. Remarkably, 62.5% of students believe they have never participated in an actual, significant environmental initiative. This indicates obstacles, presumably institutional, perhaps logistical, that restrict access, not just a number. Furthermore, students' drive to first get active may be lowered when they feel excluded or underappreciated. Equity theory alerts us to just such a sense of imbalance, giving without getting anything in return.

Another key takeaway is that while many students appreciate what they're learning, nearly 34% still feel the content misses the mark. It might be too abstract or not connected to what matters to them. On the flip side, those who take part in hands-on projects—planting a garden, organizing a campaign, joining a clean-up—tend to report stronger motivation and a clearer sense of achievement.

And let's not forget the digital world. Social media, according to 37.7% of students, is one of the most effective tools for environmental awareness. These platforms shine a spotlight on both individual and group efforts, making actions visible and meaningful. They offer instant feedback, inspire others, and give students a voice. When used well, they can transform passive concern into active participation.

All in all, one message comes through loud and clear: environmental education works best when it's lived, not just taught. When students are given the chance to act, reflect, and be seen, their engagement deepens, and with it, the potential for lasting change grows.

6 Recommendations

The initial step in more effectively aligning students' efforts with the benefits they receive is to emphasize practical, collaborative solutions. Incorporating activities such as field excursions or school-based environmental clubs into the curriculum can provide students with engaging and rewarding experiences. In addition to boosting their motivation and sense of accomplishment, these activities offer tangible knowledge grounded in consistent and meaningful actions.

It is equally important to ensure that these opportunities are accessible to all, particularly in underprivileged areas. To encourage the adoption of more complex behaviors, such as switching to sustainable transportation or reducing plastic consumption, it is essential to provide robust logistical support and visible incentives.

It is important that students feel their contributions are valued. A certificate, a word of support, or a small gift can mean a lot to a student. Not only do you want to thank them, you also want to show that you value their time, energy, and thoughts. When people are pushed in that way, it makes them want to do even more.

And now, social media gives us something strong that is easy to use. It can make a quiet classroom project into something bigger, something that is seen, talked about, and honored if it is used correctly. It lets them show off their work, get quick feedback, and feel like they're part of a bigger movement. That sense of visibility, recognition, and connectedness helps them stay strong a lot of the time.

In the end, the findings suggest that environmental education is most effective when students feel that their efforts are fairly acknowledged and rewarded. This sense of balance between what they give and what they receive is essential for lasting engagement.

7 Conclusion

In conclusion, this study sought to examine the impact of environmental education on pupils' awareness and commitment to environmentally friendly practices. The results obtained from students at Ben M'Sik High School in Casablanca show a notable awareness of simple actions, such as saving resources or sorting waste. However, they also highlight the significance of ongoing efforts to encourage individuals to adopt more complex and environmentally sustainable behaviors, which are essential for achieving a profound green transition.

The assertion that "environmental education works best when it's lived, not just taught" is strongly supported by multiple layers of evidence. Students consistently highlight the effectiveness of experiential learning, field outings, eco-clubs, and hands-on projects as the spaces where engagement and meaningful change occur. These are not abstract preferences; they reflect real behavioral shifts such as water conservation and waste sorting, indicating that action-based learning leads to concrete, observable outcomes. Moreover, the application of equity theory reinforces the idea that motivation is driven by a sense of balance between effort and reward: when students feel their actions matter and are acknowledged, their commitment deepens.

Conversely, the data also show that where this connection is missing (e.g., passive lectures or lack of recognition), motivation falters. The fact that 62.5% of students report never participating in a major environmental initiative points to institutional and structural barriers that limit lived experience and, by extension, impact. Finally, the role of social media, highlighted by over a third of respondents, further confirms that visibility, feedback, and active contribution are central to fostering lasting environmental engagement. Taken together, these insights affirm that the conclusion is not only well grounded but captures the essential dynamic revealed by the research: students learn best when they do, reflect, and feel their actions make a difference.

References

Ardoin, N. M., & Bowers, A. W. (2020). Early childhood environmental education: A systematic review of the research literature. *Educational Research Review, 31*, 100353. https://doi.org/10.1016/j.edurev.2020.100353

Armstrong, A. K., Krasny, M. E., & Schuldt, J. P. (2018). *Communicating climate change: A guide for educators* (p. 144). Cornell University Press. https://library.oapen.org/handle/20.500.12657/62154

Boulianne, S., Lalancette, M., & Ilkiw, D. (2020). "School strike 4 climate": Social media and the international youth protest on climate change. *Media and Communication, 8*(2), 208–218. https://doi.org/10.17645/mac.v8i2.2768

Davis, L., Fähnrich, B., Nepote, A. C., Riedlinger, M., & Trench, B. (2018). Environmental communication and science communication conversations, connections and collaborations. *Environmental Communication, 12*(4), 431–437. https://doi.org/10.1080/17524032.2018.1436082

Eriksson, A. (2022). *Dire et faire la "Transition Écologique": Des dissonances à la résonance: Une recherche-action avec les acteurs associatifs de l'éducation à l'environnement en Occitanie* (Doctoral dissertation, Université Paul Sabatier-Toulouse III). https://theses.fr/2022TOU30335

Girault, Y. (2018). De la prise en compte des problèmes socio-écologiques à l'évolution des principaux courants de recherche en éducation relative à l'environnement dans la francophonie. *Éducation relative à l'environnement. Regards-Recherches-Réflexions, 14*(2), 2017–2018. https://doi.org/10.4000/ere.2727

Kollmuss, A., & Agyeman, J. (2002). Mind the gap: Why do people act environmentally and what are the barriers to pro-environmental behavior? *Environmental Education Research, 8*(3), 239–260. https://doi.org/10.1080/13504620220145401

Lafitte, J. (2014). Les controverses environnementales. Entre conflit et consensus. *Éducation relative à l'environnement, 12*, 81–103. https://doi.org/10.4000/ere.453

Mallick, R., & Bajpai, S. P. (2019). Impact of social media on environmental awareness. In *Environmental awareness and the role of social media* (pp. 140–149). IGI Global. https://doi.org/10.4018/978-1-5225-5291-8.ch007

Mansoureddine, A., & Allaoui, A. (2024). La communication interne: outil de réussite d'un établissement scolaire. Revue Internationale du Chercheur.

Matteau, K. (2018). *Modification des comportements citoyens: pertinence de l'utilisation des réseaux sociaux en communication environnementale, essai de maîtrise en Environnement et développement durable*. Centre universitaire de formation en environnement et développement durable (CUFE) Université de Sherbrooke. https://core.ac.uk/download/199227484.pdf

Mitchell, C., Guy, R., Fournel, A., Treille, A., & Koning, M. D. (2019). Affective learning for effective communication. Emotion and prosody in theatre improvisation. *TIPA. Travaux interdisciplinaires sur la parole et le langage*, (35). https://doi.org/10.4000/tipa.3297

Moser, S. C. (2010). Communicating climate change: History, challenges, process and future directions. *Wiley Interdisciplinary Reviews: Climate Change, 1*(1), 31–53. https://doi.org/10.1002/wcc.11

Rouchouse, J. (2017). Les gestes pédagogiques en classe de Français Langue Étrangère/Seconde. *Synergies France, 11*, 31–41. https://www.gerflint.fr/Base/France11/rouchouse.pdf

Sauriol, G. R. (2016). Analyse Des Messages Véhiculés Par Des Médias Et Des Organisations D'intérêt Public Et Privé Au Sujet Des Néonicotinoïdes: Communiquer Pour Le Bien Commun. http://hdl.handle.net/11143/8207

Seydi, D. F. B. (2024). Impact des contextes multilingues sur l'apprentissage du français chez les préscolaires Guinéens. *ALTRALANG Journal, 6*(1), 25–40.

Sukma, E., Ramadhan, S., & Indriyani, V. (2020, March). Integration of environmental education in elementary schools. *Journal of Physics: Conference Series, 1481*(1), 012136. IOP Publishing. https://doi.org/10.1088/1742-6596/1481/1/012136

Van Steenberghe, É. (2011). Pour en lire plus: Environnement et santé: Passer de la conscience à l'acte. *Éducation relative à l'environnement. Regards-Recherches-Réflexions, 9.* https://doi.org/10.4000/ere.1701

Yalcin, T., Nistor, C., & Pehlivan, E. (2020, March). Sustainability influencers: Between marketers and educators. In '*Sustainability influencers: Between marketers and educators' Taylan Yalcin, Cristina Nistor and Ekin Pehlivan Business Forum* (Vol. 28, No. 1). https://papers.ssrn.com/sol3/papers.cfm?abstract_id=3800316

Yoshimura-Rank, M. (2013). The effects of messages on environmental behavior. *De Paul Discoveries, 2*(1), 9. https://via.library.depaul.edu/depaul-disc/vol2/iss1/9

Zandvliet, D. (2019). Ecological education via "islands of discourse": Teacher education at the intersection of culture and environment. *Journal of Outdoor and Environmental Education, 22*, 145–157. https://doi.org/10.1007/s42322-019-00037-3

CHAPTER 15

Policy and Global Perspectives in Supporting Green Education

Rachna Rathore, Priyanka Kaushik, and Saurabh Pratap Singh Rathore

1 Introduction

Inexperienced training, or environmental training, is increasingly visible as a critical component in the pursuit of sustainable development. The primary objectives of this pedagogical method are to enhance college students' understanding of environmental problems and to encourage the implementation of sustainable practices. Its objective is to equip the upcoming generation with the facts and talents essential to efficaciously address problematic ecological concerns (Orr, 1992). Thinking about India's speedy industrialization and urbanization, it's vital to combine

R. Rathore (✉)
Department of Mathematics, Jiwaji University, Gwalior, India

P. Kaushik
Department of Computer Science and Engineering, NIMS University, Jaipur, Rajasthan, India

S. P. S. Rathore
School of Management & Commerce, Vikrant University, Gwalior, India

K. Hirech (ed.), *Environmental Education and Sustainable Development*, https://doi.org/10.1007/978-3-032-05760-0_15

green training within the country's widespread instructional framework. India has numerous environmentally demanding situations, consisting of deforestation, excessive air and water pollution, and a significant reduction in biodiversity. The population of over 1.4 billion exacerbates those problems, putting tremendous pressure on the nation's natural assets. A sturdy foundation in environmental schooling is important for fostering a sustainable and ecologically conscious tradition. India has initiated many initiatives to promote ecologically sustainable training. The National Green Corps (NGC) program, administered by the Ministry of Environment, Forestry, and Climate Change, seeks to build a nationwide network of eco-club members, engaging schools in a variety of environmental conservation initiatives. The Centre for Science and Environment's Green Colleges Programme is a significant attempt that encourages colleges to undertake sustainable practices and assesses their environmental performance.

Green education, also known as environmental education, is gradually emerging as an essential key in the pursuit of sustainable development. It does not intend only to present students with knowledge about those ecological challenges, it also intends to develop values, attitudes, and achievements that sustain the stewardship of the environment and sustainability over time. Thus, environmental education in this sense is a preventative and transformative strategy, preventing ecological destruction through awareness and then transforming the environment through behavioral change and civic responsibility. With the speed of industrialization and urbanization in India it has become timely and necessary to incorporate green education in the curriculum of the nation as well. Environmental problems including deforestation, air and water pollution, and loss of biodiversity are compounded by the population of more than 1.4 billion. These increasing pressures on natural resources require a comprehensive educational solution (Fuso Nerini et al., 2024).

As such, a robust basis in environmental education is a must for the development of ecologically literate citizens who can make informed, sustainable choices. India has made great efforts to integrate sustainability into education. The National Green Corps (NGC), launched by the Ministry of Environment, Forest and Climate Change, engages students in experiential environmental conservation via eco-clubs. Likewise, the Green Schools Programme of the Centre for Science and Environment encourages institutions to practice and monitor sustainable development at the school level, fostering systemic environmental awareness across campuses (Kondratenko et al., 2017).

2 Concept of Green Education in Formal Curricula

Green schooling promotes environmental recognition and responsible consumption, equipping individuals with information and capabilities to deal with ecologically demanding situations. It emphasizes self-reliance, duty, and empathy, and the usage of nature as a coaching tool. Through fostering proactive answers and network engagement, green education aims to cultivate an environmentally aware technology for a sustainable future. When applied in a balanced way, these three pillars support learners from primary through higher education in developing environmental awareness and understanding its challenges, while also offering adaptable solutions for practical application. As a core pillar, environmental education helps students identify and interpret changes in their immediate surroundings. Beyond motivating individuals to protect and improve the environment, it strengthens comprehension of the ecological, physical, and social processes that define environmental systems. Accordingly, embedding green education within academic institutions requires that teaching and learning activities consistently foster environmental knowledge, skills, and responsible attitudes. A wholesome ecosystem is crucial for existence on this planet, but rapid environmental adjustments, driven in large part via human activities, threaten sustainability. Issues like pollution, weather trade, deforestation, and species loss are escalating. because the twenty-first century, worldwide efforts have targeted on environmental protection to save you an inhospitable international for future generations. Challenges such as pollution, climate change, deforestation, and biodiversity loss are intensifying. Since the start of the twenty-first century, global initiatives have focused on environmental protection to prevent the planet from becoming uninhabitable for future generations.

Green education helps to develop environmental awareness and responsible consumption by giving people the knowledge, skills, and values that are necessary to comprehend and tackle the dilemma of complex ecological systems. It teaches core concepts like independence, accountability, compassion, and independent thinking. Nature stands as a significant pedagogical tool, an avenue through which learners may form meaningful relationships with their world. The three legs on which effective environmental education stands are awareness, responsibility, and solution-oriented learning. When they are intelligently integrated, they help students through all levels of their education from primary onward to develop a holistic awareness of the environment (J. Sachs et al., 2023). This also entails a sound knowledge of local, regional, and worldwide environmental contexts, as well as the flexibility to propose context-specific solutions.

Environmental education not only inspires people to take concrete, positive action to improve the environment, but it also enhances people's understanding of the ecology, physics, and socio-politics of our world (Independent Group of Scientists, 2023). Thus, integrating environmental education in formal curricula implies that the educational goals (through knowledge, values, and behaviors) are oriented to develop environmental literacy and sustainable action. Life on earth depends on a healthy ecosystem. But increasingly rapid environmental deterioration caused predominantly by human activity is now the greatest threat to planetary health. With the increasing of global environmental risks, green education provides the solution to the urgency to help educate both young and older generations, thus preventing further irreparable natural environment destruction as well as developing adaptation ability.

3 Green Education for Sustainability

Hands-on Learning: Experiential study is a key component of green schooling, related to activities like subject journeys and nature walks that help students connect with the environment and understand ecological issues. By integrating environmental concerns into subjects like science, social studies, and mathematics, students comprehend the interconnectedness of environmental issues and their broader impact. Inexperienced education promotes teamwork, essential thinking, and problem-solving through tasks including waste reduction campaigns or network gardens. It encourages sustainable practices like strength saving and resource control while helping students to discover creative answers to international environmental problems. This approach fosters proactive, critical thinkers. This more comprehensive viewpoint aids pupils in comprehending how their activities affect the whole world. Experiential learning lies at the heart of green education, involving hands-on activities such as field trips, nature walks, and community-based environmental initiatives.

These learning approaches promote authentic learning, providing students with opportunities to experience ecological processes firsthand while becoming more responsible for their local surroundings. Integrating such sustainable themes across the curriculum, in subjects like science, social studies, and maths, means children begin to understand how environmental problems are linked with other more general socioeconomic systems. This sensitization process can enable students to recognize that sustainability is not just a matter of a silo but the responsibility of society at large.

Green education promotes project-based collaboration, critical thinking, and problem-solving through activities such as campus waste reduction programs, school gardens, and water conservation initiatives. These participatory methods foster sustainable behaviors like energy conservation, careful consumption, and neighborhood resource management. In addition, students are motivated to seek innovative, locally relevant solutions to global environmental issues, thus becoming active agents of transformation. This holistic approach enables learners to comprehend the ways in which their daily actions impact the environment, both locally and internationally. This pedagogical transformation positions learners as global citizens with not only knowledge but also the moral responsibility to save the planet (S.H. Lytvynova, 2015).

4 Young Generation Participation in Environmental Sustainability

Youth represent a nation's greatest opportunity. They possess the capacity to become future developers, leaders, and lawmakers. They are both the current and future inhabitants of the earth and the architects of the environment in their own eras. Consequently, it is imperative to equip children for the ecological future. An effective remedy for environmental deterioration is educating children about environmental sustainability. Children have the right to understand their surroundings and engage in their formation, since they will be the future inhabitants of the earth. Children have the right to a healthy environment that promotes their mental, social, and physical development. Children's participation encompasses the initiation of sports, promoting inclusivity, and contributing to environmental safety. Young people can develop environmental literacy through various experiential learning approaches, such as gardening, hands-on activities, collaborative projects, decision-making exercises, and other practical engagements.

5 Learner Constructivism and Critical Thinking

Crucial wondering and learner constructivism are essential in inexperienced schooling, shaping how people recognize and engage with environmental issues. Learner constructivism highlights the active role of students in building knowledge through experience and reflection, enabling them to explore environmental concepts and adapt learning to their individual styles. This technique promotes inquiry-based curiosity, encouraging college students to research and query environmental subjects, fostering deeper knowledge. Critical questioning in green training includes evaluating, synthesizing, and studying records to cope with complicated environmental issues. It facilitates students investigating issues from more than one view, using genuine proof, and advocating revolutionary sustainability solutions. Through accomplishing debates and sports, college students become active, informed citizens committed to sustainable practices.

6 Including Current Critical Approaches in Green Education

It is essential to incorporate recent critical frameworks that expand the definition of sustainability beyond traditional paradigms in order to advance policy and global perspectives on green education. Deep, introspective engagement is emphasized in transformative learning, which encourages students to challenge prevailing beliefs and reinterpret their environmental identities. By empowering people to transform from passive information consumers into active sustainability agents, green education serves as a catalyst for both individual and societal change. Formal curricula frequently overlook the rich, place-based understandings of ecology provided by Indigenous Knowledge Systems (IKS). These systems represent generations of sustainable living, spiritual connections to nature, and ecological stewardship.

Epistemic diversity is made possible by integrating IKS into green education, which encourages students to learn from regional customs like intergenerational ecological storytelling, rainwater harvesting, seed conservation, and sacred groves (V. Bykov, 2008a). By acknowledging the interconnectedness of all life forms and decentering humans, a posthumanist perspective challenges anthropocentric narratives. This viewpoint urges teachers and students to recognize non-human beings as co-actors in the sustainability discussion, including plants, animals, ecosystems, and even rivers. It promotes empathy and concern for the non-human world by bringing to light ethical issues that go beyond human welfare.

Concurrently, decolonial viewpoints challenge whose knowledge is deemed legitimate and examine the colonial causes of environmental deterioration. These methods demand the restoration of marginalized worldviews and the dismantling of Eurocentric green education models. While addressing issues of power, access, and historical injustice, decolonial environmental education encourages the use of regional languages, cultural customs, and community leadership in sustainability education. By combining these methods, green education is reframed as a process that is liberating, relational, and justice-focused rather than just a technical or behavioral intervention. The relevance, inclusivity, and transformative potential of environmental education policies in both local and global contexts are reinforced by these critical viewpoints.

7 Importance of Green Education

Inexperienced training encourages proactive moves to address environmental issues by means of promoting sustainable practices, community engagement, and advocacy for policy alternate. It highlights the worldwide nature of environmental challenges, fostering a sense of duty and worldwide citizenship. Joining forces with numerous businesses, inexperienced training empowers people to seriously determine environmental information and make knowledgeable choices. It plays a key function in growing environmentally aware individuals able to address ecological challenges. With the aid of raising attention and cultivating stewardship, green schooling equips humans with the expertise and skills to make a contribution to a sustainable and resilient future, specifically within the context of climate alternate (Juntunen, M. K. & Aksela, M. K., 2014).

Key reasons for the need for Green Education:

- Climate change awareness: Educating people about the impacts of climate change is essential for motivating action and developing effective solutions.
- Environmental literacy: Green training allows comprehension of complicated ecological systems and the interrelation of environmental demanding situations.

- Sustainable practices: Via obtaining knowledge of sustainable practices, people might also diminish their ecological effect in ordinary activities.
- Vital wondering abilities: Green education fosters vital examination of environmental problems and advocates for brand new resolutions.
- Civic Responsibility: It cultivates a feeling of responsibility to engage in environmental conservation initiatives within communities.
- Future workforce development: Green skills are becoming more essential in the labor market as sectors transition toward sustainability.
- Health and well-being: Engaging with nature via environmental education may beneficially influence mental and physical well-being.
- Interdisciplinary learning: Green education amalgamates several disciplines such as science, social studies, and economics to provide a comprehensive grasp of environmental challenges.
- Empowering youth: Involving youth in environmental issues motivates them to become proactive agents of change.

8 Methodology

Design of study: The majority of the data for this analysis came from a review of studies published between the years 2000 and 2017. Thorough evaluation was conducted on the publications to ensure that they had undergone peer review and address the following important research issues:

Achieving sustainable development goals and the need for environmentally conscious education. Strategies for advancing sustainable development goals via environmental education.

To ensure that the results were based on credible and authentic research, it was crucial that they were peer-reviewed. For this work, the narrative literature review will be used primarily. Being based on a synthesis of several data sources, the technique worked well for figuring out how well green education worked to accomplish the Sustainable Development Goals. Reviewing existing literature helps fill in knowledge gaps that need more research and shows how new findings might be used in everyday practice. In general, people put greater stock in review conclusions than they do in those of individual studies. Because it lessens the possibility of bias in the findings, the review was the method used in this research. The review also saved time and money compared to starting from scratch with a new set of experiments.

Selection of studies: The Sustainable Development Goals are a vast area of study, which is recognized in this study. The review used a number of databases, including Google Scholar, JSTOR, Emerald Group Publishing, ERIC, ResearchGate, Web of Science, and Emerald Insight, to verify that the data sources were relevant to the questions it sought to answer. Because of their narrow emphasis, these databases were considered ideal for our research. We are particularly interested in the literature on green education and its relationship to the Sustainable Development Goals (SDGs) (C.O. Ekwueme et al., 2016). Since these databases have been useful for scholars doing reviews on green education, a manual search was conducted on Google Scholar to enhance the search's comprehensiveness. Using the keyword "Green Education," the first search yielded a large number of research articles evenly dispersed among the databases that were requested. So, when the focus shifted from green education to its relationship to the SDGs, the study honed its inquiry by employing more precise terminology like sustainability literacy, cleaner production, education for sustainability, nature-based learning, and environment.

Going back to the previous topic, the 2019 Character Strengthening Program(CSP) stands apart by emphasizing the need for developing environmentally conscious individuals via eco-school branding and peer collaboration. In its first year, 2018, the intervening program aimed to implement environmental education through a green school initiative, making an effort to improve school grounds and raise students' environmental awareness. The aim for 2019 was that the program would establish a green school board and continue environmental character development.

Selection and screening: A total of 180 materials were found across several databases when the search keywords from the preceding section were used. The findings were further reviewed to remove any duplicates or publications that did not correspond to the main issues of the study. Additionally, any papers about green education that did not fall within the scope of this research were also removed. To do this, we looked at abstracts, keywords used by writers, and the titles of published works. After going through the selection and screening process, 30 papers were selected for further investigation. For in-depth evaluation and analysis, 25 articles were selected.

We looked at 25 studies that met three criteria: first, they had to be original articles or books; second, we needed to know if they contained search terms in the title or abstract; and third, we wanted to know whether they described or analyzed green education in relation to the Sustainable Development Goals. A prospective approach was used to include studies, papers, or reports that were published before 15 December 2020. This acknowledges that past studies were outliers and that the review only covered the years 2015–2020. As a result, a two-tiered eligibility screening process was introduced, enabling the more rigorous application of multiple filters. The first filter was a quick review of the titles and abstracts to weed out unrelated publications, and the second filter involved a thorough evaluation of the articles' actual content. Using conservatism at each stage allowed us to reject just those items that did not meet our inclusion criteria. If there were any questions about certain objects, they might go to the next round for further investigation. Further information was requested, if possible, to determine whether the research should be included or eliminated if eligibility was still uncertain after full text screening. In order to provide a synthesis of outcomes, all chosen publications were collected and evaluated.

With a narrative literature review serving as the main methodological framework and elements of systematic review techniques as a supplement, this study employs a qualitative research design. This methodology facilitates the investigation of theoretical frameworks, policy patterns, and pedagogical approaches related to green education in the context of the Sustainable Development Goals (SDGs). The process of data collection entailed a systematic and specific search strategy across academic databases like Google Scholar, JSTOR, ERIC, Emerald Insight, ResearchGate, and Web of Science. The search had specific keywords such as "green education," "sustainability literacy," "education for sustainable development," "environmental education," "nature-based learning," and "ecopedagogy." Truncations and boolean operators were employed to widen the coverage and narrow the results. There were 180 sources that were originally identified. A purposive sampling method was used to identify studies that touched on the intersection of green education and SDGs, transformative learning, and education policy. Studies were considered if they satisfied at least three essential inclusion criteria: (1) publication in a peer-reviewed journal or established academic publication, (2) relevance to the topic of environmental education and sustainability, and (3) publication date during the specific review period of 2015–2020. Nonetheless, foundational studies prior to this time frame were considered in order to establish conceptual foundations.

For strict screening to be guaranteed, a two-stage filtration procedure was employed. The initial step was the screening of abstracts and keywords for the removal of irrelevant or duplicate records. The second step involved full-text examination to assess against research questions and theoretical backgrounds. Of the initial set, 30 good-quality studies were identified, of which 25 were utilized for final synthesis and interpretation. Qualitative thematic coding methodology was used for data analysis. Themes emerged based on multiple readings and comparative analysis of the literature (V. Yu. Bykov, 2008b). Emergent categories including pedagogical models, student participation, institutional policy structures, Indigenous knowledge blending, and curriculum development were employed for structuring findings. Comparative and contrastive approaches were utilized where appropriate to triangulate data among studies, contexts, and levels of education.

Validity and credibility of the findings from this study were enhanced by a number of factors. To begin with, the application of cross-disciplinary, peer-reviewed sources boosts academic rigor and diminishes bias. Second, the research engages literature from various geographical contexts and learning environments, providing a wide and expansive perspective on practice and policy. Third, blending transformative, Indigenous, decolonial, and posthumanist theoretical lenses adds depth to the interpretation, rendering the conclusions better aligned with current critical discourse in sustainability education. Finally, open documentation of search strategies, inclusion criteria, and analysis procedures provides assurance of replication of the study. Although the literature review approach is confined to direct empirical observation, an integration of the current empirical evidence affirms the formulation of informed and credible knowledge, and hence presents strong theoretical and policy recommendations for promoting green education at global, national, and institutional levels.

9 Result

The articles that were chosen for this synthesis all pointed to green education as a way to teach students about the need for preserving our planet for future generations. It offers schools practical lesson plans that integrate STEM (science, technology, engineering, and mathematics) concepts with social systems. According to J. Huckle and Wals (2015), green education has the potential to equip people with the necessary information, skills, values, and attitudes to support a future that is sustainable. Biodiversity, climate change, catastrophe risk reduction, sustainable consumption, and poverty alleviation are all key issues connected to sustainable development (Kopnina, 2012). According to the reviewed study shown in Fig. 15.1, in order to promote sustainable growth in the classroom, green education must include active learning strategies that encourage student participation.

Figure 15.1 illustrates how green education integrates STEM (science, technology, engineering, and mathematics) with social systems to prepare individuals for a sustainable future. The framework emphasizes the need for teaching topics like biodiversity, climate change, and disaster risk reduction, while encouraging reforms in pedagogical approaches. Ultimately, green education aims to impart the critical knowledge, skills, values, and attitudes necessary to enable students to actively engage in local and global decision-making, fostering sustainable development.

Thus, it necessitates substantial revisions to contemporary pedagogical practices in order to foster abilities such as critical and creative thinking, as well as collective decision-making. Because it is essential to building a

Fig. 15.1 Conceptual framework of Green Education for sustainable development

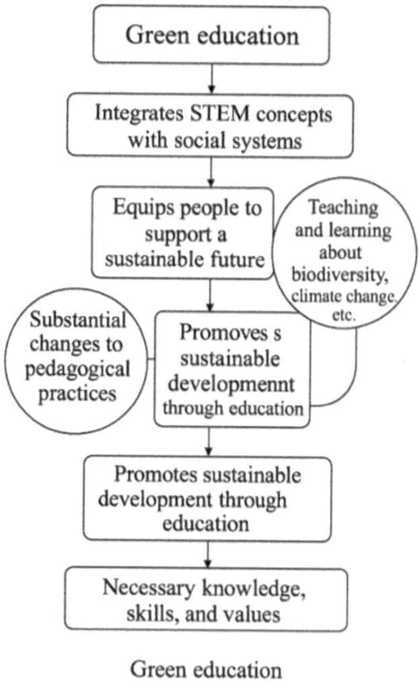

sustainable society, green education is obviously desirable in all kinds of learning environments, including official and informal schools and training programs. According to Education, it encompasses learning that secures and improves quality of life for both present and future generations. To improve people's present and future quality of life without compromising the future of our planet, sustainable development advocates for green education as a means by which students can acquire the core values, information, and abilities necessary to actively engage in decision-making processes on a local and global scale.

Figure 15.2 shows that no strategy was effective for environmentally conscious teaching. There is a lot of discussion about how students' comprehension, agency, and cognition may be enhanced by the adoption of more active, immersive, and interactive learning approaches (Kras, 2021). In collaborative learning, critical reflection, participatory learning, and creative thinking are three basic pedagogical ideas that underpin many popular approaches to bringing green education concepts into the classroom. Critical reflection encompasses both time-honored methods of

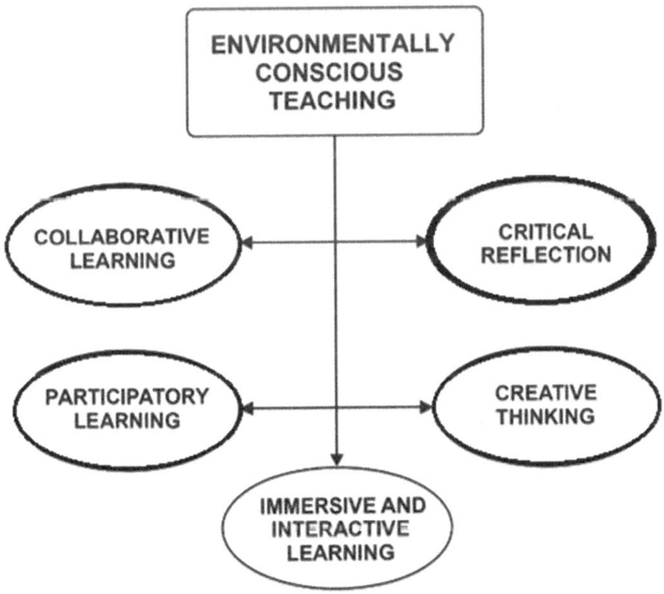

Fig. 15.2 Green educational level

instruction and more modern alternatives, such as reflective narratives, learning diaries, and online discussion forums.

To educate students to think systemically, teachers might use project-based learning, current events, and global case studies, and engaging peer or group learning, experience learning, action research, and debate are highly valued in participatory learning since they aid in decision-making and the development of case studies in collaboration with community groups and businesses (Bajaj & Chiu, 2009).

10 The Role of Green Education in Achieving Sustainable Development Goals

The review concluded that by providing resources and a curriculum, green education encourages critical and holistic thinking about environmental problems and their solutions on a global scale. Secondly, green schools are advantageous for everyone involved: teachers, students, parents, and the community as a whole. They protect health, boost student performance, save money, reduce carbon emissions, save water, and provide a unique learning environment (Nabukalu et al., 2019). Green education and learning spaces protect health by providing a great learning environment with good indoor air quality. According to reports, incorporating green building features such as non-VOC paints and carpets, enhanced ventilation, and natural daylighting makes schools healthier environments for both students and staff. By fostering the health and well-being of students, communities, and educators via the encouragement of healthy lifestyle choices, green schools and education contribute to the achievement of SDG 3, which seeks to achieve universal health coverage, end needless deaths, and decrease infant mortality (Jickling & Wals, 2008).

As shown in Table 15.1, green education supports the achievement of SDG 3 and SDG 4 by improving student performance through high-quality, inclusive, and equitable education. It enhances classroom environments by reducing air pollution, providing natural light, and using non-toxic materials, which boosts student focus and performance. Despite progress, gaps remain in global education access, with millions of children lacking preschool learning and many facing poor educational quality. Girls represent 60% of the world's illiterate youth. Green education addresses these challenges by offering an innovative alternative to traditional

Table 15.1 Green education's contributions to the attainment of sustainable development objectives

Roles of green education	*Related SDGs*
Provides unique learning opportunities	SDG 4 (Quality Education)
Creates green jobs	SDG 8 (Decent Work and Economic Growth)
Reduces student absenteeism	SDG 4 (Quality Education)
Enhances teacher retention	SDG 8 (Decent Work and Economic Growth)
Decreases water usage	SDG 6 (Clean Water and Sanitation)
Reduces carbon emissions	SDG 13 (Climate Action)
Reduces energy costs	SDG 7 (Affordable and Clean Energy)
Promotes quality education	SDG 4 (Quality Education)
Encourages holistic thinking	SDG 3 (Good Health and Well-being)

methods, improving academic outcomes and fostering lifelong learning opportunities.

The study emphasized the importance of green education in achieving Sustainable Development Goal 7, which focuses on reducing carbon emissions through cleaner production and improving access to energy. Since the 2002 World Summit on Sustainable Development, increasing emphasis has been placed on producing goods in more environmentally sustainable ways. (Goldman et al., 2018). In manufacturing, the adoption of cleaner energy reduces the consumption of water, energy, and raw materials, while at the same time minimizing the generation of hazardous and environmentally harmful substances. As a consequence of this, the range of those chemicals that end up being detrimental to our air, water, and land is reduced. The purpose of cleaner production is to reduce the poor impacts that purchased items have on human health, safety, and the environment at each level of the production process, from the extraction of raw materials to the final disposal of these substances. Thinking about one's surroundings is an crucial aspect of cleaner production, which incorporates the manufacturing of products and the provision of offerings. Cleaner manufacturing is a method that takes a holistic and linked approach to protecting our surroundings. It does this by means of addressing issues related to sustainable improvement at some point in the manufacturing procedure. Manufacturing that is less harmful to the environment is one approach that has been shown to be helpful in assisting the energy commercial enterprise function more efficaciously.Highlight the significance of cleaner manufacturing practices in promoting waste reduction, recycling, and environmental sustainability within organizations.

They argue that sustainable development requires multifaceted approaches that address environmental waste and sustainability issues, contributing to socioeconomic success. Khalili highlights that capacity building is crucial for fostering long-term social patterns and achieving SDGs. The study provides a framework for higher education leaders to invest in training programs that develop human capital. It also underscores the importance of incorporating resource management into course designs to promote sustainable development and long-term economic growth (Khalili et al., 2015).

Manufacturing in a more environmentally friendly way helps boost productivity, which in turn improves people's quality of life and makes us more competitive. In this way, cleaner manufacturing may be positioned as a tool for corporate sustainability, helping with the continuous pursuit of operational efficiency that is good for the environment by making the most efficient use of natural resources and reducing waste. Therefore, cleaner manufacturing is seen as a "win-win" strategy as it has the potential to safeguard the environment, employees, and customers while simultaneously increasing efficiency, profitability, competitiveness, and productivity (Fischer et al., 2022).

Green education supports SDGs 3, 7, 12, 13, 6, and 8 by fostering environmentally friendly jobs that promote economic growth without harming the environment. Examples include green building materials, architecture, and construction. It also helps youth employment, with initiatives like the Green School in Bali praised by Ban Ki-moon.

As far as the value of environmentally conscious education goes, the Green School Bali is among the greatest. The integration of local neighbors, workers, educators, and families is fundamental to its importance. The diversity of the community, its members' identities, attitudes, and ideals, along with its languages, traditions, and beliefs, all contribute to the many multicultural benefits that exist (Schnack, 2017). Tolerance, broader perspectives, and global knowledge are all nurtured by the school's varied student body. The school is a great place to learn, have fun, get involved, and explore new things. It is committed, in part, to creating a long-term space where people may feel comfortable talking about and sharing ideas related to global citizenship, new educational paradigms, and their connections to one another. The school's volunteer program aims to provide opportunities for learning and sharing to everyone in the local and worldwide communities in March 2015 (Abdomen) Volunteers, interns, and workers get practical experience while learning job-related skills and the importance of cross-cultural communication and teamwork. In addition, the location provides a thorough approach to work and a safe living environment marked by regular check-ins. They rely on them to keep their campus and academic lives in check. The school is now a first-rate place to study due to the passion, energy, and dedication of the interns and volunteers that work there. By sharing with the community, they help envision the future of education.

11 Strategies Used by Green Education in Fostering Sustainable Development Goals

Green education uses nature-based learning to achieve the SDGs by promoting research for sustainable innovation and economic development as shown in Table 15.2. It emphasizes nature-based solutions, such as urban green spaces and blue-green infrastructure, to address societal challenges and enhance well-being. An example is Ogun State's 2013 "green advocacy" campaign in Nigeria, aiming to mitigate climate change impacts. The program encourages decisions to defend the surroundings and its resources for future generations. It highlights the significance of schooling and ecological focus in achieving useful resource conservation, pollution reduction, and maintaining the earth's herbal stability.

Green schooling specializes in networking and education to increase the potential of education and sell sustainable improvement. A notable example is ReNature in Malta, which promotes collaboration between businesses and policymakers to support evidence-based environmental decision-making. The project emphasizes nature-based solutions to enhance urban inexperienced infrastructure, surroundings offerings, and biodiversity (Miller et al., 2022). Malta has become a leader in research and innovation for the SDGs, testing technological and coverage answers. Training and capability-building are important to achieving sustainability aims, making sure that businesses check overall performance and put in place realistic, scientifically-based solutions (Müller-Christ et al., 2014). The study identifies networks, partnerships, and evaluation as essential strategies for advancing green education. Table 15.2 lists the strategies that green education uses to advance the Sustainable Development Goals (SDGs).

Table 15.2 Strategies for advancing the Sustainable Development Goals (SDGs) via green education

Strategies	*Description*
Land use planning and green infrastructure	Using solutions derived from nature as Sustainable Development Goals, with an emphasis on green infrastructure development, ecosystem services, biodiversity, and urban land use planning
Research and innovation	Supporting studies that aim to develop and implement long-term solutions to environmental problems and the Sustainable Development Goals (SDGs), with a focus on finding new and interesting ways to tackle these problems
Consultations and community involvement	Engaging communities and stakeholders, asking for their feedback, and ensuring the relevance and effectiveness of environmental initiatives all contribute to building a sense of ownership and commitment to sustainability projects
Monitoring and evaluation	Using monitoring and evaluation to determine the impact of sustainability initiatives to guarantee that progress is in accordance with the SDGs and that measures provide substantial results
Networks and partnerships	Strengthening research capacities, providing practical solutions based on scientific findings, and appreciating partnerships and networks to advance research and practical solutions
Vision-building and advocacy	Building a greener and more sustainable future through the development of a long-term vision, the promotion of environmentally friendly practices and policies, and the lighting of a shared fire
Knowledge synthesis and capacity building	With the goal of increasing capacity in areas related to nature-based solutions, educating individuals and organizations about the benefits of nature, and integrating training with the SDGs
Collaboration and networking	Developing solid partnerships and enhancing capabilities via collaborating with businesses, lawmakers, and other interested parties to develop environmental plans and policies supported by evidence
Training and capacity building	With an emphasis on assessing organizational capacity, sharing knowledge and skills that are in line with the SDGs, and identifying members' abilities to thrive in challenging environments
Promotion of research excellence	Enhancing human well-being via prioritizing high-quality research that identifies natural answers to innovation, economic development, and environmental challenges
Nature-based learning	Bringing the outside in, forging a connection with nature, and encouraging eco-conscious actions are all ways to bring nature into the classroom

12 Outcomes of Green Education for Sustainable Development

An evaluation of the focus on green education in 18 nations found that between 2005 and 2014, the primary outcome of Education for Sustainable Development (ESD) was the development of proof of concept. So, most efforts to promote eco-friendly education focused on raising people's consciousness and piquing their interest; from there, stakeholders began the long process of integrating eco-friendly practices into both classrooms and other types of learning settings. The original goals of green education, which were to prove the concept and raise awareness, changed to capacity building, experimentation, and applying best practices after 2014. This change was driven by a desire to integrate green education into both formal and informal curricula and to evaluate its impact on sustainability.

The focus of green education changed after the first decade of increasing awareness and researching the viability of ideas, because most of the selected studies published after 2014 focused on its consequences. Green education is crucial in assisting civilizations in their transition to sustainability. In order to accelerate the achievement of the SDGs on education, stakeholders are employing green education in a three-pronged approach that begins with the creation of common and universal concepts of sustainability. Having a shared definition of sustainability makes it much easier to identify the many skills that green education recipients will need in order to achieve it. Following their identification, it is the responsibility of educators to design pedagogies that structure learners to acquire the skills.

The use of green education as a means to encourage communities to live more sustainably has been beneficial to essential aspects of sustainable lifestyles. In the first place, green education helps with green consumerism.

Figure 15.3 highlights six key service areas essential for promoting green education in support of sustainable development. These interconnected services aim to build both institutional and individual capabilities necessary for driving environmental awareness and action. Institutional capacity strengthening focuses on improving the systems, policies, and infrastructure of educational institutions to support sustainable practices. Complementing this, individual capacity strengthening develops the skills,

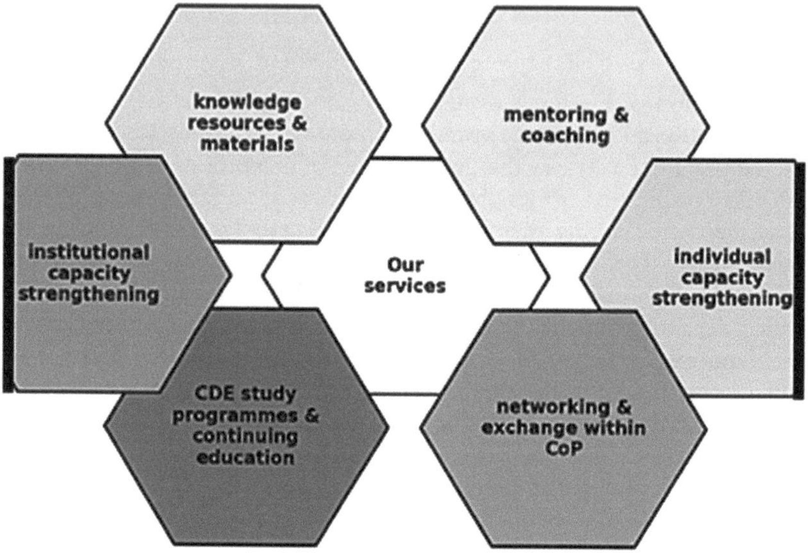

Fig. 15.3 Green education for sustainable development services

knowledge, and attitudes of learners and professionals to actively contribute to sustainability goals. Continuing and Distance Education (CDE) study programs and continuing education offer lifelong learning opportunities that integrate environmental themes, while knowledge resources and materials ensure access to relevant and up-to-date content that supports effective teaching and learning. Additionally, mentoring and coaching provide personalized support for professional growth in sustainability-related roles. Finally, networking and exchange within Communities of Practice fosters collaboration and knowledge-sharing among educators, institutions, and stakeholders. Together, these services create a comprehensive and collaborative framework for implementing green education as a tool for sustainable development.

Green education plays a key role in achieving carbon neutrality by raising environmental awareness, encouraging individuals to reduce pollution, alter consumption habits, and support cleaner production. It also familiarizes people with renewable energy technologies, aiding the transition to cleaner power sources. As a result, green education promotes reduced carbon emissions, economic growth, and greater focus on sustainability. By applying sustainability research, it fosters eco-conscious communities and strengthens efforts against climate change (Liu et al., 2021). Students in green education programs gain knowledge and values about environmental preservation, equipping them to engage in meaningful environmental actions and contribute to sustainable practices.

13 Challenges and Barriers Identified in Implementing Green Education for Sustainable Development

To achieve sustainable development, green education must be implemented, despite the fact that many parts of it, especially environmental education, have a negative effect on people's optimism about the environment. Despite numerous positive effects, one study found that green education misrepresented environmental circumstances and global ecology. Also, when Edsand and Broich adjusted the effect based on students' socioeconomic position, they discovered that only those with higher socioeconomic status were prompted to take appropriate action by parts of green education. These obstacles make it harder for green education to play a significant role in achieving SDGs 13, 4, and 10. In order to successfully raise environmental consciousness, the results indicated that educational activities need go beyond formal and informal settings.

Second, ESD is still not included in programs that help educators grow in their career, even in the decade since the conclusion of the concept's awareness-raising phase. Despite the importance of teacher education in achieving the SDGs (particularly SDG 4), research on the difficulties of green education implementation in Qatar found that teacher education curricula lacked several components of Education for Sustainable Development. Teachers also complain about a lack of resources, a lack of materials for the classroom, and an incomplete curriculum that does not adequately address ESD. The incorporation of environmental education into formal school curricula is a problem in many nations, including Qatar (V. Kioupi, N. & Voulvoulis, N., 2019). Environmental education integrates social justice, equality, and stewardship, encouraging responsible decisions for the planet. Outdoor learning spaces, combined with technologies such as online resources and virtual field trips, enhance student engagement. Green education supports SDGs 3, 4, 7, 8, 12, and 13 by way of fostering sustainability, lowering emissions, and promoting green jobs. However, demanding situations like curriculum incompatibility and restricted sources prevent its implementation. Research indicates greater instructor aid, more resources, and integrating Education for Sustainable Development (ESD) into curricula to overcome those barriers.

14 Conclusion

Environmental schooling integrates social justice, equality, and environmental stewardship, emphasizing accountable selection-making for the planet. Outdoor learning spaces such as gardens offer experiential opportunities, while technology enhances education through online resources and virtual field trips. Joining forces with local corporations allows students to interact in actual-international green practices. Green education is important for selling sustainability in classrooms, fostering participatory strategies, and supporting SDGs 3, 4, 7, and 8 through enhancing fitness, education, and sustainable process opportunities. Green colleges make a contribution to SDGs 12 and 13 by decreasing emissions and selling eco-performance. Eco-conscious education also contributes to water conservation, teacher retention, and innovative learning, supporting multiple SDGs and requiring targeted campaigns and research for effective implementation.

References

Bajaj, M., & Chiu, B. (2009). Education for Sustainable Development as peace education. *Peace & Change, 34*(4), 441–455. https://doi.org/10.1111/j.1468-0130.2009.00593.x

Bykov, V. (2008a). Teoretyko-metodolohichni zasady modeliuvannia navchalnoho seredovyshcha pedahohichnykh system vidkrytoi osvity. (Theoretical and methodological principles of modeling the educational environment of pedagogical systems of open education.). *Naukovi zapysky. Seriia: Pedahohichni nauky, 77*(1), 3–12.

Bykov, V. Y. (2008b). *Modeli orhanizatsiinykh system vidkrytoi osvity (Models of organizational systems of open education)*. Atika.

Ekwueme, C. O., Ekon, E. E., & Ezenwa-Nebife, D. C. (2016). Education for sustainability through academic freedom. *Global Journal of Educational Research, 15*(1), 23. https://doi.org/10.4314/gjedr.v15i1.3

Fischer, D., King, J., Rieckmann, M., Barth, M., Büssing, A., Hemmer, I., & Lindau Bank, D. (2022). Teacher education for Sustainable Development: A review of an emerging research field. *Journal of Teacher Education, 73*(5), 509–524. https://doi.org/10.1177/00224871221105784

Fuso Nerini, F., Mazzucato, M., Rockström, J., van Asselt, H., Hall, J. W., Matos, S., Persson, Å., Sovacool, B., Vinuesa, R., & Sachs, J. (2024 Jun). Extending the sustainable development goals to 2050 – A road map. *Nature., 630*(8017), 555–558. https://doi.org/10.1038/d41586-024-01754-6

Goldman, D., Ayalon, O., Baum, D., & Weiss, B. (2018). Influence of 'green school certification' on students' environmental literacy and adoption of sustainable practice by schools. *Journal of Cleaner Production, 183*, 1300–1313. https://doi.org/10.1016/j.jclepro.2018.02.176

Huckle, J., & Wals, A. E. J. (2015). The UN decade of education for sustainable development: Business as usual in the end. *Environmental Education Research, 21*(3), 491–505. https://doi.org/10.1080/13504622.2015.1011084

Independent Group of Scientists. (2023). *Times of crisis, times of change: Science for accelerating transformations to sustainable development.* United Nations.

Jickling, B., & Wals, A. E. J. (2008). Globalization and environmental education: Looking beyond sustainable development. *Journal of Curriculum Studies, 40*(1), 1–21. https://doi.org/10.1080/00220270701684667

Juntunen, M. K., & Aksela, M. K. (2014). Education for sustainable development in chemistry: Challenges, possibilities, and pedagogical models in Finland and elsewhere. *Chemistry Education Research and Practice, 15*(4), 488–500. https://doi.org/10.1039/c4rp00128a

Khalili, N. R., Duecker, S., Ashton, W., & Chavez, F. (2015). From cleaner production to sustainable development: The role of academia. *Journal of Cleaner Production, 96*, 30–43. https://doi.org/10.1016/j.jclepro.2014.01.099

Kioupi, V., & Voulvoulis, N. (2019). Education for sustainable development: A systemic framework for connecting the SDGs to educational outcomes. *Sustainability, 11*(21), 6104. https://doi.org/10.3390/su11216104

Kondratenko, Y., Korobko, V., Korobko, O., Kondratenko, G., & Kozlov, O. (2017). Green-IT approach to design and optimization of thermoacoustic waste heat utilization plant based on soft computing. In *Green IT engineering: Components, networks and systems implementation* (pp. 287–311). Springer International Publishing.

Kopnina, H. (2012). Education for sustainable development (ESD): The turn away from 'environment' in environmental education? *Environmental Education Research, 18*(5), 699–717. https://doi.org/10.1080/13504622.2012.658028

Kras, N. (2021). Nature-based learning at an urban community college: A case study at the Central Park Zoo. *Community College Journal of Research and Practice, 46*(6), 452–456. https://doi.org/10.1080/10668926.2021.1931557

Liu, H. Y., Jay, M., & Chen, X. (2021). The role of nature-based solutions for improving environmental quality, health, and well-being. Sustainability. *Times, 13*(19). https://doi.org/10.3390/su131910950

Lytvynova, S. H. (Ed.). (2015). *Modeliuvannia y intehratsiia servisiv khmaro orientovanoho navchalnoho seredovyshcha (Modeling and integration of cloudbased learning environment services)*. Komprynt.

Miller, N. C., Kumar, S., Pearce, K. L., & Baldock, K. L. (2022). Primary school educators' perspectives and experiences of nature-based play and learning and its benefits, barriers, and enablers: A qualitative descriptive study. *International Journal of Environmental Research and Public Health, 19*(6), 3179. https://doi.org/10.3390/ijerph19063179

Müller-Christ, G., Sterling, S., van Dam-Mieras, R., Adomßent, M., Fischer, D., & Rieckmann, M. (2014). The role of campus, curriculum, and community in higher education for sustainable development – A conference report. *Journal of Cleaner Production, 62*, 134–137. https://doi.org/10.1016/j.jclepro.2013.02.029

Nabukalu, J. B., Asamani, J. A., & Nabyonga-Orem, J. (2019). Monitoring sustainable development goals 3: Assessing the readiness of low- and middle-income countries. *International Journal of Health Policy and Management, 9*(7), 297–308. https://doi.org/10.15171/ijhpm.2019.134

Orr, D. W. (1992). Ecological literacy: Education and the transition to a postmodern world.

Sachs, J. D., Lafortune, G., Fuller, G., & Drumm, E. (2023). *Sustainable development report 2023: Implementing the SDG stimulus.* Dublin University Press.

Schnack, K. (2017). Action competence and environmental education. *Environmental Education Research, 23*(7), 918–930.

CHAPTER 16

Future Prospects and Advancing Research in Environmental Education and Sustainable Development

Priyanka Kaushik, Saurabh Pratap Singh Rathore, and Richa Sharma

1 Introduction

There are far too many inconsistencies in contemporary education, and practically everyone who has ever dealt with education in any way is aware of them (as are all socialized persons). Furthermore, we can seldom find someone who is unable to find another "recipe" for escaping the appalling condition in which the global educational system is "preserved." Many articles nowadays discuss the necessity for a fundamental shift in the approach to global education, but helpful recommendations are

P. Kaushik (✉)
Department of Computer Science and Engineering, NIMS University, Jaipur, Rajasthan, India

S. P. S. Rathore
School of Management & Commerce, Vikrant University, Gwalior, India

R. Sharma
Indian Council of Social Science Research, Special Call Division, New Delhi, Delhi, India

K. Hirech (ed.), *Environmental Education and Sustainable Development*, https://doi.org/10.1007/978-3-032-05760-0_16

sometimes limited to "patching up the holes" under the pretence of modernization. In the meantime, education has to develop into a really inventive process and be a key component of the new society that will help humanity survive the impending worldwide apocalypses, particularly the anthropoecological disaster. Rapidly shifting socio-natural dynamics force the educational process to adapt to new demands in order to provide students with innovative, advanced training that meets the demands of both the educational process and society (including strategic ones). It is evident that in order to meet the most pressing needs of both the present and, more importantly, the future generations, education must be futurized after being modernized to conform to the demands of modernity. Anticipating new opportunities and the most important demands of social development, not only those that arise right now, is already crucial. On the other hand, the historically established state of education across the world including, of course, Russian education indicates a certain commitment on the part of official educational organizations to a solid orientation toward the past, to the transmission and mastery of somewhat antiquated material during training. Even while the authors' propensity for some innovations can still be seen in the official papers of education modernization, the chosen state educational standards continue to emphasize information and ideals that are already out of date.

According to Brazil (2018), environmental education is defined as the processes by which individuals and communities develop social values, knowledge, skills, attitudes, and competencies aimed at the preservation of the environment and its shared use, which are vital to a healthy quality of life and its sustainability. Another definition of environmental education can be found in the Dictionary of Meanings (2019). It is important to realize that environmental education is a field of study that teaches people about environmental issues and how to help fight them by protecting natural resources and avoiding pollution. The goal of environmental training is to protect nature and its surroundings, which are helpful resources. From the perspectives of social, moral, monetary, ecological, and environmental factors, it seeks to pick out the distinct answers. This activity raises focus in society about the things that damage the environment, consisting of trash, soil erosion, deforestation, and pollution of the air and water. Given that it is a worldwide issue, similar concerns are found in many countries nowadays.

2 Foundations of Environmental Education

The Historical Development of Environmental Education: In recent years, environmental training has gone through enormous transformation. As environmental issues including pollution, biodiversity depletion, and climate change garnered global interest in the 1960s and 1970s, the movement intensified. The importance of training in addressing environmental challenges was emphasized by landmark events including the 1972 United Nations Conference on the Human Environment, commonly called the Stockholm Conference. The evolving framework of environmental education has led to the integration of sustainability principles into educational institutions worldwide.

Principles of Environmental Education: The ideas of environmental education awareness on enhancing environmental awareness and cultivating environmental understanding and capabilities embody fostering active engagement in environmental conservation, growing a comprehensive know-how of ecological systems, and refining essential thinking and problem-solving skills. The goal of environmental schooling is to nurture knowledgeable people capable of making accountable decisions to promote sustainability, at the same time as recognizing the interconnections among human activities and natural ecosystems.

The Role of Environmental Education in Enhancing Ecological Awareness: Environmental training is essential for reinforcing ecological attention via informing individuals about the effect of their activities on the surroundings. It assists pupils in comprehending the significance of biodiversity, resource conservation, and stopping environmental degradation. Educating the public and college students about environmental issues fosters an experience of stewardship and responsibility for maintaining the planet's fitness for future generations.

3 AI for Education

Artificial intelligence is transforming education by improving access and personalizing learning experiences to match individual preferences, abilities, and learning styles. AI-driven systems leverage data analysis and machine learning to personalize content delivery, thereby enhancing learner engagement and improving knowledge retention. Moreover, AI supports the introduction of immersive academic environments, integrating virtual reality (VR), augmented reality (AR), and gamification to simulate real-world scenarios. Those innovations make environmental training more engaging and available, motivating college students to participate in sustainability efforts via interactive simulations and modules.

4 Data Analytics

Data analytics improves environmental education via presenting insights into learner conduct, results, and academic effectiveness. It allows educators to make informed choices, refine coaching strategies, and customize interventions. Real-time tracking of pupil engagement, which includes time spent on modules and final touch charges, allows educators to address demanding situations promptly. Information analytics enables the evaluation of learning outcomes by comparing pre- and post-assessments, thereby providing evidence of program effectiveness. Moreover, it permits educators to discover traits across demographic companies, adjusting strategies to better meet diverse student needs and enhance typical learning.

5 Environmental Alertness

Environmental consciousness is vital for motivating movement closer to sustainability and fostering a collective understanding of the challenges occurring around our globe. An important step in schooling is enhancing public cognizance of environmental issues to nurture eco-aware citizenship and encourage advantageous exchange (Batool et al., 2023). AI-driven educational projects utilize immersive era, supply pertinent and appealing content, and promote a deeper knowledge of the interconnections within environmental systems to enhance environmental awareness. One of AI's number one advantages in enhancing environmental awareness is its ability to deliver customized and contextually applicable instructional content material. AI-driven systems may additionally tailor records on environmental problems to resonate with each person's personal interests, perspectives, and experiences through the analysis of user facts and choices (Chisom et al., 2024). Artificial intelligence allows educators to deliver content material that captivates students and fosters profound engagement with environmental matters, whether this be customized hints, interactive multimedia sources, or meticulously selected articles. Moreover, AI facilitates the development of engaging and impactful learning experiences through immersive technologies such as virtual reality (VR) and augmented reality (AR). Utilizing virtual and augmented reality programs, youth are able to be immersed in ecosystems threatened by pollutants, climate change, or habitat destruction, providing them with a direct experience of the results of environmental degradation. Those immersive studies elicit emotional responses that catalyze action and behavioral change while enhancing understanding of complicated environmental issues.

6 Technology Linkage to Environmental Education of SDGs

The Sustainable Development Goals (SDGs) were created to improve global welfare (Andriyani, 2021). The United Nations instituted the SDGs, a worldwide framework of 17 goals, to assist governments in achieving equitable, inclusive, and sustainable development by 2030. Anggraeni and Yanti (2024) assert that environmental education is a crucial device for promoting environmentally responsible behavior and facilitating the success of the Sustainable Development Goals, in particular on the subject of weather patterns and environmental sustainability. Engaging students in sustainability-focused projects while incorporating digital technologies helps cultivate environmental awareness, ethical values, and responsibility—key dimensions for supporting the achievement of the Sustainable Development Goals (SDGs). Information technology plays a vital role in environmental education, particularly in fostering global collaboration among schools. It strengthens students' ecological awareness while promoting global dialogue and collaboration on environmental issues. Through the use of information technology, learners can compare ecological perspectives across countries, participate in collaborative workshops, and develop innovative solutions for environmental conservation. This era of environmental education supports the transition in scientific education from a traditional model to one which emphasizes citizen empowerment. It aims to enhance public participation in technoscientific decision-making.

The third segment will provide a comprehensive analysis of the current use of AI in sustainable development by assessing the viewpoints of AI specialists in the relevant sectors and presenting insights on certain scenarios. This is essential due to the intrinsic interconnection of the Sustainable Development Goals (SDGs) and the diverse range of entities involved in sustainable development. Consequently, we are endeavoring to determine how various firms are employing AI to engage, identify, and ultimately foster sustainable growth (Abulibdeh et al., 2024). The proposed hypothesis posits that the most effectively classified political actors will not only exert a greater influence on advancements toward the SDGs but will also encompass the potential unpredictability of various causes and additional attributes, including actors, industries, non-governmental organizations, the scientific and technological community, and local producers. They will also function as conduits for AI's most advantageous impacts on sustainable development, facilitating the attainment of the SDGs. This should be the principal effect of the later formed pyramid, wherein players can influence or assign rankings to more advantageous positions and playing styles than other participants within the dynamic. In this context, AI can maintain a significant level of sustained growth by fostering social collaboration among participants (Fan et al., 2023).

Due to the many applications of AI, different use cases will exhibit varying domains, capabilities, constraints, and risk profiles. We will analyze some scenarios that illustrate the application of AI for sustainable growth and overall benefit. These are but a few cases, although they illustrate a broad spectrum of potential applications, capabilities, and results. Multiple examples of deep learning applications in medical imaging have been extensively documented in textbooks (Babalola & Olawuyi, 2021), a summary that illustrates the aforementioned trend. The sector of medical imaging and diagnostics has had a surge of successful company activity since 2015. Reports indicate that imaging and diagnostics comprised around one-third of healthcare AI start-ups that secured funding post-January 2015. Moreover, by 2026, medical images are projected to constitute at least 90% of all medical data, as indicated by experts at International Business Machines Corporation, a pioneer in the computer industry. It will serve as the paramount source of healthcare data owing to its extensive volume. Deep learning has been employed in this domain for cancer development monitoring, blood flow measurement and visualization, medical interpretation, and the treatment of diabetic retinopathy. This is apparent when examining the correlation between AI and extensive data, especially as previously shown with the emergence of big data. The use of AI is analyzed in each situation related to a corporation, particularly focusing on three prominent entities in the global technology sector: IBM, Google, and the Samsung Group, as well as a start-up. In contemplating these scenarios, another prominent individual in the IT business that was recalled is Argus, a global effort of SAP Labs China. A group developed the "Argus" approach, which uses machine learning to detect lung cancer signs in CT scans. Subsequently, Argus enhances detection precision while facilitating the testing of a greater number of patients in a reduced timeframe. The identical components are present again: artificial intelligence, the UN-designated "Trade and Industry" sector's contribution to fulfilling the Sustainable Development Goals for "Nutrition, Health, and Well-Being," and, most importantly, the stakeholders in this domain. An illustrative instance of a company employing AI responsibly to address potential effects on sustainable development is the SAP organization.

The last category of sustainability objectives is the environment; upon closer inspection, three of these goals (SDG 13, 14, and 15) are related to climate measures, life on land, and life undersea. According to PwC and Stanford Woods Institute for the Environment (2018), there are 25 (93%) goals for environmental organizations where AI may be used as an enabler. The capacity to evaluate vastly networked datasets for cooperative curatorial work may be one of artificial intelligence's advantages. SDG 13 on climate change demonstrates how developments in artificial intelligence are assisting in the understanding and modeling of climate change and its possible impacts. Because low-carbon energy systems employ renewable resources and increase energy efficiency, they are crucial in the battle against climate change. AI may potentially help build these systems. Artificial intelligence can also enhance the health of ecosystems. We can avoid and drastically reduce marine pollution by using AI-based automated oil spill detection systems. Another illustration that calls for halting desertification and restoring deteriorated soils and areas is 15.3. Artificial neural networks and certain techniques can enhance the categorization of land cover types from satellite photographs, possibly processing a significant number of images rapidly (Mary & Binu, 2024). By identifying the underlying causes, these artificial intelligence (AI) technologies can reverse trends or stop more desertification by detecting large-scale patterns of desertification and provide useful information for planning, decision-making, and environmental oversight (Egbemhenghe et al., 2023). However, as previously mentioned, the high energy requirements of AI applications, particularly when employing carbon-free energy sources, may hinder efforts to achieve SDG 13 on climate change. Furthermore, even though there are more and more examples of how artificial intelligence (AI) is being used to support biodiversity conservation and monitoring, it is predicted that there will be a greater chance of resource overexploitation as more information about AI in the environment becomes available. The existing shortcomings in AI research will be taken into consideration when we look more closely at these misuses of AI, even though they haven't been well documented yet.

7 Transformative Frameworks for Environmental Education in the Context of Sustainability

Environmental education needs to be rethought using a strong theoretical framework that incorporates modern critical viewpoints in order to critically address the complex ecological and educational issues of the twenty-first century. Conventional approaches to environmental education frequently concentrate on changing behavior and increasing knowledge, but they are unable to address more fundamental structural, cultural, and epistemological issues. A more complex, morally sound, and forward-looking basis is offered by the integration of transformative learning theory, Indigenous knowledge systems (IKS), posthumanism, and decolonial viewpoints.

Transformative Learning Perspective: Critical reflection and paradigm transformations in learners' worldviews are key components of transformative learning. According to this perspective, environmental education is about creating profound changes in awareness rather than just imparting knowledge. This allows students to challenge anthropocentric presumptions and take part in sustainable actions that are based on ecological ethics. Students are urged to reconsider how they relate to institutions, the environment, and future generations (S. Jorgenson et al., 2019).

Indigenous Knowledge Systems (IKS): The predominance of Western epistemologies in environmental discourse is contested by IKS. Indigenous viewpoints provide relational understandings of land, water, and non-human beings, rooted in centuries of ecological expertise. They place a strong emphasis on respect, reciprocity, and stewardship, all of which are in line with sustainability objectives. By recognizing different ways of knowing and being, incorporating IKS not only enhances curriculum material but also decolonizes education (Y. Tian et al., 2024).

Posthumanism and Eco-centric Ethics: The human-centered perspective that underlies environmental deterioration is criticized by posthumanist ideology. According to this theory, education has to shift toward an "eco-centric" ethics that values the agency of non-human beings since people are intertwined with larger ecological systems (Han & Liang, 2023). With an emphasis on interdependence and group survival, this reorientation forces a reassessment of educational objectives.

Decolonial and Social Justice Perspectives: Systemic injustices, cultural erasures, and colonial histories are all closely related to environmental degradation. A decolonial paradigm examines how marginalized people are disproportionately affected by environmental damages and calls for socially, culturally, and politically sensitive education. Students who receive decolonial environmental education are better equipped to question prevailing structures and imagine different, more sustainable futures (Tao & Tao, 2024).

8 AI Applications in Environmental Monitor and Protection

Artificial intelligence (AI) is a transformational force in the complex dance of Earth's ecosystems, where each beat and movement adds to the harmonic balance of nature. The lucidity of this reasoning is enhanced by actual international examples that clearly illustrate the palpable effects of AI in protecting ecosystems (Y. Junjia et al., 2023).

AI is rising as a protector of biodiversity and an added boost behind sustainable environmental practices, from detecting illicit logging operations inside the Amazon rainforest to monitoring elephant poaching routes in Africa. Tracking the natural world is crucial to comprehending the complicated patterns that characterize the lives of many animals. This richness is frequently no longer captured by way of traditional techniques like manual monitoring and sparse observational facts. The emergence of artificial intelligence (AI) has converted wildlife tracking by giving us previously unheard of powers to observe and understand animal activities. Machine learning-driven AI structures integrate a variety of information sources, such as GPS information, drone pictures, and digicam lure recordings.

Machine learning algorithms, trained on massive datasets, help with wildlife monitoring by reading species' precise behaviors and actions. This enables real-time data interpretation and a deeper knowledge of animal conduct. AI is particularly beneficial in monitoring African elephant poaching routes, for which traditional strategies are much less powerful due to vast regions and complex migration patterns. By studying records on movement patterns, habitats, and environmental changes, AI predicts poaching hotspots, permitting conservationists to respond unexpectedly and save animals from poaching. This facts-driven technique complements conservation strategies and enables extra proactive, targeted interventions to defend wildlife.

9 Implement Virtual Reality in Environmental Knowledge

The method of including more context or statistics from the retrieval process to already-existing material is known as augmentation. Virtual reality is a form of media that creates a digital arena that allows the viewer to enjoy the activities or conditions shown within the media, which will increase environmental ethics. Environmental training relies heavily on digital truth technology. Via simulation and visualization, AR and VR technology may additionally assist in explaining standards in environmental schooling. By way of improving the educational efficacy of complex topics like sustainability and weather exchange, AR and VR are useful resources for college students' comprehension. Without ever leaving the classroom, virtual and augmented technology provides students with immersive reviews that help them realize how human activity influences the surroundings. Moreover, it is believed that augmented reality and interactive infographics can improve environmental cognizance and literacy.

So that it will foster the ethical features of the surroundings and create ideal environmental preservation behaviors, technology may additionally expand knowledge of and improve interplay with nature. Virtual generation can consequently inspire experienced environmental behavior and increase pupil involvement. Real and virtual imagery are combined to create simulated settings with virtual reality technology. A highly lifelike three-dimensional world is available in which users can observe and interact with things. Learner involvement and engagement in the learning process are successfully increased when virtual reality is used in environmental education.

Mobile applications and augmented reality are two examples of how this technology is being used in environmental education through project-based learning, as shown in Fig. 16.1. In one instance, students at a school created an application that customers could download by scanning QR

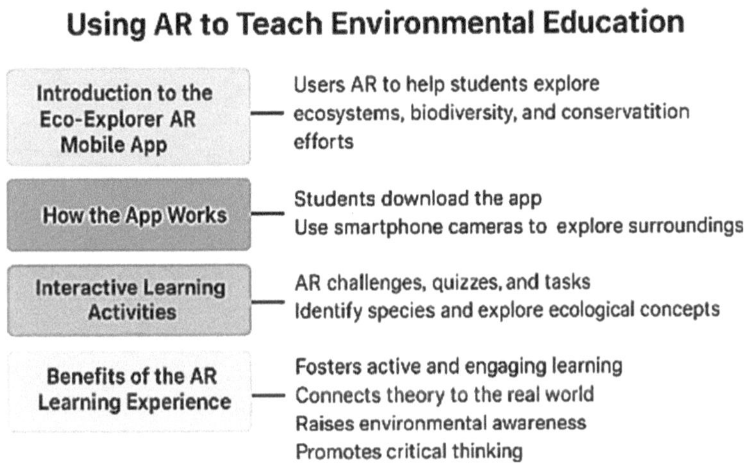

Fig. 16.1 Enhancing environmental education with augmented reality tools

codes placed along a promenade along a river. The purpose of the app was to notify users about the route. Additionally, the app was made to work with augmented reality elements that could overlay neighbourhoods from the past on top of neighbourhoods from the present, enabling users to "visit" precolonial eras and investigate Indigenous peoples' customary land use.

One of the cornerstones of Industry 4.0 is the Internet of Things (IoT), whose ideas and methods are practical in the current technological age. Furthermore, the science of the Internet of Things is expanding. Optimizing life via the use of smart sensors and smart gadgets that collaborate over the Internet is incredibly promising. Environmental education is more successful when digital technologies and the Internet of Things are used. Students can carry out practical experiments and environmental studies. Ultimately, data from IoT devices raises awareness of environmental concerns by illuminating ecosystem interconnections and the effects of human activity. These tools foster a sense of connection between students and their surroundings and boost their involvement in environmental project-based learning. Projects involving the design, construction, and programming of environmental protection prototypes make use of IoT. Science, computer science, physics, and mathematics are all used in these initiatives.

Sensors and programs are examples of the IoT era which are used to promote sensory engagement and raise expertise on plant needs. For instance, the Virtual Boggarts project connects temperature and humidity data with hands-on activities, enabling students to investigate microclimates using Microbit sensors. The NFC Shield and Tag were utilized with the Arduino Uno to generate digital Boggart characters and accompanying audio narratives. This makes students more aware of non-human reactions and environmental consequences while enabling them to keep an eye on the necessities of flora and the environment. In addition, initiatives like a sensor-equipped buoy for monitoring water quality in the Sarno River and a robotic arm for detecting indoor air pollution illustrate how low-cost solutions can provide real-time data to help authorities respond effectively to pollution challenges. In light of environmental problems, this academic experience indicates how a sensible, transdisciplinary study method can cope with real-international issues and equip students to become educated and engaged citizens.

10 Integrating Environmental Education With Global Challenges and the SDGs

In order to solve pressing global issues including resource depletion, ecological inequity, and climate change, environmental education is essential. It is becoming a key factor in the global shift toward sustainability rather than a supplemental field. In order to develop environmentally conscious individuals, educational systems must become more in line with the Sustainable Development Goals (SDGs) of the UN.

10.1 Alignment with Sustainable Development Goals (SDGs)

SDG 4: Quality Education: Students who receive modern environmental education gain vital skills that promote sustainable development. Virtual reality (VR), artificial intelligence (AI), and adaptive learning platforms make education more inclusive, dynamic, and able to meet the demands of a wide range of learners throughout the world (G. Cebrián et al., 2020).

SDG 13: Climate Action: Climate education increases scientific knowledge and public awareness of extreme weather patterns, global warming, and adaptation measures. The way that students learn about and address climate-related issues is being completely transformed by technological tools like satellite images, AI models, and Internet of Things-based climate sensors (M. Bian et al., 2020).

SDG 15: Life on Land: Environmental education raises awareness of biodiversity and protects ecosystems. Students can learn about land use, deforestation, and conservation techniques firsthand using data analytics and geospatial intelligence, which strengthens their bond with the natural world (Y. Li et al., 2023).

SDG 10: Reduced Inequalities: By combining equity frameworks with environmental education, under-represented and marginalized people are guaranteed equal access to information about sustainability. Indigenous Knowledge Systems (IKS) and regional environmental customs are included in instructional materials to make them inclusive and culturally sensitive (Z. Shen et al., 2023).

10.2 Promoting Climate Literacy and Environmental Justice

In order to address social justice, historical injustices, and policy-level ramifications, true climate education must go beyond fundamental environmental understanding, and should:

- Examine how vulnerable people are disproportionately impacted by environmental degradation
- Promote moral consideration of waste, consumption, and environmental stewardship
- Encourage youth-led climate action and civic involvement
- Encourage a range of viewpoints, particularly from marginalized and Indigenous speakers (L. Ladachart et al., 2020)

Technology-Driven Transformations in Environmental Learning: The delivery of environmental education is changing as a result of technological advancement.

- Environmental simulations and customized learning pathways are offered by artificial intelligence (AI).
- Real-time ecosystem monitoring and pollutant tracking are supported by IoT devices.

Immersion experiences produced by augmented and virtual reality technologies enhance knowledge of environmental systems and human effect (B. Herman et al., 2020). These resources increase the impact, context, and experiential nature of environmental education while equipping students to address sustainability issues in the real world.

10.3 A Roadmap for Transformative and Inclusive Sustainability Education

Environmental education needs to be transformatively inclusive and technologically advanced in order to effectively support global sustainability. In addition to guaranteeing academic relevance, aligning it with the SDGs gives students the knowledge they need to become knowledgeable change agents. We provide the groundwork for a sustainable and just future for everybody by integrating ecological ethics, environmental justice, and interdisciplinary knowledge into educational frameworks (P. Glavič, 2020).

11 Challenge in Bridge Environmental Education and Sustainable Growth

Integrating sustainability into educational institutions faces several challenges, including rigid curricula, a shortage of trained educators, financial constraints, outdated resources, and inadequate infrastructure. Political and social limitations, which include resistance to trade and absence of popularity for sustainability packages, additionally hinder development. Marginalized businesses may also opt out of sustainability training due to inequitable entry opportunities. In developing countries, monetary increase regularly takes precedence over environmental worries, similarly obstructing sustainability schooling. Political instability and vulnerable environmental rules also can put off efforts to sell sustainability in schooling.

Environmental training policies and their real implementation can reveal great divergence. Notwithstanding the approval of legislation encouraging sustainability education in numerous nations, sensible implementation frequently remains deficient. Colleges often face difficulties in effectively implementing these policies due to limited funding, insufficient teacher training, and lack of proper oversight (Zywiolek et al., 2022). Moreover, whereas country-wide projects may also comprise sustainability objectives, there are frequently no incentives or enforcement measures to ensure local adherence. Such gaps between policy and practice can lead to slow progress and limited impact on student learning outcomes and environmental awareness.

12 Methodology

In their study on the impact of environmental education on students' awareness, attitudes, and behaviors, the researchers identified television, newspapers, parents, teachers, and schools as the primary sources of environmental knowledge. Researcher indicate that senior college students mostly acquired knowledge of ecological conservation through literature and media, followed by computer networks, lectures, and coursework. García-Saiz, Palazuelos, and Zorrilla (2013) indicate that college students mostly obtain knowledge on water conservation via media and lectures delivered by instructors. Dong et al. (2014) executed environmental education at the Sihcao Wildlife Reserve, a designated conservation area, and discovered that college students' principal sources of knowledge were field excursions, media, and lectures. Thus, college students' comprehension of environmental issues may be substantially enhanced by environmental education initiatives conducted by instructors or educational institutions. This research hypothesis is derived from the aforementioned literature for this study.

H1: There are strong relationships between environmental education and environmental knowledge.

Researchers observed a significant correlation between students' perceptions of water and their overall environmental understanding. At public vocational training institutes, a strong relationship was identified between students' environmental views and their knowledge. Findings further indicated that students with greater environmental knowledge demonstrated more positive attitudes toward environmental issues. Similarly, a robust positive correlation was found between environmental knowledge and environmental sensitivity among high school teachers. Additional studies reported that after four to eight weeks of environmental courses or programs, students showed a strong link between knowledge and attitudes. These results suggest that educational interventions can effectively enhance both environmental knowledge and attitudes. Moreover, research conducted in Taiwan highlighted a positive correlation between social values and environmental knowledge, awareness, cosmic beliefs, and cognition of environmental challenges among educators. Based on these findings, this study formulates the following research hypothesis (Islam (2024)).

H2: Environmental awareness has significant relationships with environmental attitudes.

Joo-Nagata et al. (2017) assert that the primary aim of environmental education is to cultivate environmental literacy to enhance essential environmental knowledge, promote problem-solving skills, and elevate attitudes and participation in environmental initiatives. To cultivate college students' attitudes toward environmental protection, enhance environmental attitude education, and apply it to sustain environmental quality in daily life, they posited that the primary aim of environmental education should be to provide college students with experiential learning opportunities regarding the environment and to instruct them on the essential dynamics of the human–environment relationship. This study presents the following research hypothesis based on the aforementioned literature.

H3: There are significant relationships between environmental mindset and environmental education.

Sample and measure display: A quasi-experimental study with 186 college students as the research subjects administered environmental education using artificial intelligence teaching systems for 16 weeks, totalling three hours each week for an aggregate of 48 hours. Statistical software is employed to examine the study data, and several hypotheses are assessed.

Reliability and validity test: Validity refers to a measurement instrument's ability to accurately assess the intended research issues. It includes construct, criterion-related, and content validity. In this study, the questionnaire items are based on domestic and international students, with pretesting to confirm content validity. A linear structural equation model analyzes environmental education, knowledge, and attitudes. The model fit is acceptable, indicating positive predictive validity. Reliability is evaluated using item-to-total correlation coefficients, with values above 0.5 confirming construct validity.

Assessments of validity and reliability precede this. A higher Cronbach's α signifies more reliability. Cronbach's α values ranging from 0.7 to 0.98 indicate great dependability; however, values below 0.35 should be excluded. A Cronbach's α score over 0.7 signifies strong dependability.

Furthermore, items failing to match the criteria must be discarded; item-to-total correlation coefficients should exceed 0.4. The official questionnaire of the study was developed in alignment with the specified requirements, and the assessed Cronbach's α values vary from 0.70 to 0.80, indicating acceptable reliability.

Furthermore, discriminant validity would employ the correlation coefficients of alternative dimensions and the square root of the average variance extracted from each dimension to assess differentiation among them. The model exhibits discriminant validity when the former is superior. Consequently, the mean variance of "environmental education," "environmental knowledge," and "environmental attitude" is squared, and the outcomes are juxtaposed with the dimension correlation coefficients. This study model's dimensions demonstrate discriminant validity.

13 Research Design and Type

This study adopts a quantitative and quasi-experimental research design to evaluate the relationship between environmental education and its impact on environmental knowledge, awareness, attitudes, and behaviors among college students. The study is grounded in empirical literature suggesting that educational interventions significantly affect students' environmental literacy, awareness, and actions (E. Waltner et al., 2020).

Research Hypotheses: Based on the synthesis of prior literature, the following hypotheses are proposed:

- *H1: There are strong relationships between environmental education and environmental knowledge.*
- *H2: Environmental awareness has significant relationships with environmental attitudes.*
- *H3: There are significant relationships between environmental mindset and environmental education.*

Sampling and Participants: A quasi-experimental sampling method was used to recruit 186 college students from diverse academic disciplines to ensure a representative cohort. The selection criteria included enrollment in undergraduate environmental education programs. The participants were exposed to a structured intervention using artificial intelligence (AI)-based teaching systems for 16 weeks, comprising three hours per week, amounting to a total of 48 instructional hours.

Data Collection Methods: Data was gathered using a pre-validated structured questionnaire, adapted and contextualized from national and international literature. The questionnaire encompassed sections on environmental education, environmental knowledge, environmental attitudes, and behavioral intentions. All items were measured on a five-point Likert scale (1 = strongly disagree to 5 = strongly agree) (Tao & Tao, 2024).

Additionally, field observations, attendance tracking, and AI system-generated analytics were used to complement quantitative self-reported responses, ensuring a triangulation of findings.

Data Analysis Methods: Data analysis was conducted using Statistical Software Packages (e.g., SPSS and AMOS). The following methods were used:

- Descriptive statistics to summarize participant demographics and item-level responses
- Correlation analysis to examine the inter-relationships among constructs
- Linear Structural Equation Modeling (SEM) to test the conceptual model and proposed hypotheses
- Confirmatory Factor Analysis (CFA) to validate the construct structure of the instrument

13.1 Reliability and Validity of Data

Reliability Analysis: Reliability was evaluated using Cronbach's α, with values ranging between 0.70 and 0.80, confirming high internal consistency. Items with item-to-total correlation coefficients below 0.40 were excluded from the final model, in accordance with best practices.

Validity Analysis: Content validity was ensured by aligning questionnaire items with previously validated tools and having them reviewed by subject matter experts.

- Construct validity was examined using CFA, where factor loadings exceeded the recommended threshold of 0.60.
- Discriminant validity was assessed by comparing the square root of average variance extracted (AVE) for each construct with the inter-construct correlations. The AVE square roots exceeded the correlations, indicating satisfactory discriminant validity.

- Predictive validity was supported by the SEM model fit indices (e.g., CFI, RMSEA, and χ^2/df) all falling within acceptable ranges.

Justification of Methodological Rigor: This methodology is justified based on its alignment with empirical standards in environmental education research. The use of quasi-experimental design allows for control over intervention exposure, while SEM provides robust hypothesis testing. The use of AI-based teaching systems enhances consistency and objectivity in instructional delivery. The multi-step validation process, combined with stringent reliability checks, confirms the credibility of the findings and supports the generalizability of the conclusions.

14 Result

14.1 *LISREL Model Assessment Display*

In order to compute numerous factors and multiple causal routes, the LISREL (linear structural relations) model combines simultaneous equations from econometrics with factor and path analysis from classical statistics. Preliminary fit criteria, overall model fit, and internal structure fit might all be used to assess the model's fit (Submit Ramírez Suárez et al., 2023). Table 16.1 shows the research data.. The following explains the preliminary fit criteria, the model's internal structure fit, and the overall model fit.

Figure 16.2 illustrates the LISREL (linear structural relations) model used to assess the relationships among Environmental Education, Environmental Knowledge, and Environmental Attitude. It demonstrates significant and positive causal paths: H1 (0.866) links Environmental Education to Environmental Knowledge, H2 (0.883) connects Environmental Knowledge to Environmental Attitude, and H3 (0.871) ties Environmental Education directly to Environmental Attitude. The model fit is validated through key indices: $\chi^2/df = 1.462$ (acceptable threshold < 3), RMR = 0.007, GFI = 0.934, and AGFI = 0.904, indicating strong overall model fit and internal structure reliability.

H1, H2, and H3 are supported by the positive and significant correlations that Environmental Education presents with Environmental Knowledge (0.866), Environmental Knowledge and Environmental Attitude (0.883), and Environmental Education and Environmental Attitude (0.871), according to the model's internal structure fit.

Table 16.1 The total linear structural relation model's analysis result

Evaluation item	*Parameter/evaluation standard*	*Result*	*T*
Preliminary fit criteria	Environmental education	0.645	9.23
	Affection idea	0.667	9.67
	Technical concept	0.634	10.26
	Nature knowledge	0.689	7.89
	Problem knowledge	0.621	7.67
	Action strategy	0.623	8.11
	Environmental attitude	0.756	8.67
	Environmental belief	0.645	11.03
	Environmental value	0.656	10.78
	Environmental ethics	0.778	13.36
Fit of internal structure of model	Environmental education → Environmental knowledge	0.878	22.49
	Environmental knowledge → Environmental attitude	0.844	27.21
	Environmental education → Environmental attitude	0.866	24.78
Overall model fit	Squared chi-square value/degree of freedom	1.423	
	Fit index	0.956	
	Adjusted fit index	0.978	
	Residual root mean square	0.009	
	Standardized fit index	0.934	

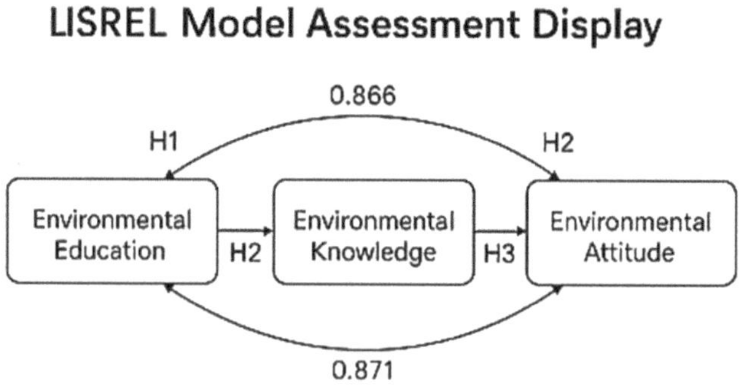

Fig. 16.2 LISREL model assessment display for environmental education and attitudes

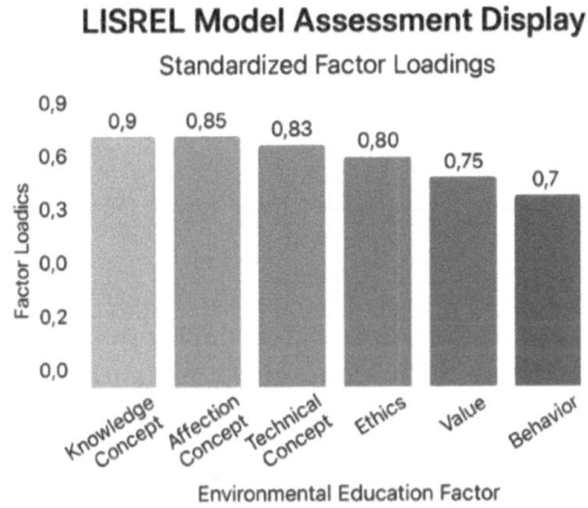

Fig. 16.3 LISREL standardized factor loadings of environmental education components

The criteria $\chi2/Df = 1.462$, which are less than the standard 3, and RMR = 0.007 demonstrate the suitability of $\chi2/DF$ and RMR in terms of overall model fit. Furthermore, chi-square cannot be used to directly assess fit since it is sensitive to sample size. The model exhibits improved fit indices as the overall model fit standards GFI = 0.934 and AGFI = 0.904 are higher than the usual 0.9 (the closer GFI and AGFI to 1 reflecting the greater model fit).

Figure 16.3 illustrates the standardized factor loadings derived from the LISREL model to assess various dimensions of Environmental Education. Each bar represents a specific component, indicating its relative contribution to the structural model. The Knowledge Concept shows the highest factor loading (0.90), suggesting it is the most influential variable in shaping environmental education outcomes. This is followed by Affection Concept (0.85), Technical Concept (0.83), and Ethics (0.80), all of which demonstrate strong influence. Moderate loadings are observed for Value (0.75) and Awareness (0.70), while Behavior (0.65), though still significant, contributes comparatively less. The gradient color bars visually highlight the variance in loading strengths, supporting the internal structural fit and validating the model's effectiveness in representing environmental education dimensions

Table 16.1 presents the analysis results of the linear structural relations (LISREL) model, divided into three main sections: preliminary fit criteria, internal structure fit, and overall model fit.

Table 16.2 Test of hypothesis

Research hypothesis	*Correlation*	*Empirical result*	*P*	*Result*
H-1	Positive	0.834	0	Supported
H-2	Positive	0.824	0	Supported
H=3	Positive	0.867	0	Supported

- Under preliminary fit criteria, the model evaluates the relationship between environmental education and two latent variables: environmental knowledge and environmental attitude. Each component (e.g., knowledge idea, affection idea, technical concept) shows a significant loading (ranging from 0.621 to 0.778) with high *t-values*, all significant at the 0.01 level, indicating strong validity of each measured item in contributing to the construct.
- In the fit of internal structure, the relationships between constructs are assessed. For instance, environmental education strongly influences environmental knowledge (0.878), and environmental knowledge influences environmental attitude (0.844), both with high *t-values* (> 22), confirming a robust model structure.
- The overall model fit statistics confirm the model's adequacy. Values such as a chi-square/df ratio of 1.423, a high fit index (0.956), adjusted fit index (0.978), low residual root mean square (0.009), and standardized fit index (0.934) suggest excellent model fit.

Table 16.2 outlines the hypothesis testing results. Three hypotheses (Hl, H2, H3) were tested, showing strong positive correlations (0.834, 0.824, and 0.867 respectively) with p-values of 0.00, indicating statistical significance. All hypotheses were supported, confirming the proposed positive relationships among environmental education, knowledge, and attitude.

In summary, the LISREL model is statistically robust, with well-supported constructs, high factor loadings, strong internal structural relationships, and excellent overall model fit.

14.2 Findings

Current understanding can no longer adequately meet the constantly evolving environmental difficulties associated with diverse environmental issues and varying conservation efforts. Educators are encouraged to familiarize themselves with literature and data on environmental education, participate in relevant training, and connect with organizations or individuals engaged in environmental protection prior to teaching, in order to augment their professional knowledge and introduce courses to enrich the lessons in a timely manner. To ensure college students acquire systematic environmental education, grasp ecological principles, recognize the origins of environmental pollution and its effects on life, and cultivate the mindset and behaviors necessary for environmental protection, it is advisable to integrate environmental education into formal curricula. To facilitate college students' comprehension of environmental education materials and enhance their learning engagement, educators should prioritize individual variations among students and utilize diverse teaching methodologies in conjunction with artificial intelligence teaching systems. The teaching module indicates that the knowledge structure within domain knowledge is essential for artificial intelligence teaching systems to provide optimal flexibility in rebuilding material knowledge (A. Telešienė et al., 2021).

Consequently, it is advisable to standardize the resources and delineate their overall resource attributes, educational merits, and significances. In this case, the artificial intelligence teaching system could easily access any materials that fulfilled the same criteria. This encompassed understanding ideas and their interconnections with other concepts. Alongside tackling the challenge of expediting the development of AI teaching systems, it would achieve the objectives of content dissemination and repeated use.

15 Conclusion

The study reveals that environmental education, enhanced by artificial intelligence (AI) teaching systems, significantly improves college students' environmental knowledge, attitudes, and awareness. AI systems align teaching content with students' knowledge, offering personalized teaching strategies and fostering a deeper understanding of environmental issues. These systems help alter college students' environmental concerns, encouraging them to consider the growing sustainability of the sector. The interactive nature of AI-based training allows students to actively participate in discussions and express their concerns, moving beyond textbook data. Through improving engagement and stimulating creativity, AI-based environmental training motivates students to participate more actively and expand a robust interest in environmental problems.

References

Abulibdeh, A., Zaidan, E., & Abulibdeh, R. (2024). Navigating the confluence of artificial intelligence and education for sustainable development in the era of industry 4.0: Challenges, opportunities, and ethical dimensions. *Journal of Cleaner Production, 437*(140527), 140527. https://doi.org/10.1016/j.jclepro.2023.140527

Andriyani, S. (2021). Pencarian informasi mengenai Sustainable Development Goals: Reduced inequalities. *Electronic, and Telecommunication, 2*, 1–10. https://doi.org/10.52435/complete.v1i1.103

Anggraeni, I., & Yanti, I. (2024). Pengembangan ecoliteracy melalui program bank sampah: Sebagai upaya mewujudkan Sustainable Development Goals (SDGs). *Jurnal Anak Bangsa, 3*(2), 158–169. https://doi.org/10.46306/jas.v3i2.63

Babalola, A., & Olawuyi, D. S. (2021). Advancing environmental education for sustainable development in higher education in Nigeria: Current challenges and future direction. *Sustainability, 13*(19), 10808. https://doi.org/10.3390/su131910808

Batool, K., Zhao, Z.-Y., Irfan, M., & Żywiołek, J. (2023). Assessing the role of sustainable strategies in alleviating energy poverty: An environmental sustainability paradigm. *Environmental Science and Pollution Research International, 30*(25), 67109–67130. https://doi.org/10.1007/s11356-023-27076-0

Bian, M., Yu, H., & Zhang, Y. (2020). Research trends and prospects of green innovation in China: An analysis based on bibliometrics and knowledge graph. *Sci. Technol. Manag. Res., 16*, 236–243.

Cebrián, G., Junyent, M., & Mulà, I. (2020). Competencies in education for sustainable development: Emerging teaching and research developments. *Sustainability, 12*, 579.

Chisom, O. N., Biu, P. W., Umoh, A. A., Obaedo, B. O., Adegbite, A. O., & Abatan, A. (2024). Reviewing the role of AI in environmental monitoring and conservation: A data-driven revolution for our planet. *World Journal of Advanced Research and Reviews, 21*(1), 161–171. https://doi.org/10.30574/wjarr.2024.21.1.2720

Dong, Y., Zhang, X., Dai, J., & Guan, H. (2014). HYVI: A HYbrid VIrtualization solution balancing performance and manageability. *IEEE Transactions on Parallel and Distributed Systems: A Publication of the IEEE Computer Society, 25*(9), 2332–2341. https://doi.org/10.1109/tpds.2013.239

Egbemhenghe, A. U., Ojeyemi, T., Iwuozor, K. O., Emenike, E. C., Ogunsanya, T. I., Anidiobi, S. U., & Adeniyi, A. G. (2023). Revolutionizing water treatment, conservation, and management: Harnessing the power of AI-driven ChatGPT solutions. *Environmental Challenges, 13*(100782), 100782. https://doi.org/10.1016/j.envc.2023.100782

Fan, Z., Yan, Z., & Wen, S. (2023). Deep learning and artificial intelligence in sustainability: A review of SDGs, renewable energy, and environmental health. *Sustainability, 15*(18), 13493. https://doi.org/10.3390/su151813493

García-Saiz, D., Palazuelos, C., & Zorrilla, M. (2013). Data mining and social network analysis in the educational field: An application for non-expert users. In *Educational Data Mining* (pp. 411–439). Springer International Publishing. https://doi.org/10.1007/978-3-319-02738-8_15

Glavič, P. (2020). Identifying key issues of education for sustainable development. *Sustainability, 12*, 6500.

Han, Y., & Liang, Y. (2023). Scientific knowledge map study of therapeutic landscapes and community open spaces: Visual analysis with CiteSpace. *Sustainability, 15*, 15066.

Herman, B. M., Zeidler, D. L., & Newton, M. E. (2020). Students' emotive reasoning through place-based environmental socioscientific issues. *Research in Science Education, 50*, 2081–2109.

Islam, M. M. (2024). Utilizing AI for social good: Tackling global issues and fostering inclusive solutions. *Journal of Artificial Intelligence General Science (JAIGS) ISSN:3006-4023, 3*(1), 341–362. https://doi.org/10.60087/jaigs.v3i1.124

Joo-Nagata, J., Martinez Abad, F., García-Bermejo Giner, J., & García-Peñalvo, F. J. (2017). Augmented reality and pedestrian navigation through its implementation in m-learning and e-learning: Evaluation of an educational program in Chile. *Computers & Education, 111*, 1–17. https://doi.org/10.1016/j.compedu.2017.04.003s

Jorgenson, S. N., Stephens, J. C., & White, B. (2019). Environmental education in transition: A critical review of recent research on climate change and energy education. *Journal of Environmental Education, 50*(3), 1–18. https://doi.org/10.1080/00958964.2019.1604478

Junjia, Y., Alias, A. H., Haron, N. A., & Bakar, N. A. (2023). A bibliometric review on Safety Risk assessment of construction based on CiteSpace software and WOS database. *Sustainability, 15*, 11803.

Ladachart, L., PooThawee, M., & Ladachart, L. (2020). Toward a place-based learning progression for haze pollution in the northern region of Thailand. *Cultural Studies of Science Education, 15*, 991–1017.

Li, Y., Du, Q., Zhang, J., Jiang, Y., Zhou, J., & Ye, Z. (2023). Visualizing the intellectual landscape and evolution of transportation system resilience: A bibliometric analysis in CiteSpace. *Developments in the Built Environment, 14*, 100149.

Mary, K. J. A., & Binu, V. P. (2024). Helping hand of AI for achieving SDG. *Journal of Applied Science, Engineering, Technology and Management, 2*(01), 29–34. https://doi.org/10.61779/jasetm.v2i1.6

Shen, Z., Ji, W., Yu, S., Cheng, G., Yuan, Q., Han, Z., Liu, H., & Yang, T. (2023). Mapping the knowledge of traffic collision Reconstruction: A scientometric analysis in CiteSpace, VOSviewer, and SciMAT. *Science & Justice, 63*, 19–37.

Submit Ramírez Suárez, V., Acosta-Castellanos, P. M., Castro Ortegon, Y. A., & Queiruga-Dios, A. (2023). Current state of environmental education and education for sustainable development in primary and secondary (K-12). Boyacá, Colombia. https://doi.org/10.3390/su151310139

Tao, Y., & Tao, Y. (2024). Integrating aesthetic education in quality education: A bibliometric analysis of sustainable development perspectives. *Sustainability, 16*, 855.

Telešienė, A., Pauw, J. B., Goldman, D., & Hansmann, R. (2021). Evaluating an educational intervention designed to foster environmental citizenship among undergraduate university students. *Sustainability, 13*, 8219.

Tian, Y., Jin, Y., Zhao, Y., Du, Y., Shen, S., & An, J. (2024). Analysis of knowledge graph: Hotspots and future trends in environmental education research. *Sustainability, 16*(6), 2378. https://doi.org/10.3390/su16062378

Waltner, E., Scharenberg, K., Hörsch, C., & Riess, W. (2020). What teachers think and know about education for sustainable development and how they implement it in class. *Sustainability, 12*, 1690.

Zywiolek, J., Sarkar, A., & Sial, M. S. (2022). Biometrics as a method of employee control. 2022 16th International Conference on Ubiquitous Information Management and Communication (IMCOM). IEEE. https://doi.org/10.1109/imcom53663.2022.9721809

CHAPTER 17

Interdisciplinary Education for Social Sustainability: Smoking Prevention from an Educational Perspective

Guillermina Ávila García

1 Introduction

This proposal stems from the concept of social sustainability, which stems from the sustainable development defined by the World Commission on Environment and Development in 1987, which states "development that meets the needs of the present without considering the capacity of future generations to meet their own needs" (cited in Cuesta et al., 2024). Barron et al. (2023, p. 30) defines social sustainability as occurring when all people feel part of the development process and believe that they and their descendants will benefit from it.

The Economic Commission for Latin America and the Caribbean (CEPAL, 2019) specifically states in target 3.a of SDG 3: Strengthen the implementation of the World Health Organization Framework Convention

G. Á. García (✉)
Instituto Politécnico Nacional, Agricultura, Miguel Hidalgo, Mexico City, México
e-mail: gavilag@ipn.mx

K. Hirech (ed.), *Environmental Education and Sustainable Development*, https://doi.org/10.1007/978-3-032-05760-0_17

on Tobacco Control in all countries, as appropriate; with indicator 3.a.1 representing the prevalence of current tobacco use among those aged 15 years and older (ages adjusted). This research justifies the use of this objective, considering the context of students aged 15 to 18, and according to a survey, 73% of them have smoked or vaped.

Philip Morris International (PMI) states that the majority of carbon dioxide (CO_2) emissions occur in the tobacco cultivation phase, followed by the manufacturing phase, and finally, the distribution and logistics phase. Philip Morris International (PMI, 2023) states that maximizing the benefits of smoke-free products involves developing and marketing science-based anti-smoking alternatives, making them available in 100 markets (at least half of which are low- and middle-income markets), and in that sense, implementing robust youth access prevention programs in indirect retail channels.

In Mexico, in 2023, the Addiction Prevention classroom strategy was implemented, and a teacher's guide was distributed throughout the country. It indicates that the average age for starting to use both legal and illegal drugs is 13 years old. The guide's objective is to promote knowledge, reflection, and awareness among students about the serious consequences of drug use, among which smoking and vaping stand out, which are problems that primarily affect the student population (Government of Mexico, 2024).

The Government of Mexico (2024) highlights the problems of vaping use, the effects, and the main irreversible damage to health, such as cardiovascular diseases, serious respiratory diseases, and cancer (OMS, 2021). The Federal Commission for the Protection of Sanitary Risks (COFEPRIS) points out that there is a belief that the use of vapers does not entail serious health risks because the percentage of nicotine they introduce into the body is lower.

These national and international documents provide the opportunity to shape and guide work toward interdisciplinary education that integrates the different knowledge derived from the learning units to foster understanding and awareness of both the environment and health.

2 Environmental Education and Interdisciplinarity

2.1 *Interdisciplinary Education*

Before discussing interdisciplinarity, let's consider the concept of a "comprehensive institutional approach," which UNESCO considers the recommended strategy for implementing Education for Sustainable Development (ESD) (Kohl et al., 2022). ESD is an educational agenda that dates to Agenda 21 (the program resulting from the United Nations Earth Summit held in Rio de Janeiro, 1992) and consists of a series of ongoing efforts to transform educational practices and systems to better address today's complex sustainability issues. From this perspective, this chapter seeks to demonstrate the transformation of education through educational practice, where "students are allowed to live what they learn and learn what they live" (UNESCO, 2020). It is precisely in this sense that the institutional approach to sustainability is considered a good strategy for achieving this transformation.

On the other hand, sustainability science aims to understand the complex and dynamic interactions between natural and human systems to transform them sustainably (Clark & Dickson, 2003; Jerneck et al., 2011; Kates et al., 2001; Komiyama & Takeuchi, 2006; Mochizuki & Yarime, 2015). Thus, ESD can be considered in three stages that involve progressive levels of learning: first-order, second-order, and third-order learning, which corresponds to education about, education for, and education as sustainability (Mochizuki & Yarime, 2015).

Along the same lines, Didham et al. (2024) indicate that interdisciplinary education for sustainable development is considered a valuable mechanism for improving the quality and overall relevance of education and learning systems; seeking to integrate knowledge and methods from different disciplines to create syntheses of approaches; likewise, Ajiga et al. (2025) indicate the need for interdisciplinary models for the challenges posed by sustainability that require an integrated approach that leverages knowledge from various fields, considering interdisciplinary curricular models as essential to promote a comprehensive understanding (George, 2023).

Furthermore, Olvitt et al. (2024) point out that interdisciplinary and transdisciplinary learning can equip students with a nuanced understanding of global sustainability concerns, while cultivating creativity, agency, and skills such as critical thinking.

This information is justified by these principles of sustainability, initially considering students' concerns and interests as part of the diagnosis, thereby generating opportunities that combine physics learning with the knowledge and application of social sustainability.

The classroom project proposal focuses precisely on this connection and integration of knowledge from various learning units that high school students take, having as its central axis the study of Physics, the detection of the problem, and later the understanding of it, resulting in a range of strategies that are carried out in the classroom toward social sustainability. Below, the study plan and program are described, which integrates the competence to be developed focusing particularly on the social sustainability that was worked on in the project.

Mathematics focused towards the study of Differential Calculus, which establishes the core competency: *Solve problems related to the variation of functions, based on the concept of the derivative, in theoretical and real situations of their academic, social, and global environment*. This learning unit is related to sustainability in the integration of research on the current problem in Mexico, which is studied from variational information, such as the functions that describe the situation of decrease or increase in smoking in young people.

Physics II: Solve problems of mechanics, conservation of energy, thermodynamics and hydrodynamics from the application of the laws and principles to phenomena that occur in their personal or social environment. This learning unit establishes its importance in the experiment based on Pascal's Principle which states that when a change in pressure occurs at any point in a confined fluid, an equal change in pressure occurs at all points in the fluid (Bauer & Westfall, 2011).

Chemistry II: Solve qualitative and quantitative aspects of chemical changes, using a language specific to the field and with a focus on Science - Technology - Society and Environment that applies in personal, academic, and work contexts. In this section, students investigated the harmful effects on health of traditional cigarette consumption combined with the very frequent use of vaping among young people, considering national surveys on drug use, including alcohol and tobacco.

Technical Drawing II: Apply methods of graphic representation in volumes showing its external and internal structure, using instruments and software, in accordance with current official engineering standards. It is precisely in this learning unit where students show the designs of healthy lungs and those that are contaminated by smoking.

English III: Communicates experiences, events, aspirations, achievements, and opinions in English, using the relevant structures, vocabulary, and expressions, orally and in writing, in formal and informal social contexts. An interview with a psychiatrist took place. The interview was given in English for both the students and the psychiatrist, considering the topic of addictions in general.

The classroom project is developed progressively and considers the accumulation of knowledge and skills acquired by students. Each semester, a different perspective is addressed. It is important to note that in the first two semesters, students developed new ideas, and it wasn't until the third semester that they developed a more solid outline, opting for mental health or addiction prevention. In the fourth semester, the idea of working specifically on addiction prevention focused on smoking took shape.

It can be said that this classroom project is the initial phase, and work will continue over the next two semesters to achieve consolidation and a final product. However, this phase shows significant progress compared to the initial research and surveys, as well as experiments aimed at addressing smoking or vaping addiction.

2.2 *Smoking and Climate Change*

The World Health Organization [OMS] (OMS, 2021) states that tobacco directly harms our health through consumption and exposure to secondhand smoke and its negative impact on the environment; in addition, tobacco production and consumption also contribute to global warming by releasing 80 million tons of carbon dioxide (CO_2) into the environment each year, which is equivalent to driving 17 million gasoline-powered vehicles each year. Furthermore, tobacco use pollutes the air we breathe, while leaving nicotine residues and other harmful chemicals on indoor surfaces, thus exposing non-smokers to the toxins in secondhand smoke (OMS, 2021).

Also, the Pan American Health Organization [OPS] (OPS, 2022) points out that cigarette consumption generates various types of environmental costs, such as chemicals, metals involved in manufacturing, energy used, numerous chemical additives, among others.

3 Methodology

The Classroom Project (PA), as a methodological and collaborative strategy, considers various thematic axes. In this investigation, only those that comprise this work are included: Sustainable Development and Mental Health. These topics are proposed to the students in an assembly, and by mutual agreement, the students choose the topic to be developed. In this case, the interest in social sustainability is illustrated, focusing on mental health, addictions, and smoking.

4 Project-Based Learning (PBL)

In this inquiry, the application of the Classroom Project as part of the pedagogical approach for sustainable teaching and learning highlights the advantage of experiential methods based precisely on the project for sustainable practices in education; in this way problem-based learning immerses students in the real challenges of sustainability and interdisciplinary collaboration, while Project-Based Learning allows them to assume leadership roles empowering students to become proactive advocates for sustainability (Singha & Singha, 2024).

Project-Based Learning (PBL) is part of active teaching-learning strategies. It also directs students' goals toward finding solutions to problems or challenges, implementing research strategies, knowledge, and skills that will be part of their new learning.

In accordance with Cobo and Valdivia (2016), the following stages are considered, considering some adjustments that are integrated as part of the classroom project (which are indicated in italics, as a distinction from the stages of PBL) that is developed from the institutional part:

- Assembly with the students to detect the problem.
- Project approach and organization. In this case, the students are those who put on the table the problems they observe in their school community or even in their group.
- The organization is carried out by the students, considering the learning units where they consider they have more affinity.
- Research on the topic. The information is collected, the teacher acts as a guide, and the students, according to the learning unit that corresponds to them, begin the research process in different sources supported by the teacher in charge of the subject.

- Definition of the objectives and work plan. In this case and in accordance with the PA, a research protocol is delivered where the work plan, objectives, and schedule of activities are defined.
- Project implementation. Each classroom session presents progress and monitors the work, difficulties, and opportunities to make the necessary adjustments.
- Presentation and evaluation of the results through a presentation where communication skills are put into practice, evaluation criteria are established by the Directorate of Upper Secondary Education (DEMS) of the IPN, where the evaluators are external to the educational institution, in this case CECyT 11.

With respect to the evaluation, checklists and rubrics were used to evaluate teamwork, research, experimentation, and oral presentation of the results of the research for each learning unit, as well as the final report of all the data and information collection.

Twenty-five high school students from CECyT 11 of the IPN participated, considering the consent of the students, their ages are between 15 and 18 years old, including both genders: female and male. The students belong to a social and urban context, living within the metropolitan area of Mexico City.

5 Results

In the survey of the topic/problem to be worked on, 92% of the students were inclined to give attention to the problem of smoking, arguing that on the way from home to school and vice versa they see that many students are vaping, just as they see how several teachers smoke in the vicinity of the school. Eight percent of the students thought that there are also suicide problems and that it is an area of opportunity to address mental health problems. Finally, the core of the topic arose around the problem of smoking in the student population.

During the student assembly in accompaniment with the teacher (author of this work), they had a democratic space, 100% of the students were actively involved in the selection of the research topic, the students expressed concern about the problem of smoking, especially in the vicinity of the school by teachers, which for them implied that it is a show of authority, so it should be avoided.

<<Some teachers are our figure of admiration, and I believe that some colleagues can be induced to smoke by these teachers...>>

<<I personally do not like that adults want to impose something on you, when they themselves are not capable of controlling their habits, I say this about the teachers who smoke at the entrance of the school...>>

<<I have relatives in Asia, and they tell us that they smoke in rooms so that the smoke produced by tobacco does not affect non-smokers. It would be good if they implemented it in Mexico, because in my case I do not smoke, but I must tolerate those who do and above all even the bad breath that these smokers and vapers emit>>.

The students deliberated on the specific aspects of smoking and what line their research would follow, considering the following aspects:

- The causes and consequences of tobacco consumption in young people.
- Impact of smoking on physical and mental health.
- Science in smoking.
- The environment and smoking.
- The transversality of the learning units and the study of smoking.
- How to design experiments that demonstrate the harmful effects of tobacco.

Organization of the project, the students distributed tasks and responsibilities and the different learning units for the assignment of roles for the coordination and communication of the project, in this way the autonomy and responsibility of the students was promoted, which is precisely what PBL seeks.

From the initial work, the students who were part of the classroom project carried out the application of a survey (originally in Spanish) to 250 students from different semesters of CECyT 11, to find out how often they smoke or vape. The English translation of the complete survey form can be found in the Appendix of this chapter, obtaining the following results:

- Have you ever smoked cigarettes? Yes: 112, No: 138.
- How often do you currently smoke? Rarely: 74; daily: 38; I don't smoke: 138.

- Do you use other tobacco or nicotine products, such as vapes, electronic cigarettes, or smokeless tobacco? Yes: 96; Sometimes: 16; No: 138.
- What are the main reasons why you smoke or have smoked? (You can mark several options) I don't smoke: 138; Stress and anxiety: 84; Curiosity: 35; Pleasure: 38; Addiction: 8
- Do you think smoking helps you manage stress or anxiety? Yes: 77; I don't know: 12; No: 161
- Would you like to quit smoking? Yes: 97; I don't smoke: 138; No: 15.
- Do your friends or close relatives smoke? Yes: 184; No: 66.

This is where the importance of carrying out research to show the damage caused by smoking and vaping in the final presentation comes from.

The objectives were defined, with respect to the classroom project as shown:

5.1 *Prevention of Smoking and Vaping Addiction at CECyT 11*

The Challenge: To significantly raise awareness among CECyT 11 students regarding the dangers of smoking and vaping, ultimately preventing addiction through targeted information campaigns led by specialists.

The Need: Recognizing the observed, experienced, or closely faced issue of nicotine addiction among CECyT 11 students, this initiative aims to provide crucial information on reducing and overcoming vaping and cigarette consumption, as well as directing individuals to specialized support.

Expected Social Impact: A healthier CECyT 11 community with reduced rates of smoking and vaping, leading to fewer negative health consequences and a greater focus on positive recreational activities.

Integrated Approach: This project strategically connects internal resources (teachers from Industrial Processes and students from group 4IV3) with external expertise (addiction specialists, doctors, and psychologists). Awareness will be disseminated through targeted campaigns and conferences for students.

Evidence of Learning: Students will engage in an experiment to understand the impact of smoking/vaping on lung health. They will also analyze the evolution of consumption through graphical representations and explore the chemical processes and organ damage via informative posters. Visual understanding will be enhanced through sketches of affected lungs, and English language skills will be utilized to create impactful posters and brochures explaining the risks.

Learning Unit Integration: This project is embedded within the curriculum through activities in:

- **Physics II:** Analyzing dynamics, properties of matter, and thermodynamics through the Pascal and Bernoulli experiment (linking to respiratory function).
- **Differential Calculus:** Graphing and analyzing the rate of change in smoking/vaping prevalence.
- **Chemistry II:** Investigating the chemical processes of addiction and its effects on the body.
- **Technical Drawing:** Creating visual representations of lung damage caused by smoking/vaping.
- **English III:** Developing informational materials (posters and brochures) on the dangers of smoking/vaping.

Note: The specific achievements within each learning unit are detailed in Table 17.1.

Finally, the students evaluated the work of the classroom project and the impact on the student community applied to 250 students who visited the exhibition and interacted with it, in addition to the two external evaluators from other schools, shown in Table 17.2.

The external evaluators talked with the students, considering the most outstanding and most frequent questions. The following nomenclature was applied to distinguish the interventions of both students and evaluating teachers: E1 = Evaluator 1, E2 = Evaluator 2, …; in the case of students, the nomenclature of Est1 = Student 1, Est2 = Student 2 and so on was considered:

- *E1: What were your main achievements when developing this project?*
- *Est1: We realized the importance of disseminating information about the serious consequences of smoking or vaping.*
- *E2: Which of the activities that you are showing in this exhibition can help students reduce or eradicate the habit of smoking?*
- *Est5: I really liked the levitation experiment; it goes in two directions; the first is that you learn something about science, the second most important thing is that you have fun. I think that if we do fun things in our lives, we stay away from addictions.*

Table 17.1 Achievements obtained by students (organization and delivery of work)

Activity carried out	*Proportion of the sample of 25 students who carried out the classroom project*	
	Complies	*Does not comply*
Research corresponding to the learning unit	88%	12%
Participation and collaboration with team members	88%	12%
Relates the concepts of tobacco addiction (cigarettes or vaping) with social sustainability	76%	24%
Establishes and forms a responsible criterion about tobacco/vaping consumption	96%	4%
Raising awareness through experimental activities	100%	0
Shares information with colleagues at the institution to raise awareness about tobacco or vaping consumption	100%	0
Mathematics: it is an example of how the growth of smoking and vaping among the population can be modeled.	60%	40%
Physics: in the application of the Bernoulli experiment, applying the concepts of pressure and air flow in a real situation, where students put into practice the importance of breathing.	84%	16%
Chemistry: understanding the toxic components of cigarettes and their effects on the body.	100%	0%
Technical drawing, visualization of the effects of smoking on the human body, especially on the lungs.	72%	28%
English: the interview with an expert and the transmission of this allows to have first-hand information about the effects of smoking and vaping in the student community.	64%	36%
Exhibition and interaction with partners from the educational institution.	100%	0%
Exhibition and interaction with teachers from the educational institution.	100%	0%

Source: Own elaboration

- *E1: What would you say to a classmate to get them to stay away from smoking?*
- *Est2: We invite them to our science club (because we want to make this project a science club) so that they don't think about cigarettes.*

Table 17.2 Assessment of the classroom project

The project allows me to acquire new knowledge on how to apply knowledge of scientific subjects for the benefit of health and at the same time social sustainability.	**Always**	**24**
	Often	1
	Sometimes	0
	Rarely	0
Each activity developed for the classroom project allowed me to significantly apply the knowledge acquired from the different learning units.	**Always**	3
	Often	22
	Sometimes	0
	Rarely	0
I identified strategies that allow me to support other colleagues, and, above all, I learned that well-applied science can solve social problems.	**Always**	7
	Often	8
	Sometimes	0
	Rarely	0

Source: Own elaboration

- *Est3: We are telling them to look at our sketches of what the lung would look like and the consequences such as cancer.*
- *Est10: Invite them to therapy. Our next classroom project will bring experts to the school and help lines so that they can be supported in their addiction.*
- *E2: Do you know anyone who has quit or reduced their cigarette or vaping habit because of the strategy of the project they implemented?*
- *Est7: No, we don't know yet. But this is a project that considers that it will have further scope in the future.*

The checklist used by the evaluators is shown in Annex. Most of the established criteria are evaluated in a similar way. Some discrepancies that arose were due to the external environment of the institution itself. In general, both evaluators consider that if the project is continued, it has great expectations, in addition to being a relevant topic due to the high percentage of smokers at the institutional level, coupled with the fact that the strategy is addressed in the classroom: prevention of addictions, implemented by the federal government of Mexico and in accordance with SDG 3 toward social sustainability.

6 Conclusion

The classroom project implemented at the IPN high school level makes a valuable contribution to the comprehensive education of high school students. In parallel, work was done on Sustainable Development Goal (SDG): 3, which aims to ensure an education oriented toward sustainable development and sustainable lifestyles. The integration of the PBL methodology transcends knowledge across different learning units, enriching meaningful learning for students and contributing to the strategy implemented by the Mexican federal government, by being agents of change that involves promoting healthy lifestyles and preventing disease.

The Classroom Project-based methodology, interdisciplinarity, and PBL represent an innovative and effective pedagogical model, given the evaluations recorded by external evaluators with no direct knowledge of the project. These evaluations provide suggestions that will undoubtedly allow for improvements in this proposal in subsequent semesters, in addition to addressing a complex problem aligned with the 2030 Agenda.

Finally, these results obtained during the investigation stimulate reflection on the importance of implementing educational projects aimed at integrating the largest number of learning units and, first and foremost, viewing science from a different perspective, contributing to the construction of a more sustainable future for all.

Appendix: Smoking and Vaping Questionnaire

This questionnaire is intended for informational purposes to identify some of the addictions experienced by high school students and is in accordance with the National Strategy for the Prevention of Addictions.

Objective:

To understand the prevalence, risk factors, and motivations associated with tobacco use (cigarettes and other products) among high school students at the Scientific and Technological Studies Center No. 11 of the National Polytechnic Institute, to identify areas of intervention and design effective prevention strategies for social sustainability.

This questionnaire is part of the research being conducted at the IPN in conjunction with the Secretariat of Research and Graduate Studies.

Section 1: Demographic Information (Optional)

Please answer the following questions truthfully. You do not need to provide your name.

Name:
Age:
Gender:
Educational Level:

Section 2: Smoking Habits

Please answer using the answer that best matches your habits.

Have you ever smoked cigarettes?

☐ Yes
☐ No

If you answered "Yes" to the previous question, how often do you smoke?

☐ Daily
☐ Several times a week
☐ Occasionally
☐ Rarely
☐ I no longer smoke

Do you use other tobacco or nicotine products, such as vapes, e-cigarettes, or smokeless tobacco?

☐ Yes
☐ No

If you answered "Yes" to the previous question, how often do you use them?

☐ Daily
☐ Several times a week
☐ Occasionally
☐ Rarely
☐ I no longer smoke

Section 3: Motivations for Smoking

What are the main reasons you smoke or have smoked? (You can mark multiple options)

Social pressure
☐ Stress
☐ Anxiety
☐ Curiosity
☐ Pleasure
☐ Addiction
☐ Influence of friends or family
☐ Other

Do you think smoking helps you manage stress or anxiety?

☐ Yes

☐ No

Would you like to quit smoking?

☐ Yes

☐ No

Section 4: Environmental influence

Do your friends or close family members smoke?

☐ Yes

☐ No

Are you exposed to tobacco smoke in public places or at home?

☐ Yes

☐ No

Do you think advertising for tobacco products influences your decision to smoke or not to smoke?

☐ Yes

☐ No

☐ Not sure

Thank you for your participation. The information collected is for research purposes only and is treated confidential

References

Ajiga, D. I., Hamza, O., Eweje, A., Kokogho, E., & Odio, P. E. (2025). Developing interdisciplinary curriculum models for sustainability in higher education: A focus on critical thinking and problem solving. *Journal of Public Administration and Social Welfare Research*, 10(1), 87–102. https://doi.org/10.56201/jpaswr.v10.no1.2025.pg87.102

Bauer, W., & Westfall, G. D. (2011). *Física para ingeniería y ciencias* (2nd ed., Vol. 1). McGraw-Hill. https://archive.org/details/vol-1-bauer-wolfgang-westfall-gary-d-fisica-para-ingenieria-y-ciencias.-volumen-1/page/n11/mode/2up?view=theater

Barron, P., Cord, L., Cuesta, J., Espinoza, S. A., Larson, G., & Woolcock, M. (2023). *Social sustainability in development: Meeting the challenges of the 21st century*. New Frontiers of Social Policy – World Bank Group. https://doi.org/10.1596/978-1-4648-1946-9

Clark, W. C., & Dickson, N. M. (2003). Sustainability science: The emerging research program. *Proceedings of the National Academy of Sciences, 100*(14), 8059–8061. https://doi.org/10.1073/pnas.1231333100

Cobo, G., & Valdivia, S. (2016). Aprendizaje basado en proyectos. In *Proyecto de Innovación Educativa y Desarrollo Curricular, 11*. Instituto de Docencia Universitaria. http://repositorio.pucp.edu.pe/index/handle/123456789/170374

Comisión Económica para América Latina y el Caribe [CEPAL]. (2019). *La Agenda 2030 para el Desarrollo Sostenible y los Objetivos de Desarrollo Sostenible: Una oportunidad para América Latina y el Caribe*. Naciones Unidas.

Cuesta, J., Madrigal, L., & Pecorari, N. (2024). Una perspectiva global de la sostenibilidad social y la pobreza. *ICE, Revista De Economía, 934*. https://doi.org/10.32796/ice.2024.934.7733

Didham, R. J., Fujii, H., & Torkar, G. (2024). Exploring interdisciplinary approaches to education for sustainable development. *Nordic Journal of Comparative and International Education (NJCIE), 8*(2). https://doi.org/10.7577/njcie.5877

George, A. S. (2023). Preparing students for an AI-driven world: Rethinking curriculum and pedagogy in the age of artificial intelligence. *Partners Universal Innovative Research Publication, 1*(2), 112–136. https://doi.org/10.5281/zenodo.10245675

Gobierno de México. (2024). *Guía para docentes: Estrategia en el aula. Prevención de adicciones.* Government of Mexico. (2024, December 27). CSP teaching guide [PDF]. https://www.gob.mx/cms/uploads/attachment/file/964368/Guia_docentes_CSP_2024_271224_WEB.pdf

Jerneck, A., Olsson, L., Ness, B., Anderberg, S., Baier, M., Clark, E., Hickler, T., Hornborg, A., Kronsell, A., Lövbrand, E., & Persson, J. (2011). Structuring sustainability science. *Sustainability Science, 6*, 69–82. https://doi.org/10.1007/s11625-010-0117-x

Kates, R. W., Clark, W. C., Corell, R., Hall, J. M., Jaeger, C. C., Lowe, I., McCarthy, J. J., Schellnhuber, H. J., Bolin, B., Dickson, N. M., Faucheux, S., Gallopin, G. C., Grubler, A., Huntley, B., Jager, J., Jodha, N. S., Kasperson, R. E., Mabogunje, A., Matson, P., Mooney, H., Moore, B., O'Riordan, T., & Svedin, U. (2001). Sustainability science. *Science, 292*(5517), 641–642. https://doi.org/10.1126/science.1059386

Kohl, K., Hopkins, C., Barth, M., Michelsen, G., Dlouha, J., Razak, D. A., Bin Sanusi, Z. A., & Toman, I. (2022). A whole-institution approach towards sustainability: A crucial aspect of higher education's individual and collective engagement with the SDGs and beyond. *International Journal of Sustainability in Higher Education, 23*(2), 218–236. https://doi.org/10.1108/IJSHE-10-2020-0398

Komiyama, H., & Takeuchi, K. (2006). Sustainability science: Building a new discipline. *Sustainability Science, 1*, 1–6. https://doi.org/10.1007/s11625-006-0007-4

Mochizuki, Y., & Yarime, M. (2015). Education for sustainable development and sustainability science: Re-purposing higher education and research. In W. Leal Filho, L. Brandli, O. Kuznetsova, & A. Paço (Eds.), *Routledge handbook of higher education for sustainable development* (pp. 11–24). Routledge. https://doi.org/10.4324/9781315852249

Olvitt, L., Davies, M., Ebrahim, A. I., & Cockburn, J. (2024). Transdisciplinary curriculum design for sustainability transitions: A reflective dialogue. *South African Journal of Science*, *120*(9/10). https://doi.org/10.17159/sajs.2024/17884

Organización Panamericana de la Salud [OPS]. (2022). *El tabaco y su impacto ambiental: panorama general.* https://iris.paho.org/bitstream/handle/10665.2/56065/9789275325681_spa.pdf#:~:text=Los%20cigarrillos%20contaminan%20nuestro%20aire%2C%20como%20han,de%20las%20que%20se%20desechan%20millones%20de

Organización Mundial de la Salud [OMS]. (2021). *Informe OMS sobre la epidemia mundial de tabaquismo, 2021: Abordar los productos nuevos y emergentes*, Ginebra, 2021. https://www.who.int/es/publications/i/item/9789240032095

Organización Mundial de la Salud [OMS]. (2022). *El tabaco: un veneno para nuestro planeta.* https://iris.who.int/bitstream/handle/10665/373168/9789240055117-spa.pdf

Philip Morris International [PMI]. (2023). *What does our sustainable future look like.* https://www.pmi.com/sustainability/our-approach/sustainable-future

Singha, R., & Singha, S. (2024). Application of experiential, inquiry-based, problem-based, and project-based learning in sustainable education. In S. C. Shukla & R. Singha (Eds.), *Teaching and learning for a sustainable future: Innovative strategies and best practices* (pp. 109–128). IGI Global. https://doi.org/10.4018/978-1-6684-9859-0.ch006

UNESCO. (2020). *Education for sustainable development: A roadmap.* https://doi.org/10.54675/YFRE1448

CHAPTER 18

Interdependence Approach to Sustainability Education in Gulf Cooperation Countries

Raed Awashreh and *Almi Hassiba*

1 Introduction

The GCC nations, Bahrain, Kuwait, Oman, Qatar, Saudi Arabia, and the United Arab Emirates, have experienced rapid economic growth, largely due to their rich reserves of oil and gas. This development has led to significant improvements in infrastructure, rising standards of living, and greater global influence. However, the swift pace of growth has also exacerbated environmental degradation, driven the overuse of natural resources, and highlighted an urgent need for educational programs that embed sustainability principles. As a result, education for sustainability is increasingly viewed as a critical mechanism for aligning economic progress with environmental stewardship and social well-being in these transitioning economies (Cong & Ren, 2023).

R. Awashreh (✉)
Government and Society, United Arab Emirates University, Al Ain, United Arab Emirates

A. Hassiba
Faculty of Economics and Management, Annaba University, Annaba, Algeria
e-mail: hassiba.almi@univ-annaba.dz

K. Hirech (ed.), *Environmental Education and Sustainable Development*, https://doi.org/10.1007/978-3-032-05760-0_18

In alignment with the United Nations Sustainable Development Goals (SDGs), particularly those focused on clean water, affordable energy, sustainable consumption, climate action, and ecosystem conservation, education plays a pivotal role in preparing younger generations to address these interconnected challenges (Shayan et al., 2022). Although scholars such as Amin et al. (2023) have called for robust sustainability curricula in the GCC, implementation remains fragmented and inconsistent. Often, sustainability education is treated as a stand-alone initiative rather than integrated across disciplines and institutional structures, limiting its long-term impact.

To address these shortcomings, this chapter proposes the use of an interdependence model that emphasizes the interconnectedness of environmental, economic, and social systems. This model fosters a holistic understanding of sustainability, encouraging students to think beyond silos and engage in systemic problem-solving (White et al., 2024). By prioritizing long-term thinking, interdisciplinary collaboration, and stakeholder engagement, the interdependence model aligns with emerging global frameworks while remaining regionally relevant. It provides a comprehensive educational structure that supports both critical reflection and action, helping learners develop the capacity to address sustainability challenges across local, national, and international contexts (Awashreh & Al Ghunaimi, 2025a). Moreover, the model promotes systems thinking, enabling students to understand how decisions in one domain such as resource management can impact social equity, economic stability, and environmental resilience. Rather than relying on isolated, subject-based approaches, the interdependence model advocates for an integrated curriculum that facilitates cooperative problem-solving and meaningful participation in policy and community decision-making processes (Weber et al., 2021).

Given the urgency of environmental and sustainability challenges in the GCC, this study explores how the interdependence model can enhance the effectiveness of sustainability education and contribute to long-term socioeconomic and ecological resilience in the region. Specifically, the study addresses the following research questions:

1. How can the interdependence model be effectively integrated into GCC educational systems?
2. What are the major obstacles to its implementation, and how can these be mitigated?

3. In what ways can this model contribute to the broader goal of sustainable development in GCC countries?

By answering these questions, the chapter aims to offer actionable insights and policy recommendations for embedding sustainability education into the curricula of GCC institutions. To achieve these goals, the chapter proceeds as follows: Sect. 2 reviews the relevant literature, Sect. 3 presents the methodology, Sect. 4 discusses results and insights, and Sects. 5 and 6 offer conclusions and implications for future research and policy development.

2 Context of Sustainability in GCC Countries

The GCC consists of Bahrain, Kuwait, Oman, Qatar, Saudi Arabia, and the United Arab Emirates, all of which have undergone very rapid economic growth due to their natural endowment in oil and gas. This growth was accompanied by several benefits, including the establishment of infrastructure, the rise in living standards, and increased global influence. However, it also resulted in large-scale environmental challenges, overconsumption of natural resources, and a growing demand for educational strategies that will guarantee sustainability. Against this backdrop, and in the context of these countries trying to diversify their economies and decrease reliance on fossil fuels, sustainability education remains a key tool in shaping a future that balances economic progress with environmental preservation and social well being (International Monetary Fund, Middle East and Central Asia Dept., 2024).

Nevertheless, most schooling systems, even at their best, are too often limited to fragmented efforts that impart training in sustainability in a comprehensive manner, often splitting pieces of sustainability from one another within particular systems. Therefore, now might be an opportunity to bring the conversation more in line with an interdependence model, which is engaged with the discernment of environmental, economic, and social systems as necessarily interdependent. This model builds a holistic understanding of particular sustainability challenges, by considering long-term impacts hidden in strategies impacting sectors and generations (Leal Filho et al., 2018). Moreover, the interdependence model for sustainability education promotes economic prosperity, environmental health, and social justice as interdependent, facilitating cross-disciplinary, cross-sector problem-solving. It promotes systems thinking to connect local-to-global impacts. Consequently, shifting from single sustainability courses to an interdisciplinary curriculum enhances critical engagement and decision-making for a sustainable future (Gras-Velazquez & Fronza, 2020).

In light of the sustainability concerns of the GCC, this study examines how an interdependence strategy may assist in enhancing education toward sustainable environmental and socio-economic outcomes. Specifically, integration strategies, potential barriers, and contributions to sustainable development are analyzed in this research, with pragmatic conclusions and policy recommendations for the integration of sustainability in GCC curricula (Al-Saidi, 2022).

The GCC countries stand at the crossroads in their journey toward sustainability, being beset with pressing challenges hitherto produced by rapid urbanization, industrialization, and resource-intensive development: environmental degradation, social inequality, and dependence of the economy on fossil fuels. Environmental degradation **is** manifest in the overexploitation of water resources, air pollution, and the loss of biodiversity, while the social challenges take the form of unequal access to resources, job opportunities, and education. Economically, over-reliance on oil and gas revenues comes at the price of diversified development and susceptibility to heavy volatility in world energy markets (ElMassah & Hassanein, 2023). Nevertheless, GCC countries have increasingly awakened to the concept of sustainable development, as evidenced by recent initiatives such as Saudi Arabia Vision 2030, Qatar National Vision 2030, and Oman Vision 2040, all of which intend to address important sustainability challenges concerning the environment, economy, and society. In these national strategies, the focus is basically on renewable energy, better resource management, and inclusive economic growth. **In this regard,** education has a pivotal role to play by raising awareness about sustainability, inculcating skills, and bringing behavioral changes (Al-Saidi, 2022).

Therefore, this has made integration with national curricula all the more urgent for immediate attention, given that future generations can be better equipped with knowledge and skills related to such challenges. Schools and universities in the GCC have integrated principles of sustainability into their programs, helping students take a multidisciplinary approach toward learning and apply sustainable solutions to existing problems (Winston, 2021). Furthermore, there is an urgent need for collaboration between the government, private sector, and academia to advance the cause of sustainability in the right direction and link education reforms with wider imperatives of national development (Onyeaka & Akinsemolu, 2024). However, the region must overcome the existing status of policy fragmentation, lack of funding for sustainability research, and poor community engagement if it is to progress further on sustainability. To this end, the GCC can create a full regulatory framework that will help move the countries toward sustainability and resiliency by 2025 through investment in green technologies, thereby instilling a culture of sustainability in future generations while they are young.

The GCC countries continue to be fast-growing economies, spurred by the wealth generated from oil and gas. Improved infrastructure and better living standards have coincided with degraded environments, overconsumption of resources, and a need for educational strategies focused on sustainability (World Bank, 2024). Accordingly, there is a clear call for education for sustainability to be inherently embedded in balancing economic progress with environmental preservation and social well-being as these countries seek to diversify their economies away from heavy reliance on fossil fuels.

In practice, despite all efforts to embed sustainability within education, it tends to be fragmentary, with sustainability issues taught in isolation. Hence, a shift toward an interdependence approach is now needed, wherein environmental, economic, and social systems are grasped to be inextricably interrelated. The interdependence approach allows for a holistic understanding of challenges for sustainability to be developed; it builds long-term strategies considering impacts across different sectors and generations (Khodadadi, 2025). Furthermore, the interdependence approach supports the view that an education model should be oriented to underline the linkages between economic growth, environmental health, and social equity. It fosters systems thinking so that students realize the global repercussions of local decisions. Additionally, the sustainability approach within every subject provides a foundation for developing enablers toward issues that build resilience for sustainability (Evans et al., 2017). The interdependence approach assists by extending the scope of sustainability education within the GCC, its effective implementation, adoption barriers, and contribution to the attainment of the SDGs are researched in this study. It is expected that the results will provide actionable insights and policy recommendations toward integrating sustainability into GCC curricula, thus equipping the future generation with the means to tackle regional challenges (Alhammadi et al., 2024).

The GCC region stands at an important juncture, having to grapple with challenges regarding environmental degradation, social inequality, and economic dependency on fossil fuel resources. National strategies on sustainable development for renewable energy, resource management, and inclusive economic growth have been highlighted in the Vision 2030 for Saudi Arabia, Qatar National Vision 2030, and Oman Vision 2040. Education also plays a crucial role by raising awareness and building capacities regarding sustainability. Schools and universities are increasingly including sustainability in their programs (Elrahmani et al., 2021; Elgendy, 2024).

Nevertheless, the GCC still grapples with challenges such as policy fragmentation, very low funding for research, and poor community engagement. To address these issues, an enhanced regulatory framework, investment in green technologies, and the instigation of a culture of sustainability are essential for enabling the Gulf region to move toward a resilient future. This chapter discusses how an interdependence approach can help overcome these barriers toward sustainable development (World Bank Group, 2021).

The interdependence approach places systems thinking at its core, where students are made to consider all challenges to sustainability as interlinked. For example, only through the application of systems thinking can better management of connected resources be established to understand why there is always a connection linking water scarcity with energy use. Social inequalities influencing access to clean energy, for instance, are being tackled through equitable policies. Students and educators apply systems thinking to study the ripple effects of human actions and make better, more responsible decisions (Iacovidou et al., 2021).

For the integration of multidisciplinary knowledge, the key to sustainability education in the GCC will involve incorporating knowledge from different disciplines to understand sustainability challenges from a wider perspective (Al-Hail et al., 2024). This includes:

- Environmental Science: Ecosystems, renewable energy, sustainable resource management, and related ecological dimensions.
- Economics: Economics of sustainability, green technologies, the circular economy, thereby facilitating students' financial assessment of sustainability options.
- Social Sciences: Addressing social equity, the impact of development on communities, and the role of cultural values in sustainability,

which helps students appreciate the human dimensions of sustainable development.

Such a multidisciplinary approach allows students to fathom the different dimensions of sustainability and how they interrelate. This model of holistic education equips them with the competencies needed to develop innovative solutions to the complexities of sustainability challenges both in the GCC region and beyond (Annelin & Boström, 2024).

3 Methodology

This study adopts a qualitative approach based on secondary data analysis, which integrates a comprehensive review of research papers, reports, policy documents, and existing literature with the specific aim of analyzing the interdependence model for integrating sustainability education within the GCC context. The interdependence model was selected because it uniquely addresses the interconnectedness of environmental, economic, and social systems a critical perspective given the complex sustainability challenges facing the GCC countries (Leal Filho et al., 2018; Gras-Velazquez & Fronza, 2020). This framework facilitates a systemic understanding necessary for interdisciplinary educational reforms, unlike other models which tend to isolate components of sustainability.

Secondary data were deliberately chosen to enable a broad and rich analysis of existing knowledge, allowing for the identification of trends, patterns, and insights from diverse, authoritative sources without the resource intensiveness of primary data collection (Martins et al., 2018). **Specifically,** the study conducts a critical content analysis of policy documents, government reports, and scholarly articles focused on sustainability education in the GCC. It systematically explores the current status of sustainability education, government strategies, and international benchmarks. **Moreover,** data were sourced from reports and publications by prominent international organizations such as the United Nations, World Bank, and International Energy Agency, providing a reliable basis for comparison and benchmarking of GCC countries' performance within a global context.

To operationalize the qualitative analysis, the study utilized thematic coding procedures facilitated by qualitative data analysis software (e.g., NVivo). Initial open coding was applied to identify recurring ideas and concepts within the documents. These codes were then grouped into categories representing broader themes such as barriers to integration, benefits of the interdependence model, and interdisciplinary education approaches. Axial coding was subsequently employed to explore relationships between these themes and subthemes.

To ensure reliability and validity of the thematic analysis, a triangulation process was conducted by comparing findings across multiple sources, and analysts independently reviewed the coding framework to enhance consistency. Discrepancies were discussed and resolved to finalize the thematic categories (Mokski et al., 2022). In addition to thematic analysis, comparative analysis was applied to identify differences and similarities across GCC countries' policies and educational practices. This step helped pinpoint areas requiring improvement and highlighted strategies to avoid common pitfalls in sustainability education implementation.

Overall, the study acknowledges existing gaps in sustainability education, the challenges faced by educators, and the potential for interdisciplinary learning environments. It further examines the roles of government bodies, academic institutions, and the private sector in advancing sustainability education in the region. Ultimately, the study's findings aim to inform evidence-based policy development and support the GCC region's transition toward a more sustainable and integrated educational framework. Table 18.1 shows the steps taken in the qualitative data analysis process.

Table 18.1 Summary of qualitative data analysis process

Step	*Description*
Data collection	Review of policy documents, reports, scholarly articles
Open coding	Identification of initial codes from text segments
Axial coding	Grouping codes into thematic categories and exploring relationships
Triangulation	Cross-validation of themes through multiple data sources and analysts
Comparative analysis	Comparing policies and practices across GCC countries
Interpretation	Synthesizing findings to generate insights and policy recommendations

Source: Authors work

4 Systems Thinking in Sustainability Education

Systems thinking is an essential approach in education for sustainability that allows learners to discover how environmental, social, and economic systems are interconnected (Table 18.2). Instead of solving problems in isolation, it encourages an exploration of interdependencies within broader ecological and human systems, enabling holistic problem-solving (Marcos-Sánchez et al., 2022). This is particularly critical for the GCC region, where rapid economic development and environmental fragility intersect in unique ways.

Table 18.2 Key aspects of systems thinking in sustainability education

Aspect	*Description*
Interconnectedness	Recognizing how environmental, social, and economic factors interrelate within sustainability challenges. In the GCC, this means understanding how oil revenue, climate adaptation, and youth employment are interdependent.
Cause-and-effect relationships	Identifying how actions in one area impact others, both positively and negatively. For example, rapid urbanization in the Gulf affects coastal ecosystems, air quality, and public health.
Feedback loops	Understanding reinforcing and balancing loops, such as how deforestation leads to soil erosion and further land degradation. In the GCC, this can include how poor waste management increases public health costs and environmental burdens.
Trade-offs and decision making	Evaluating sustainability choices in terms of benefits and unintended consequences, such as biofuel policies leading to deforestation. Desalination, while solving water scarcity, can cause marine ecosystem stress through brine discharge.
Application to real-world issues	Applying theoretical knowledge to practical challenges like urban sustainability, climate change, and food systems. Students can explore smart city solutions in Riyadh or Masdar City as models.
Multidisciplinary integration	Incorporating knowledge from environmental science, economics, and social sciences to understand sustainability comprehensively. This supports GCC-wide educational reform integrating STEM with environmental ethics.
Stakeholder involvement	Engaging communities, businesses, and governments to address sustainability challenges collaboratively. Green initiatives in UAE schools involving NGOs exemplify this approach.
Long-term perspective	Encouraging forward-thinking approaches to sustainability that consider future generations. This aligns with Vision 2030 and national strategies across GCC states.

Source: Authors work

For instance, a student studying water scarcity in the GCC may initially attribute the problem solely to limited rainfall. However, through systems thinking, they begin to link water scarcity to energy-intensive desalination, inefficient agricultural practices, population growth, and the broader implications of climate change. This approach highlights how interventions in one domain such as increased energy consumption can yield unintended consequences in another, underscoring the need for integrated solutions (Karaarslan Semiz, 2020).

Moreover, systems thinking encompasses the identification of reinforcing feedback loops. A region-specific example includes over-extraction of groundwater in arid GCC regions, which reduces aquifer levels, leading to land subsidence and increased salinity. These feedback loops guide students in recognizing self-perpetuating environmental degradation and identifying points of intervention (Puspitasari et al., 2024).

Additionally, sustainability challenges in the GCC are deeply intertwined. For example, water and energy are co-dependent: water is essential for cooling in power generation and oil refining, while energy powers desalination plants crucial for drinking water. Understanding such linkages allows learners to consider holistic approaches such as solar-powered desalination as viable alternatives that support both environmental and economic sustainability goals (Olabi et al., 2023).

There are also trade-offs that must be critically analyzed. For instance, expanding food production in the desert might improve food security but increase water stress and greenhouse gas emissions. While global examples like agroforestry are informative, in the GCC, sustainable agriculture could include hydroponics or precision irrigation using AI technologies. Conversely, policies promoting biofuels may appear beneficial but can result in food insecurity or loss of biodiversity due to land-use changes, an issue applicable to palm-based biofuel investments in Gulf agricultural experiments (Jeswani et al., 2020).

Furthermore, students can apply systems thinking to GCC-specific urban challenges. Learning how green public transportation initiatives reduce air pollution, lower carbon emissions, and improve social mobility helps students connect environmental, economic, and equity outcomes. Educators must therefore play a key role by integrating project-based learning, simulations, and modeling tools that reflect the GCC's socio-ecological context (Karaarslan Semiz, 2020). These pedagogical tools not only enhance knowledge of global challenges but also strengthen students' capacity to develop localized, sustainable solutions.

GCC sustainability education must therefore be inherently interdisciplinary, integrating environmental science, economics, engineering, and policy analysis. Students should learn to apply sustainability practices such as smart irrigation, AI-based waste management, and renewable energy deployment, while simultaneously addressing economic growth, resource efficiency, and climate adaptation (Amin et al., 2023; D'Amato & Korhonen, 2021). Such integrative learning fosters critical thinking and aligns with the region's long-term strategic sustainability objectives (Shayan et al., 2022).

Equally important is experiential learning that connects education with real-world sustainability efforts. Internships with clean-tech companies, collaborative policy labs, and NGO-led community projects allow students to engage with GCC sustainability transitions firsthand. This prepares future professionals for green job markets and supports responsible consumption, carbon neutrality, and innovation (Abo-Khalil, 2024).

To mainstream systems thinking, sustainability must be institutionalized within GCC education systems (Awashreh & Al Ghunaimi, 2025b). This includes embedding environmental ethics and sustainability modules at all levels, encouraging student participation in policymaking, and promoting civic engagement through environmental clubs and local green initiatives (Kioupi & Voulvoulis, 2019). Such efforts help transform passive learners into active agents of change.

Ultimately, a systems-thinking-based model empowers students to become sustainability leaders in their organizations and communities. By cultivating a mindset that prioritizes long term environmental stewardship, socio-economic equity, and technological resilience, students will be equipped to navigate the GCC's complex future. Activities such as strategic foresight, problem-based scenarios, and climate simulations nurture adaptive leadership essential for achieving national visions and the UN Sustainable Development Goals (Awashreh, 2025; Bourmistrov & Åmo, 2022).

5 Strategies for Implementing the Interdependence Approach in GCC Education

To effectively embed the interdependence model of sustainability education within GCC institutions, a strategic and phased approach is essential. This section outlines four key, high-impact strategies supported by real-world practices and aligned with the region's socio-political context. First, curriculum transformation is fundamental. GCC universities should gradually integrate sustainability principles across disciplines, enabling students to connect environmental, economic, and social dimensions of development. This involves developing interdisciplinary modules that link subjects such as environmental science, economics, and public policy, using systems thinking and project-based learning. Such experiential learning fosters critical thinking and active problem-solving, while also connecting learners with local community, business, and government initiatives (George, 2022; Biasutti et al., 2016; Christou et al., 2024). However, this process requires coordinated curriculum committees, faculty training, and accreditation reforms highlighting the need for institutional support and phased implementation.

Second, leveraging digital technologies can significantly enhance the reach and quality of sustainability education. Tools such as AI, big data analytics, and virtual platforms can facilitate interactive, collaborative learning across GCC institutions. For example, AI-powered simulations can help students model climate or energy policy outcomes, while online platforms support regional knowledge-sharing (Annelin & Boström, 2024). Nevertheless, challenges such as digital infrastructure gaps and faculty digital readiness must be addressed through investment and professional development. Third, universities must lead by example through campus sustainability initiatives. Implementing renewable energy systems, green-certified buildings (e.g., LEED), and sustainable transportation sets a tangible standard for students. Additionally, resource efficiency practices such as composting, recycling, and water conservation can institutionalize sustainability norms (Green et al., 2016; Cole et al., 2023; Barman et al., 2023; Rodríguez-Guerreiro et al., 2024). Yet, such infrastructure transitions must be tailored to financial and climatic constraints of each GCC country and supported by strong leadership and monitoring mechanisms.

Fourth, student engagement is essential for cultural transformation. Universities should empower students to lead sustainability initiatives by providing grants, mentorship, and physical space for projects targeting local issues such as energy use, food security, and waste reduction. Embedding sustainability concepts into diverse disciplines will enhance students' ability to tackle global challenges post-graduation (Uzorka et al., 2024; Marcone, 2022). Building this bottom-up momentum requires a shift in institutional culture, from passive instruction to active learning communities. Finally, regional and international collaboration is a strategic imperative. GCC universities should form regional consortia to address shared challenges, such as desertification and water scarcity, through joint research and policy proposals. Moreover, partnerships with global institutions can enable knowledge transfer and contextual adaptation of successful sustainability models, accelerating progress (Zaidan et al., 2019; Al-Sarihi & Mansouri, 2022; Faccia et al., 2023). However, political and institutional coordination barriers may require long-term diplomatic and educational strategies.

At a theoretical level, this transformation aligns with social learning theory and constructivist pedagogy, which emphasize learner agency and real-world application. This implies a paradigmatic shift from passive information delivery to active, interdependent learning environments where sustainability becomes embedded in institutional identity and behavior over time.

6 Conclusion

In summary, the GCC countries face complex and interconnected sustainability challenges that demand strategic, context-sensitive solutions. The adoption of the interdependence model for sustainability education presents an opportunity to develop future leaders capable of navigating environmental, economic, and social systems in an integrated and adaptive manner. This model fosters deep learning, critical thinking, and systems-based approaches that enable students to anticipate long-term and intergenerational outcomes of their actions. Furthermore, it encourages cross-sector collaboration, positioning universities as innovation hubs for sustainability research across disciplines including environmental science, economics, engineering, social policy, and governance.

However, despite its potential, the implementation of this model faces several structural and contextual limitations. A major obstacle lies in the traditional, discipline-siloed organization of many universities, which limits the development of interdisciplinary programs essential for sustainability education. To overcome this, institutional reforms are needed to embed interdisciplinary collaboration within both curricula and research agendas.

Additionally, significant disparities exist in institutional capacity across GCC universities. While some institutions have taken initial steps to integrate sustainability into their academic frameworks, others especially smaller or underfunded universities lack the financial, infrastructural, and human resources to do so effectively. This uneven capacity presents a critical barrier to region-wide adoption of sustainability education.

To address these gaps, policymakers and university leaders should prioritize the development of sustainability education in alignment with local environmental, economic, and social needs. This includes tailoring programs to regional priorities such as water scarcity, energy efficiency, and desertification, and working closely with government and industry partners to ensure relevance and applicability.

Based on the findings, three key recommendations are proposed:

1. Develop region-specific interdisciplinary curricula that integrate sustainability with real-world challenges and promote applied, project-based learning.
2. Establish collaborative platforms between universities, governments, and industries to support joint research, faculty development, and policy innovation.
3. Implement phased infrastructure investment to support green campuses, digital sustainability tools, and faculty training in under-resourced institutions.

This study acknowledges certain methodological limitations. The exclusive reliance on secondary data may not capture all contextual nuances or institutional practices across the GCC. Future research should include primary data collection, such as stakeholder interviews and case studies, to deepen understanding of institutional barriers and implementation pathways. Longitudinal studies of graduate outcomes and program effectiveness are also recommended to inform continuous improvement.

In conclusion, while challenges remain, the interdependence approach offers a valuable roadmap for embedding sustainability in GCC higher education. By aligning educational frameworks with local priorities and fostering inclusive, interdisciplinary learning environments, universities can play a pivotal role in advancing sustainable development across the region.

References

Abo-Khalil, A. G. (2024). Integrating sustainability into higher education: Challenges and opportunities for universities worldwide. Heliyon, 10(9), e29946. https://doi.org/10.1016/j.heliyon.2024.e29946

Al-Hail, M., Zguir, M. F., & Koç, M. (2024). Exploring digital learning opportunities and challenges in higher education institutes: Stakeholder analysis on the use of social media for effective sustainability of learning–teaching–assessment in a university setting in Qatar. *Sustainability, 16*(15), 6413. https://doi.org/10.3390/su16156413

Alhammadi, A., Alsyouf, I., Semeraro, C., & Obaideen, K. (2024). The role of industry 4.0 in advancing sustainability development: A focus review in the United Arab Emirates. *Cleaner Engineering and Technology, 18*, 100708. https://doi.org/10.1016/j.clet.2023.100708

Al-Saidi, M. (2022). Disentangling the SDGs agenda in the GCC region: Priority targets and core areas for environmental action. *Frontiers in Environmental Science, 10*, 1025337. https://doi.org/10.3389/fenvs.2022.1025337

Al-Sarihi, A., & Mansouri, N. (2022). Renewable energy development in the Gulf Cooperation Council countries: Status, barriers, and policy options. *Energies, 15*(5), 1923. https://doi.org/10.3390/en15051923

Amin, H., Zaman, A., & Tok, E. (2023). Education for sustainable development and global citizenship education in the GCC: A systematic literature review. *Globalisation, Societies and Education*, 1–16. https://doi.org/10.1080/14767724.2023.2265846

Annelin, A., & Boström, G.-O. (2024, July 10). Interdisciplinary perspectives on sustainability in higher education: A sustainability competence support model. *Frontiers in Sustainability, 5*. https://doi.org/10.3389/frsus.2024.1416498

Awashreh, R. (2025). Leadership in the pursuit of net-zero emissions: Overcoming barriers and embracing strategic innovation. Arab Gulf Journal of Scientific Research. Advance online publication. https://doi.org/10.1108/AGJSR-02-2025-0030

Awashreh, R., & Al Ghunaimi, H. (2025a). Bridging cultural gaps: Enhancing student motivation and academic integrity in Oman's universities. *Futures Learning and Studies, 7*(2). https://doi.org/10.30564/fls.v7i2.8300

Awashreh, R., & Al Ghunaimi, H. (2025b). Rethinking the PhD: Aligning higher education with market needs and practical skills. *Forum for Linguistic Studies, 7*(2), 254–264. https://doi.org/10.30564/fls.v7i2.7942

Barman, P., Dutta, L., Bordoloi, S., Kalita, A., Buragohain, P., Bharali, S., & Azzopardi, B. (2023). Renewable energy integration with electric vehicle technology: A review of the existing smart charging approaches. *Renewable and Sustainable Energy Reviews, 183*, 113518. https://doi.org/10.1016/j.rser.2023.113518

Biasutti, M., De Baz, T., & Alshawa, H. (2016). Assessing the infusion of sustainability principles into university curricula. *Journal of Teacher Education for Sustainability, 18*(2). https://doi.org/10.1515/jtes-2016-0012

Bourmistrov, A., & Åmo, B. W. (2022). Creativity, proactivity, and foresight. Technological Forecasting and Social Change, 174, 121215.

Christou, O., Manou, D. B., Armenia, S., Franco, E., Blouchoutzi, A., & Papathanasiou, J. (2024). Fostering a whole-institution approach to sustainability through systems thinking: An analysis of the state-of-the-art in sustainability integration in higher education institutions. *Sustainability, 16*(6), 2508. https://doi.org/10.3390/su16062508

Cole, L. B., Fallahhosseini, S., Zangori, L., & Oertli, R. T. (2023). Learnscapes for renewable energy education: An exploration of elementary student understanding of solar energy systems. *Interdisciplinary Journal of Environmental and Science Education, 19*(1), e2305. https://doi.org/10.29333/ijese/13034

Cong, Y., & Ren, C. (2023). Does natural resource dependency impede sustainable development? Exploring the non-linear consequence of economic complexity. *Resources Policy, 85*(B), 103972. https://doi.org/10.1016/j.resourpol.2023.103972

D'Amato, D., & Korhonen, J. (2021). Integrating the green economy, circular economy and bioeconomy in a strategic sustainability framework. *Ecological Economics, 188*, 107143. https://doi.org/10.1016/j.ecolecon.2021.107143

Elgendy, K. (2024, June 6). *Bracing for the sandstorm: The Gulf energy transition imperative*. Arab Center Washington DC. https://arabcenterdc.org

ElMassah, S., & Hassanein, E. A. (2023). Economic development and environmental sustainability in the GCC countries: New insights based on the economic complexity. *Sustainability, 15*(10), 7987. https://doi.org/10.3390/su15107987

Elrahmani, A., Hannun, J., Eljak, F., & Kazi, M.-K. (2021). Status of renewable energy in the GCC region and future opportunities. *Current Opinion in Chemical Engineering, 31*, 100664. https://doi.org/10.1016/j.coche.2020.100664

Evans, S., Stevenson, R. B., Lasen, M., Ferreira, J. A., & Davis, J. (2017). Approaches to embedding sustainability in teacher education: A synthesis of the literature. *Teaching and Teacher Education, 63*, 405–417. https://doi.org/10.1016/j.tate.2017.01.013

Faccia, A., Le Roux, C. L., & Pandey, V. (2023). Innovation and e-commerce models, the technology catalysts for sustainable development: The Emirate of Dubai case study. *Sustainability, 15*(4), 3419. https://doi.org/10.3390/su15043419

George, J. (2022). *Sustainability education: Are UAE schools doing enough?* Education Middle East. https://educationmiddleeast.com/features/sustainability-education-are-uae-schools-doing-enough/

Gras-Velazquez, A., & Fronza, V. (2020). Sustainability in formal education: Ways to integrate it now. *IUL Research, 1*(2), 154–175. https://doi.org/10.57568/iulres.v1i2.84

Green, A., Wilson, I., & Bothun, G. (2016). Achieving sustainability through rooftop solar electricity generation on the college campus: A case study. In *Challenges in higher education for sustainability* (pp. 299–316). Springer. https://doi.org/10.1007/978-3-319-23705-3_15

Iacovidou, E., Hahladakis, J. N., & Purnell, P. (2021). A systems thinking approach to understanding the challenges of achieving the circular economy. *Environmental Science and Pollution Research, 28*(24), 1–22. https://doi.org/10.1007/s11356-020-11725-9

International Monetary Fund, Middle East and Central Asia Dept. (2024). *Gulf Cooperation Council: Pursuing visions amid geopolitical turbulence: Economic prospects and policy challenges for the GCC countries* (Vol. 2024, Issue 066). International Monetary Fund. https://doi.org/10.5089/9798400295744.007

Jeswani, H. K., Chilvers, A., & Azapagic, A. (2020). Environmental sustainability of biofuels: A review. *Proceedings of the Royal Society A: Mathematical, Physical and Engineering Sciences, 476*(2243), 20200351. https://doi.org/10.1098/rspa.2020.0351

Karaarslan Semiz, G. (2020). Systems thinking research in science and sustainability education: A theoretical note. In *Transition to quality education*. MDPI. https://doi.org/10.3390/books978-3-03897-893-0-3

Khodadadi, A. H. (2025, January 16). *Green vs. global: Can sustainability and economic growth go hand in hand?* Future Policy & Economics. Earth.Org. https://earth.org/green-vs-global/

Kioupi, V., & Voulvoulis, N. (2019). Education for Sustainable Development: A Systemic Framework for Connecting the SDGs to Educational Outcomes. Sustainability, 11, 6104. https://doi.org/10.3390/su11216104

Leal Filho, W., Raath, S., Lazzarini, B., Vargas, V. R., de Souza, L., Anholon, R., Quelhas, O. L. G., Haddad, R., Klavins, M., & Orlovic, V. L. (2018). The role of transformation in learning and education for sustainability. *Journal of Cleaner Production, 199*, 286–295. https://doi.org/10.1016/j.jclepro.2018.07.017

Marcone, G. (2022). Humanities and social sciences in relation to sustainable development goals and STEM education. Sustainability, 14(6), 1–14. https://doi.org/10.3390/su14063279

Marcos-Sánchez, R., Ferrández, D., & Morón, C. (2022). Systems thinking for sustainability education in building and business administration and management degrees. *Sustainability, 14*(19), 11812. https://doi.org/10.3390/su141911812

Martins, F. S., Carneiro da Cunha, J. A., & Serra, F. A. R. (2018). Secondary data in research – Uses and opportunities. *Revista Ibero-Americana de Estratégia, 17*(4), 1–4. https://doi.org/10.5585/ijsm.v17i4.2723

Mokski, E., Leal Filho, W., Sehnem, S., & Guerra, J. B. S. O. d. A. (2022). Education for sustainable development in higher education institutions: An approach for effective interdisciplinarity. *International Journal of Sustainability in Higher Education, 24*(1). https://doi.org/10.1108/IJSHE-07-2021-0306

Olabi, A. G., Elsaid, K., Obaideen, K., Abdelkareem, M. A., Rezk, H., Wilberforce, T., Maghrabie, H. M., & Sayed, E. T. (2023). Renewable energy systems: Comparisons, challenges and barriers, sustainability indicators, and the contribution to UN sustainable development goals. *International Journal of Thermofluids, 20*, 100498. https://doi.org/10.1016/j.ijft.2023.100498

Onyeaka, H., & Akinsemolu, A. A. (2024). Advancing green education in the MENA region: Challenges, opportunities, and best practices. *Sustainable Development.* https://doi.org/10.1002/sd.3182

Puspitasari, R., Nurmalina, R., Hariyadi, H., & Agustian, A. (2024, September 18). Systems thinking in sustainable agriculture development: A case study of garlic production in Indonesia. *Frontiers in Sustainable Food Systems, 8.* https://doi.org/10.3389/fsufs.2024.1349024

Rodríguez-Guerreiro, M.-J., Torrijos, V., & Soto, M. (2024). A review of waste management in higher education institutions: The road to zero waste and sustainability. *Environments, 11*(12), 293. https://doi.org/10.3390/environments11120293

Shayan, F. N., Mohabbati-Kalejahi, N., Alavi, S., & Zahed, M. A. (2022). Sustainable Development Goals (SDGs) as a Framework for Corporate Social Responsibility (CSR). *Sustainability, 14*(3), 1222. https://doi.org/10.3390/su14031222

Uzorka, A. (2024). Perspectives on online education in higher education. International Journal of Technology in Education and Science (IJTES), 8(1), 111-120. https://doi.org/10.46328/ijtes.523

Weber, J. M., Lindenmeyer, C. P., Liò, P., & Lapkin, A. A. (2021). Teaching sustainability as a complex systems approach: A sustainable development goals workshop. *International Journal of Sustainability in Higher Education, 22*(8), 25–41. https://doi.org/10.1108/IJSHE-06-2020-0209

White, P. J., Ardoin, N. M., Eames, C., & Monroe, M. C. (2024). Agency in the Anthropocene: Education for planetary health. *The Lancet Planetary Health, 8*(2), e117–e123. https://doi.org/10.1016/S2542-5196(23)00271-

Winston, N. (2021). Sustainable community development: Integrating social and environmental sustainability for sustainable housing and communities. *Sustainable Development, 30*(1). https://doi.org/10.1002/sd.2238

World Bank. (2024, May 29). *New GCC economic update finds improved quality of education is critical for sustained economic growth.* World Bank. https://www.worldbank.org/en/news/press-release/2024/05/29/new-gcc-economic-update

World Bank Group. (2021). *Country private sector diagnostic: Volume I: Main report.* https://documents1.worldbank.org/curated/en/099033206292221012/pdf/IDU0eddeb4dd009380497b080a009e10329c3c21.pdf

Zaidan, E., Al-Saidi, M., & Hammad, S. H. (2019). Sustainable development in the Arab world – Is the Gulf Cooperation Council (GCC) region fit for the challenge? *Development in Practice, 29*(5), 670–681. https://doi.org/10.1080/09614524.2019.1628922

CHAPTER 19

Environmental Education Through Clubs and Climate Resilience: A Statistical Analysis of Activation Factors

Amine Hmid and Driss Abbadi

1 Introduction

In the 1960s, the primary actors in Environmental Education (EE) were ecologists and naturalist scientists, each within their respective fields, working to promote a message of respect and protection for nature (Baldi & al., 2013). The institutionalization of environmental issues strengthened with the creation of ministries of nature protection and environment in several countries, as well as the introduction of environmental studies as a new school subject. The joint efforts of naturalists, educators, and nature protection organizations culminated in the organization of the first United Nations Conference on the Human Environment, held in Stockholm in 1972 (Kiss et al., 1972).

A. Hmid (✉) • D. Abbadi
Faculty of Legal, Economic, and Social Sciences of Ain Sebâa, Hassan 2 University, Casablanca, Morocco
e-mail: amine.hmid1-etu@etu.univh2c.ma

K. Hirech (ed.), *Environmental Education and Sustainable Development*, https://doi.org/10.1007/978-3-032-05760-0_19

The first United Nations Conference on the Human Environment represents the first official historical milestone of Environmental Education (EE) on an international scale. Among the recommendations of this conference, it was emphasized that an international framework should be established to develop EE, particularly in response to the global environmental crisis. This awareness sparked a wide international debate, leading to several regional meetings. The "Belgrade Charter" defined for the first time the goal of environmental education: "To educate a world population that is aware of environmental issues and capable of working both individually and collectively to solve current problems and prevent future challenges." In 1977, the Tbilisi Conference expanded this perspective, focusing on the importance of the environment beyond its problematic aspects (Bekhat et al., 2020). One of the major challenges of contemporary education is to foster values, behaviors, and attitudes that contribute to building a more sustainable, livable, and cohesive world for present and future generations (Kimengele et al., 2025).

Environmental education (EE) is increasingly positioned as a strategic entry point for building climate resilience, especially among youth. In recent years, environmental clubs in schools and communities have emerged as participatory platforms where environmental awareness, civic engagement, and local adaptation strategies intersect (Bekhat et al., 2020; Mousavi et al., 2024). This study examines the activation factors that make these clubs effective, using a critical theoretical framework grounded in contemporary scholarship. Drawing on recent peer-reviewed articles, we articulate four complementary lenses transformative learning, Indigenous knowledge systems, posthumanist thinking, and decolonial perspectives to understand how environmental clubs contribute to climate resilience. Transformative learning emphasizes the role of critical reflection, emotional engagement, and identity development in shaping learners' environmental behavior. As shown in article of Luo and other authors on 2025, environmental education initiatives that include reflexive dialogue, problem-solving activities, and scenario analysis significantly enhance students' capacity to respond to climate challenges (Luo et al., 2025). Similarly, Hu in her article, "Environmental Education in Low-Income and Middle-Income Countries: A Systematic Review and Meta-analysis," finds that club-based participatory learning fosters agency, resilience, and social responsibility. These findings support our inclusion of variables such as critical reflection, participatory action, and collaborative problem-solving as core activation factors within environmental clubs (Hu et al., 2025).

A growing body of research recognizes the importance of Indigenous and local knowledge systems in enhancing climate resilience (Moser, 2010). Tiemann highlights how community-based education programs that integrate traditional ecological knowledge (TEK) contribute to adaptive capacity in marginalized regions (Tiemann & Ring, 2025). Eregha and other authors further demonstrate that students exposed to Indigenous environmental practices develop deeper ecological understanding and intergenerational solidarity. In the context of environmental clubs, these insights inform activation variables such as cultural relevance, intergenerational learning, and place-based knowledge, which are statistically analyzed in this study (Eregha et al., 2024). Posthumanist theories call for a shift beyond anthropocentric views, emphasizing the interconnectedness of human and non-human entities. Other studies explore the affective and sensorial dimensions of environmental learning, where students build emotional ties with ecosystems through immersive, hands-on experiences. These articles reveal that such emotional bonds can trigger long-term behavioral shifts and environmental stewardship. This aligns with our empirical investigation into the role of emotional connectedness to nature as a statistically significant factor in club activation and sustained engagement (Eregha et al., 2024; Gebrekidan, 2024). Decolonial frameworks critique dominant, Eurocentric models of education and advocate for learning approaches rooted in local needs, knowledge, and agency. Igei's study emphasizes how decolonial pedagogies empower learners to challenge structural inequalities and mobilize for environmental justice (Igei et al., 2024). Likewise, Article of Eregha illustrates how environmental clubs can serve as spaces for community-led adaptation and resistance to extractivist practices. These insights justify the inclusion of community leadership, local ownership, and justice orientation as critical dimensions in our analysis of activation factors (Eregha et al., 2024). Additional contributions, such as the articles by Ekren et al. (2025) and Igei et al. (2024), provide empirical evidence linking club-based education to enhanced environmental risk perception and proactive behavior. Similarly, the studies by Salazar et al. (2024) and Zouine et al. (2024) further reinforce the connection between educational design and climate action literacy, offering statistical validation for the activation metrics employed in this study.

Together, these perspectives offer a robust, multidimensional framework that allows us to interpret how and why environmental clubs activate youth toward climate-resilient behaviors. By situating our statistical analysis within this theoretical foundation, we aim not only to quantify patterns of participation and impact but also to critically engage with the broader epistemological and pedagogical dynamics at play in contemporary environmental education.

At the Moroccan level, the integration of environmental education into school curricula was achieved following the directives of the National Charter for Education and Training in 1999 (Cherai et al., 2017; Choukri et al., 2011).

Schools, as primary sites for shaping the foundational values of society, must integrate environmental education through concrete measures aimed at establishing effective environmental policies, especially in countries like Morocco, which is highly vulnerable to climate change and actively working to align, coordinate, and plan its policies toward sustainable development (Amaghouss et al., 2024). As a result, various themes and concepts related to the environment have been addressed in the curricula of several disciplines, primarily History-Geography and Life and Earth Sciences, across different educational levels (Hamouchi et al., 2010). Research conducted over the last decade by Moroccan scholars involved in environmental issues (Khzami & Ben-Fares, 2009; Hamouchi et al., 2011; Agorram et al., 2013; Agorram et al., 2016; Andrade et al., 2025) has shown that attempts have been made to integrate this dimension into the school programs. However, these efforts often remain theoretical, primarily leading to environmental knowledge that, although present in textbooks, is difficult to apply concretely. As a result, these lessons are limited to knowledge accumulation without effectively fostering the development of eco-responsible skills, behaviors, or attitudes.

Other studies, based on an analysis of textbooks, revealed that environmental education, supposedly integrated into the Life and Earth Sciences curriculum at the secondary level, remains mainly focused on the subject of the environment. It relies on a passive transmission of knowledge about environmental realities, often presented from an ethical perspective (Hamouchi, 2016). The scientific literature typically distinguishes three types of relationships with the environment: education about the environment, education for the environment, and education through or in the environment (Ziaka et al., 2002; Mamadou, 2025).

The strengthening of environmental integration in Morocco has also been achieved through extracurricular activities. These activities play an essential role in school life. Linked to education, they are implemented by schools to promote and develop students' knowledge and skills. These activities cover various cognitive, emotional, and psychomotor aspects and take place within educational structures in the form of educational clubs, which are responsible for organizing and supervising extracurricular activities. These clubs provide a methodological framework to energize school life, involving various stakeholders: students, educational and administrative staff, and various partners. They become spaces for the development of social and psychological skills, communication, teamwork, citizenship values, and civic behavior. Working within these educational clubs is an essential pedagogical approach, encouraging students to transfer and apply their school learning to their daily lives, while placing them at the center of reflection and pedagogical action. Environmental education finds its place in these extracurricular activities, particularly through environmental clubs. These clubs, managed by volunteer teachers who are committed to the importance of the environment, play a key role in transmitting ecological values and empowering students with respect to their environment. They promote the institutionalization of their activities into concrete actions, fostering cooperation among different school actors and allowing students to acquire contextualized knowledge about the environment.

Environmental education, particularly within school-based environmental clubs, plays a crucial role in raising awareness and engaging younger generations in the fight against environmental and climate-related challenges. In the Moroccan context, where the effects of climate change are already being felt through phenomena such as desertification, heatwaves, and prolonged droughts, the issue of climate resilience has become critical. Resilience refers to the ability of communities to anticipate, adapt to, and recover from the negative impacts of climate change.

Environmental education (EE) is no longer limited to the mere dissemination of information but seeks to encourage in-depth reflection on the human causes of environmental crises, as highlighted by Smyth et al. (2004). This holistic approach goes beyond intellectual awareness, pushing for concrete actions to reduce the ecological footprint, as suggested by Fien (1993). The study results confirm that encouraging responsible behavior is essential.

Tilbury (1995) emphasizes the importance of EE for the responsible management of natural resources and for raising students' awareness of the interdependence of ecosystem elements. In this regard, EE should be transversal, integrated into all school curricula, and not limited to a single subject, as Sterling (2001) underscores. UNESCO (2005) adds that the ultimate goal of EE is to train responsible citizens, capable of making informed decisions to preserve the planet.

Climate change, as a global and long-term disruption of the Earth's climate systems, has become a central concern in environmental education (EE). In Morocco, this urgency has been reflected in educational policy: since the 2020 school year, the integration of climate change education into environmental education frameworks has been reinforced in official texts (Déduit & Polo, 2022). This aligns with broader perspectives in the literature, where scholars such as Huckle et al. (1998) argue that EE plays a crucial role in raising awareness of pressing environmental issues like biodiversity loss and climate change. Lundholm and Plummer (2009) further highlight that such awareness is a necessary precursor to meaningful action.

Indeed, climate change education must first become an educational space where citizens are informed and made aware of both the causes and the consequences of climate change. This includes a particular focus on the abnormal increase in extreme weather events and the many forms of damage they inflict on ecosystems, biodiversity, and human societies as a whole (Agundez-Rodríguez & Sauvé, 2022). Such an approach underlines the need for education that not only informs but also fosters critical thinking and engagement.

However, the practical implementation of EE initiatives such as environmental clubs and climate-focused programs often encounters significant barriers. These include limited resources, time constraints, and inadequate funding (Wals, 2007; Rickinson et al., 2004). Moreover, the success of these initiatives relies heavily on the quality of teacher training and the availability of continuous professional support, as emphasized by Palmer (1998) and Bengtsson et al. (2009). In this context, Morocco's curricular reinforcement of climate change education represents a valuable step, but its long-term impact will likely depend on addressing these persistent structural challenges.

In this context, it becomes essential to critically examine the role of sustainable development within the educational framework and advocate for increased emphasis on Environmental and Sustainability Education (Ayotte, 2020). Understanding and implementing these distinct approaches can enhance the effectiveness of environmental education and ultimately support the broader goals of sustainable development (Nicol et al., 2000; Girault & Sauvé, 2008; Simmons et al., 2008; Mammino, 2025) .

However, to assess the effectiveness of these educational actions, it is essential to understand the activation factors that influence teachers' and students' engagement in environmental clubs. These factors can be numerous: teachers' motivations, students' perceptions, available resources, institutional support, and the clubs' ability to incorporate climate issues into their daily activities. A statistical analysis of the responses from 42 teachers, through a questionnaire designed for this purpose, would help identify the factors that either facilitate or hinder involvement in environmental education and, consequently, in building climate resilience.

Thus, the central issue of this study is to understand how much the environmental education implemented by environmental clubs contributes to strengthening the climate resilience of students and the educational community, and what are the key factors that either activate or hinder this dynamic within the institution.

2 Research Methodology

2.1 Investigation Method

In terms of research methodology, we have chosen to adopt a mixed method approach, addressing both qualitative and quantitative aspects. This decision stems from the recognition that the study envisaged requires an in-depth exploration that cannot be captured exhaustively by a single methodological perspective. The joint use of quantitative and qualitative methods has therefore enabled us to gain a better understanding of the various aspects relating to the activation of environmental clubs and their role in climate change education. The quantitative method, based mainly on a structured questionnaire, made it possible to collect figures from respondents on specific dimensions.

At the same time, the qualitative dimension made it possible to contextualize the quantitative results, to explain the trends observed, and to bring out more subtle or complex elements, such as the internal dynamics of the clubs, interpersonal relations, or subjective representations of environmental change. Thanks to this complementary approach, the mixed method has provided a richer, more nuanced reading of the results, offering both a global vision and more specific insights into the experiences and local realities. It also strengthened the validity of the conclusions, by cross-referencing statistical data with qualitative accounts. This methodological choice proved particularly relevant in meeting the objectives of the study, namely to identify the determining factors in the effective activation of environmental clubs and to put forward concrete recommendations for improving and sustaining them in the school context.

2.2 *Representative Sample*

The participant selection strategy in this study is grounded in purposive sampling, a method characterized by informed and deliberate choice rather than randomization. Unlike probabilistic approaches that seek to minimize researcher influence, purposive sampling embraces the researcher's judgment in identifying individuals whose profiles closely align with predefined research criteria. This method enables the selection of participants who are particularly relevant to the study's objectives, thereby facilitating a form of analytical generalization, even if statistical representativeness is not the goal.

Accordingly, a total of 42 teachers and school administrators were selected based on specific criteria, including their demonstrated interest in environmental education (EE) and their potential to contribute meaningfully to its implementation within their institutions. Teachers, given their direct engagement with students, are central to incorporating environmental issues into everyday learning, combining theoretical knowledge with practical application to foster environmental awareness. Administrators, on the other hand, play a pivotal role in supporting and coordinating these efforts at the organizational level, helping to cultivate a school culture that encourages environmental responsibility and long-term commitment to sustainability initiatives.

2.3 *Data Collection*

In alignment with the research objectives, the selection of data collection tools in this study was carefully considered to ensure methodological coherence. Initially, document analysis was employed to explore the topic through existing literature, including academic articles, institutional reports, books, and other relevant written sources. This approach enabled the researcher to gather credible and accessible information without necessitating direct contact with participants.

Following this preliminary phase, a structured questionnaire was developed and administered to collect primary data from the selected participants. The instrument was distributed simultaneously to a targeted group, providing a practical and systematic means of data collection. A total of 42 questionnaires were disseminated, and all were successfully completed and returned, ensuring a full response rate.

To facilitate comprehension and reduce any potential language barriers, the questionnaire was administered in French, the participants' primary language. It was delivered via Google Forms, allowing for efficient distribution and data compilation. The questionnaire consisted of 26 items, organized into four thematic sections, each addressing key dimensions related to environmental education and the role of environmental clubs in promoting ecological awareness within the school context.

- Part 1: Personal Data
- Part 2: Perceptions regarding environmental clubs
- Part 3: Extracurricular activities and environmental clubs
- Part 4: Respondents' satisfaction
- Part 5: Barriers to the activation of environmental clubs

2.4 *Data Analysis*

To analyze the collected data, various methods were used to ensure a comprehensive approach. Descriptive analysis was first used to explore participants' perceptions of environmental education and environmental clubs, revealing common patterns and recurring themes within their responses. This approach helped to illuminate prevailing social and educational attitudes toward these initiatives.

To investigate the relationship between the presence of environmental clubs and training in environmental education, the chi-square test was applied. This statistical technique allowed for an assessment of the association between these two variables, providing insight into their interdependence.

Participants' satisfaction with the activation of environmental clubs was evaluated using a Likert scale, which addressed various factors such as administrative support, involvement of stakeholders, availability of resources, and student participation. Moreover, the Weighted Average Index (WAI) was calculated to rank the obstacles identified through the questionnaire, offering a clear picture of the relative importance and impact of each challenge.

Together, these methods offered a comprehensive understanding of the factors influencing the operation and effectiveness of environmental clubs and educational efforts within the studied context.

The formula for calculating the WAI is as follows:

$$\mathbf{WAIi} = \sum(\mathbf{gij} \times \mathbf{pj}) / \mathbf{Ni}$$

- WAIi is the weighted average index for barrier *i*
- gij is the severity assigned to barrier *i* by participant *j*
- pj is the weight corresponding to the severity *gij*
- Ni is the total number of evaluations for barrier *i*

3 Results and Discussions

3.1 Sample Processing

The analysis of the observational data confirmed the completeness and validity of all 42 cases, with no missing information recorded. This completeness ensures the robustness and reliability of the subsequent statistical analyses.

Cross-tabulations involving variables such as "*Professional Status*," "*Years of Experience*," "*Level of Education*," and "*Specialization*" revealed a balanced and comprehensive distribution without any gaps.

Examining the relationship between professional status and years of experience, the sample is composed predominantly of teachers, who represent 35 of the 42 participants (approximately 83%), while administrators account for the remaining 7 (around 17%). Among administrators, none fall within the five to ten years of experience range; three have under five years, and four have over ten years of experience. In contrast, a significant majority of teachers 25 out of 35 (71%) possess more than ten years of experience, with five teachers reporting less than five years and another five between five and ten years.

Overall, the teaching cohort is characterized by considerable experience, largely concentrated among those with over a decade in the field. Administrators, although fewer in number, display a more varied distribution across experience levels.

The intersection of specialization and education level indicates that most participants hold a Master's degree, accounting for 81% (34 out of

42). Doctorate and Bachelor's degree holders are equally represented, each comprising approximately 9.5% of the sample (four participants each).

Breaking down by specialty, all eight English specialists hold Master's degrees, while the four Mathematics specialists have attained Doctorates. History-Geography participants predominantly hold Master's degrees (10 individuals), as do those specializing in Physics-Chemistry (4 participants) and Life and Earth Sciences (3 participants). Participants in pedagogical, social, or administrative fields show more diversity, with two holding Bachelor's degrees and one a Master's. Other fields such as Physical Education, Philosophy, French, and Computer Science are less represented, with one or two participants each, mainly holding Master's degrees.

In summary, the sample is marked by a high level of academic qualification, with Master's degrees prevailing across most disciplines. The Doctorate degree is notably concentrated within Mathematics, underscoring advanced specialization in that area.

3.2 *Perceptions Related to EE and CC*

At the level of social perceptions, the findings reveal a multifaceted understanding of environmental education (EE), highlighting several key dimensions that capture the breadth and complexity of this pedagogical approach.

Awareness of ecological issues: One respondent described EE as "teaching ecological issues," pointing to an emphasis on imparting knowledge about current environmental challenges. This perspective frames EE as a means to raise learners' consciousness about global problems such as climate change, pollution, and biodiversity loss, encouraging not only theoretical comprehension but also critical reflection on the causes and implications of environmental degradation.

Encouraging environmentally responsible lifestyles: The call for fostering "a more environmentally friendly lifestyle" underscores a vital aspect of EE: promoting behavioral change. Environmental education extends beyond the mere transmission of information to actively influence daily practices such as recycling, mindful consumption, and sustainable resource use thereby positioning students as engaged participants in environmental stewardship.

Promoting environmental protection and sustainable resource management: References to "raising students' awareness of environmental protection and sustainable resource management" highlight a central objective of EE. This involves cultivating an understanding of the importance of preserving natural resources through systemic, sustainable practices, preparing learners to responsibly manage the environment in the long term.

Integrating environmental concerns into the curriculum: The incorporation of environmental themes across disciplines is seen as essential for a cohesive and comprehensive educational experience. Rather than treating EE as an isolated subject, it should be embedded throughout various academic areas such as sciences, geography, and history shaping both content and pedagogy to foster a holistic perspective on environmental challenges.

Education aimed at environmental protection: Finally, the view that EE "aims to protect the environment" reflects its outcome-driven nature. The ultimate goal is to nurture informed, responsible citizens who are committed to preserving the planet through thoughtful and concrete actions.

Examples of questions included in the questionnaire:

- In your view, what are the primary objectives of environmental education?
- What do you consider to be the main goals of climate change mitigation?
- To what extent do you believe environmental education can contribute to addressing climate change?

The study's findings on the objectives of environmental education (EE) reveal a strong consensus regarding the importance of raising awareness about ecological issues. Participants consistently emphasized the need to inform students about significant environmental challenges. Beyond awareness, EE is also seen as a means to cultivate responsible behaviors and practical skills, motivating students to actively participate in environmental protection. Although somewhat less emphasized, a solid theoretical understanding of ecosystems remains a vital foundation for comprehensively addressing these challenges.

With respect to climate change (CC), the majority of respondents (64.3%) identified reduced rainfall as a primary concern, linking it directly to disruptions in water cycles. Fewer participants acknowledged the rise in average temperatures, indicating a potential gap in awareness concerning this fundamental aspect of climate change. Nonetheless, many respondents recognized the association between CC and extreme weather events, highlighting its immediate and visible impacts.

A substantial majority (92.9%) agreed that environmental education constitutes a powerful instrument in the fight against climate change. However, a minority (7.1%) expressed some skepticism, suggesting that educational initiatives must be complemented by political and economic measures to achieve meaningful progress.

Regarding environmental clubs in Moroccan schools, responses underscored their valuable role in raising environmental awareness and fostering student engagement. Yet, the effectiveness of these clubs is hindered by challenges such as insufficient resources, lack of coordination, and limited institutional support. Addressing these obstacles will be crucial to ensuring their long-term success.

3.3 *Extracurricular Activities and EC*

Collaboration on climate change education:

The results presented in Table 19.1 indicate a notable level of teacher involvement in educational projects addressing climate change, highlighting an increasing acknowledgment of the significance of environmental education and sustainable development within school curricula. Most teachers appear to focus their initiatives on practical actions such as tree planting or the management of natural resources which not only raise

Table 19.1 Chi-square tests

Test	*Value*	*Df*	*Asymptotic significance (bilateral)*	*Exact Sig. (bilateral)*	*Exact Sig. (unilateral)*
Pearson Chi-Square	9.105	1	0.003		
Continuity Correction[b]	5.457	1	0.019		
Likelihood Ratio	8.724	1	0.003		
Fisher's Exact Test				0.014	0.014
Linear-by-Linear Association	8.888	1	0.003		
Number of Valid Cases	42				

students' awareness but also actively involve them in contributing to a more sustainable future.

However, the 28.57% of negative responses reveal that a considerable portion of teachers or schools have yet to incorporate these themes into their educational activities. This gap may be attributed to logistical limitations, pedagogical difficulties, or insufficient training related to environmental education.

The chi-square test is a statistical method used to assess whether a significant association exists between two categorical variables. The result presented in Table 19.1 and Fig. 19.1 with a p-value of 0.003 falls below the conventional significance threshold of 0.05. This indicates a statistically significant relationship between the presence of environmental clubs and training in environmental education.

This finding suggests that environmental clubs and environmental education training are meaningfully connected. It may imply that receiving training in environmental education contributes to the establishment or development of environmental clubs, or alternatively, that underlying factors influence both variables simultaneously.

The results of the chi-square (Table 19.2 and Fig. 19.2) test in this case, yielding a p-value of 0.491, indicate that there is no statistically significant association between the level of engagement of environmental clubs and training in environmental education. This suggests that, within the context of this study, such training does not appear to directly influence or correlate with the degree of activity or involvement of environmental clubs.

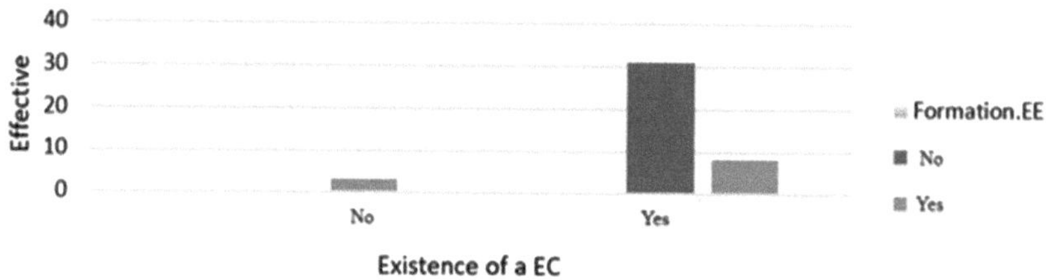

Fig. 19.1 Graphical representation of chi-square results

Table 19.2 Chi-square tests

Test	*Value*	*Df*	*Asymptotic significance (bilateral)*	*Exact Sig. (bilateral)*	*Exact Sig. (unilateral)*
Pearson Chi-square	0.474[a]	1	0.491	0.491	
Continuity correction[b]	0.113	1	0.737	0.737	
Likelihood ratio	0.479	1	0.489	0.489	
Fisher's exact test					0.726
Linear-by-linear association	0.463	1	0.496	0.496	
Number of valid cases	42				

[a]Pearson Chi-Square: This indicates the standard Pearson Chi-Square test, which assesses whether there is a statistically significant association between the two categorical variables. In this table, the Pearson Chi-Square value is 0.474 with 1 degree of freedom and a bilateral asymptotic significance of 0.491, indicating no significant association at the 0.05 level.

[b]Continuity Correction: This refers to the Yates' continuity correction applied to the Pearson Chi-Square test. It adjusts for the overestimation of chi-square values in small samples. Here, the Continuity Correction value is 0.113 with 1 degree of freedom and a bilateral asymptotic significance of 0.737, also indicating no significant association.

Thus, both superscripts "a" and "b" denote specific variations of the Chi-Square test, with "a" representing the standard Pearson test and "b" representing the continuity-corrected version for small samples.

Fig. 19.2 Graphical representation of chi-square results

3.4 *Satisfaction of Participation Related to the Activation Variables of the EC*

The results reveal a combination of strengths and areas for improvement that can guide future efforts to enhance the effectiveness of environmental clubs. With regard to administrative support, the majority of participants reported being fairly or fully satisfied, reflecting positive institutional backing. Nonetheless, a minority expressed dissatisfaction, indicating that there is still scope for strengthening this support.

Overall satisfaction with general support was also relatively high, although a small number of participants conveyed mild dissatisfaction, pointing to the need for targeted improvements. A significant concern emerged around the availability of resources: considerable dissatisfaction suggests that both material and financial support are inadequate, limiting the clubs' capacity to implement activities effectively.

Another key issue identified is the lack of recognition for the efforts of club members. Participants expressed frustration over the limited acknowledgment of their contributions, highlighting a need for greater institutional appreciation. Financial support remains a persistent challenge, with respondents indicating that insufficient funding undermines the sustainability and impact of environmental clubs.

On a more positive note, member engagement received generally favorable feedback, with most participants expressing satisfaction regarding their involvement. However, collective participation was identified as an area for further development. Enhancing collaboration among members could significantly boost the clubs' overall effectiveness.

Finally, time constraints were cited as a notable source of dissatisfaction, with some members feeling that the demands of club activities were excessively burdensome. Addressing these concerns, while building on existing strengths such as engagement and institutional support, will be essential to improving the long-term success and impact of environmental clubs.

3.5 *Barriers to the Activation of Environmental Clubs*

The results presented in Table 19.3 indicate that the primary barriers to the effective activation of environmental clubs in schools are related to time constraints, limited teaching resources, and insufficient funding. These are closely followed by challenges such as lack of training, weak coordination, and inadequate administrative support. While factors such

Table 19.3 Ranking of barriers to the activation of environmental clubs

Barrière	*WAI*
Lack of time	**4.79**
Limited educational resources	**4.64**
Lack of funding	**4.57**
Lack of training	**3.57**
Coordination issues	**3.50**
Lack of administrative support	**3.07**
Resistance to change and lack of interest	**2.71**

as lack of interest and resistance to change are perceived as less critical, they nonetheless remain relevant considerations.

These findings suggest that for environmental clubs to operate effectively and achieve their intended goals, targeted investments are required. This includes the provision of adequate material and financial resources, stronger institutional backing, and the development of specific competencies among club facilitators. Moreover, restructuring the school schedule to allocate dedicated time for environmental activities could significantly enhance club functioning.

Raising awareness about the value of environmental clubs, alongside reinforced administrative support, also emerges as essential for ensuring the long-term sustainability and impact of environmental education initiatives within schools. The analysis of the Weighted Average Index (WAI) in the table 19.3 highlights the primary obstacles hindering the activation of environmental clubs. Among the barriers, "Lack of time" received the highest WAI score (4.79), indicating that time constraints are perceived as the most significant challenge by respondents. Following closely are "Limited educational resources" (WAI = 4.64) and "Lack of funding" (WAI = 4.57), suggesting that resource limitations and financial support are also critical factors affecting the establishment and functioning of these clubs. Moderate barriers include "Lack of training" (3.57) and "Coordination issues" (3.50), reflecting that skill gaps and organizational challenges still impede effective implementation but are less severe compared to time and resource constraints. Lower-ranked barriers, such as "Lack of administrative support" (3.07) and "Resistance to change and lack of interest" (2.71), indicate that, although these factors are present, they are considered less influential in comparison to the more pressing structural and resource-related obstacles. Overall, the results suggest that initiatives aimed at promoting environmental clubs should prioritize time management strategies, resource provision, and funding support, while addressing training and coordination issues as secondary concerns to enhance club activation and sustainability.

3.6 *Analysis and Discussion of Results*

The findings of this study reveal a multifaceted understanding of Environmental Education (EE), encompassing awareness-raising, behavioral change, sustainable resource management, and curricular integration. One of the central outcomes is the strong emphasis on major ecological challenges, particularly climate change, pollution, and biodiversity loss. This aligns with Smyth et al. (2004), who argue that EE should extend beyond the transmission of factual knowledge to foster critical reflection on humanity's role in environmental degradation. Consistent with this view, the study shows that EE encourages learners not only to understand these issues theoretically but also to engage in thoughtful analysis of their causes and consequences.

A second key theme is the promotion of environmentally responsible lifestyles, reinforcing the action-oriented nature of EE. This reflects Fien's (1993) and Luo et al.'s (2025) assertion that EE must move beyond intellectual awareness to drive concrete behavioral change. In our findings, students are encouraged to adopt practical habits such as recycling and responsible consumption behaviors that align with the goal of reducing ecological footprints and promoting sustainable living.

Sustainable resource management also stands out as a key priority. As Tilbury (1995) highlights, environmental education (EE) should equip learners with a holistic understanding of ecological relationships while fostering responsible stewardship of natural resources. Our study reinforces this perspective by demonstrating that EE enables students to grasp the interconnectedness of ecosystems, thereby preparing them to manage resources both effectively and ethically. Schools, as pivotal institutions in shaping society's core values, must embed environmental education through concrete actions that support the development and implementation of robust environmental policies, particularly in countries like Morocco, which face high vulnerability to climate change and are actively striving to align, coordinate, and advance policies in line with sustainable development goals (Amaghouss et al., 2024).

Furthermore, the findings support Sterling's (2001) and Déduit and Polo's (2022) argument that EE should be integrated across the curriculum rather than treated as a discrete subject. Respondents emphasized the need for a cross-disciplinary approach, incorporating environmental themes into science, geography, history, and other subjects to foster a holistic understanding of environmental issues.

The ultimate aim of EE environmental protection is also affirmed by the study, in alignment with UNESCO (2005), which advocates for an education that empowers learners to engage actively in environmental preservation. The study illustrates that EE fosters both knowledge and action, equipping students to become informed, responsible citizens who can make environmentally conscious decisions.

In terms of social perceptions of EE's objectives, the study reveals a strong consensus on the importance of raising awareness about global environmental issues such as climate change and biodiversity loss. This aligns with Huckle et al. (1998), who argue that developing ecological awareness is a foundational step toward meaningful environmental action. While the theoretical understanding of ecosystems was mentioned less frequently, its presence indicates that a foundational ecological literacy remains important, even if immediate issues tend to dominate the discourse.

Regarding climate change, most respondents associated it with reduced rainfall, reflecting localized concerns over water scarcity especially relevant in drought-prone contexts. This finding resonates with Bennett et al. (2015), who note that climate change often manifests most visibly through disruptions in local water cycles. Conversely, fewer respondents identified rising global temperatures as a key feature of climate change, suggesting a knowledge gap. This may be explained by Gifford (2011), who argues that temperature shifts, being gradual and abstract, are less perceptible to the general public than more immediate phenomena like rainfall changes or extreme weather.

When asked about EE's effectiveness in combating climate change, a significant majority (92.9%) affirmed its importance in driving behavioral change and fostering awareness. This view supports Huckle et al.'s (1998) argument that EE is a vital component of broader efforts to address environmental challenges. However, the small percentage of skeptical responses (7.1%) underscores the perception that educational interventions alone are insufficient and must be accompanied by comprehensive political and economic strategies again echoing the concerns raised by Huckle et al.

The study also highlights the significant role played by environmental clubs in raising awareness and promoting student involvement. Nonetheless, several challenges were identified, including inadequate resources, weak coordination, and limited institutional support. These findings are consistent with previous research emphasizing the need for sustained investment, training, and institutional backing to ensure the success of school-based environmental initiatives. Overcoming these obstacles will require enhanced collaboration among educators, school leaders, and external partners, as well as a greater commitment to equipping schools with the tools and support necessary to sustain impactful environmental programs.

In conclusion, the findings of this study are broadly consistent with the existing literature, reinforcing the value of environmental education in raising awareness, promoting sustainable behaviors, and cultivating long-term engagement with ecological issues. At the same time, the results point to critical areas for improvement particularly the need to address practical constraints faced by environmental clubs and to fully integrate EE into mainstream educational practices.

4 Conclusion

Environmental education constitutes a foundational pillar of sustainable development, and its integration into formal education systems is essential for achieving the Sustainable Development Goals. In Morocco, while significant efforts have been made to promote environmental education particularly through the establishment of environmental clubs in schools various barriers continue to limit their effectiveness. These clubs are instrumental in raising awareness and fostering action on climate-related issues; however, their impact remains constrained by a lack of resources, insufficient funding, limited training, low levels of motivation, and time constraints.

The findings of this study confirm the presence of these key challenges, which hinder the successful activation and operation of environmental clubs. To enhance their effectiveness, it is imperative to strengthen training programs for teachers and facilitators, allocate adequate material and financial resources, and actively promote the motivation of all stakeholders involved. Additionally, improved financial support and more efficient management practices could significantly enhance the visibility and impact of these clubs in addressing climate change.

Several areas for improvement have been identified to ensure the long-term success of these initiatives. First, the development of partnerships with public institutions, private sector actors, and civil society organizations could help secure stable and sustainable funding. Expanding and diversifying training opportunities for teachers and club leaders is equally crucial, enabling them to manage and implement projects more effectively. Furthermore, the introduction of systematic monitoring and evaluation mechanisms would allow for continuous improvement and the refinement of club activities. Addressing time constraints is also critical: allocating dedicated time slots for environmental club activities and reducing scheduling conflicts for teachers and students would help ensure consistent participation.

Funding remains a major obstacle to the sustainability of these clubs. It is therefore essential to establish specific budgets at the institutional level and explore avenues for external financing. In parallel, the lack of adequate materials such as gardening tools, workshop equipment, and educational resources on environmental topics must be addressed. Investing in the necessary equipment would enhance the clubs' operational capacity and make their activities more engaging and impactful.

The findings of this study highlight a series of systemic barriers that hinder the effective implementation and sustainability of environmental clubs in schools. These challenges such as insufficient training, lack of funding, limited time allocation, and inadequate materials are not only operational issues but also reflect broader global challenges addressed by the United Nations' Sustainable Development Goals (SDGs), particularly in the context of climate education and environmental justice.

SDG 4: Quality Education emphasizes the importance of inclusive and equitable education and promotes lifelong learning opportunities for all. Target 4.7 specifically calls for ensuring that learners acquire the knowledge and skills needed to promote sustainable development, including through education for sustainable lifestyles, human rights, gender equality, and a culture of peace and non-violence. Strengthening environmental clubs through teacher training, structured activities, and appropriate resources directly supports this goal by equipping students with environmental literacy, critical thinking skills, and active citizenship.

SDG 13: Climate Action underlines the urgent need to combat climate change and its impacts. Environmental clubs represent practical, school-based initiatives that foster awareness, engagement, and action on climate-related issues among young people. Supporting these clubs contributes to

Target 13.3, which aims to improve education, awareness-raising, and human and institutional capacity on climate change mitigation and adaptation.

Moreover, the study touches on issues of environmental justice, as the lack of material and financial resources often disproportionately affects under-resourced schools and communities. Ensuring equitable access to high-quality environmental education is essential for addressing the uneven distribution of environmental burdens and opportunities. By advocating for institutional support, diversified partnerships, and stable funding, the study promotes a more just and inclusive approach to environmental learning.

In conclusion, this research highlights the need for a comprehensive and sustained strategy to unlock the full educational potential of environmental clubs. By aligning with key global goals and principles, namely SDG 4, SDG 13, and the broader vision of environmental justice, these initiatives can become powerful agents of change, fostering a generation that is informed, empowered, and actively engaged in building a sustainable and equitable future.

References

Agorram, B., Khzami, S., Selmaoui, S., Razouki, A., Arfaoui, M., & Caravita, S. (2013). Le concept "écosystème" dans les manuels scolaires du Maroc: Analyse comparative dans six pays méditerranéens. In *Proceedings of the 7th World Environmental Education Congress* (pp. 46–57). Marrakech.

Agorram, B., Zak, M., Selmaoui, S., & Khzami, S.-E. (2016). *An analysis of the treatment of evolution in Moroccan secondary textbooks.* The Eurasia Proceedings of Educational & Social Sciences (EPESS), 5, 286–288.

Agundez-Rodríguez, M.-H., & Sauvé, L. (2022). L'éducation relative au changement climatique: une lecture à la lumière du Pacte de Glasgow. *Éducation relative à l'environnement: Regards – Recherches – Réflexions, 17*(1). https://doi.org/10.4000/ere.8421

Amaghouss, J., Aalioua, A., & Aziz, K. (2024). Mesure et déterminants du comportement pro-environnemental des lycéens au Maroc: cas de la direction provinciale de Youssoufia. *Revue marocaine de l'évaluation et de la recherche éducative, 10.*

Améziane, N. (2018). Place of the environment in the Moroccan curricula of Life and Earth Sciences in secondary education. *The Journal of Quality in Education, 8(11), 19.* https://doi.org/10.37870/joqie.v8i11.153

Andrade, C., Stathopoulos, S., Mourato, S., Yamasaki, E. N., Paschalidou, A., Bernardo, H., Papaloizou, L., Charalambidou, I., Achilleos, S., Psistaki, K., Sarris, E., Carvalho, F., & Chaves, F. (2025). The role of interventions in enhancing indoor environmental quality in higher education institutions for student well-being and academic performance. *Current Opinion in Environmental Science & Health, 45*, 100611. https://doi.org/10.1016/j.coesh.2025.100611

Ayotte, C. (2020). *Engagement politique des jeunes dans la lutte aux changements climatiques: rôle de l'éducation relative à l'environnement*. Master's thesis, Université de Sherbrooke, CUFE. Essais et productions de fin d'études, Université de Sherbrook.

Baldi, I., Cordier, S., Coumoul, X., Elbaz, A., & Gamet-Payrastre, L. (2013). Pesticides : Effets sur la santé. Paris, France : Institut national de la santé et de la recherche médicale (INSERM).

Bekhat, B., Madrane, M., Janati-Idrissi, R., Zerhane, R., & Laafou, M. (2020). Améliorer l'enseignement de l'Éducation relative à l'Environnement dans le système éducatif marocain. *International Journal of Innovation and Applied Studies, 28*(2), 361–365.

Bengtsson, J., Kross, L., & Ostman, L. (2009). Educational strategies for sustainable development: A review. *Journal of Environmental Education, 40*(2), 91–104.

Bennett, L., Smith, J., & Albert, l. (2015). Understanding and responding to climate change: Insights from local experiences. *Environmental Education Research, 21*(1), 3–18.

Cherai, B., El Hnot, H., Idrissi, B. E. F., & Akkaoui, E. G. (2017). Les clubs d'environnement dans les établissements scolaires au Maroc: Une voie prometteuse pour l'ancrage de l'éducation à l'environnement. *European Scientific Journal, 13*(13), 337. https://doi.org/10.19044/esj.2017.v13n13p337

Choukri, H. (2011). *Le statut de l'éducation relative à l'environnement dans l'enseignement primaire marocain*. Centre de formation des inspecteurs.

Déduit, C., & Polo, C. (2022). La dimension affective de l'éducation au changement climatique: étude mixte au cycle 3 de l'enseignement primaire français. *Éducation relative à l'environnement: Regards – Recherches – Réflexions, 17*(2). https://doi.org/10.4000/ere.8768

Ekren, E., Hall, R. E., Pierdolla, E., Barnes, V., Jarzombek-Torralva, A., Morrish, D., & Martinez-Prather, K. (2025). Crime prevention through environmental design in public school career and technical education facilities: Principals' perceptions of security enhancement. *Safety Science, 185*, 106781. https://doi.org/10.1016/j.ssci.2025.106781

Eregha, P. B., Nathaniel, S. P., & Vo, X. V. (2024). COP28 roadmap: Do ICT, education, and renewable energy consumption matter for environmental quality? Quantile evidence for emerging economies. *Journal of Cleaner Production, 477*, 143744. https://doi.org/10.1016/j.jclepro.2024.143744

Fien, J. (1993). *Education for the environment: Critical curriculum theorising and environmental education*. Deakin University Press.
Gebrekidan, T. K. (2024). Environmental education in Ethiopia: History, mainstreaming in curriculum, governmental structure, and its effectiveness: A systematic review. *Heliyon, 10*(9). https://doi.org/10.1016/j.heliyon.2024.e30573
Gifford, R. (2011). The dragons of inaction: Psychological barriers that limit climate change mitigation and adaptation. *American Psychologist, 66*(4), 290–302.
Girault, Y., & Sauvé, L. (2008). L'éducation scientifique, l'éducation à l'environnement et l'éducation pour le développement durable: Croisements, enjeux et mouvances. *Aster, 46*, 7–30.
Hamouchi, A. (2016). L'éducation relative à l'environnement des manuels scolaires marocains: Conceptions et éthiques environnementales. *RADISMA, 12*, 14p.
Hamouchi, A., Essafi, K. H., & El Hajjami, A. (2010). Intégration de l'éducation relative à l'environnement dans le système éducatif marocain. *RADISMA, 5.*
Hamouchi, A. (2011). *Intégration de l'environnement et de l'éducation relative à l'environnement dans le système éducatif marocain : Cas de l'enseignement des sciences de la vie et de la terre* (Doctoral dissertation, Sidi Mohamed Ben Abdellah University, Faculty of Sciences Dhar El Mahraz–Fez). 225 p.
Hu, B., Wang, C., Xia, Z., Lu, C., & Ju, X. (2025). Environmental education in low-income and middle-income countries: A systematic review and meta-analysis. *Journal of Environmental Psychology, 104*, 102613. https://doi.org/10.1016/j.jenvp.2025.102613
Huckle, J., Sterling, S., & Barnett, J. (1998). *Education for sustainability: An agenda for action*. Earthscan.
Igei, K., Kurokawa, H., Iseki, M., Kitsuki, A., Kurita, K., Managi, S., Nakamuro, M., & Sakano, A. (2024). Synergistic effects of nudges and boosts in environmental education: Evidence from a field experiment. *Ecological Economics, 224*, 108279. https://doi.org/10.1016/j.ecolecon.2024.108279
Khzami, S., & Ben-Fares, S. (2009). L'environnement et l'éducation à l'environnement : conceptions d'enseignants du primaire et du secondaire. *Revue Africaine de didactique des sciences et des mathématiques*. (4).
Kimengele, B. P., Niyonkuru, C., Bapolisi, B. P., Nihoreye, F. J., Bahati, W. J.-M., & Muhunga, M. R. (2025). Obstacles to the integration of environmental education and sustainable development in life and earth sciences courses in the final cycle of basic education in the Congolese education system (DR Congo). *Revue Internationale du Chercheur, 6*(1).
Kiss, A. C., & Sicault, J.-D. (1972). La Conférence des Nations Unies sur l'environnement (Stockholm, 5–16 juin 1972). *Annuaire français de droit international,* 18, 603–628. https://doi.org/10.3406/afdi.1972.1717

Lundholm, C., & Plummer, R. (2009). Environmental education and sustainability: A critical review. *Environmental Education Research, 15*(2), 207–228.

Luo, L., Yang, Q., Qiao, D., Cao, Y., Ding, J., Ma, H., & Wei, J. (2025). How does environmental education affect college students' waste sorting behavior: A heterogeneity analysis based on educational background. *Journal of Environmental Management, 389*, 126064. https://doi.org/10.1016/j.jenvman.2025.126064

Mamadou, T. (2025). Place des élèves et des enseignants en classes de géographie dans la gestion de l'environnement en milieu scolaire au Sénégal. *Revue internationale de la recherche scientifique et de l'innovation, 3*(4), 817–824. https://doi.org/10.5281/zenodo.15812908

Mammino, L. (2025). Cross-bridging green chemistry education and environmental chemistry education. *Sustainable Chemistry for the Environment, 9*, 100195. https://doi.org/10.1016/j.scenv.2024.100195

Moser, S. C. (2010). Communicating climate change: History, challenges, and future directions. *Wiley Interdisciplinary Reviews: Climate Change, 1*(1), 31–39.

Mousavi, N., Ahmadi, S., Sharifian Sani, M., Irandoost, S. F., Mohammadi Gharehghani, M. A., & Abdolhai, Z. (2024). Identifying environmental education strategies for children with an emphasis on children under four years old: A qualitative study in Iran. *Heliyon, 10*(17), e37161. https://doi.org/10.1016/j.heliyon.2024.e37161

Nicol, H., Davies, B., & Grant, S. (2000). *Sustainable development education: A strategy for the future.* Cambridge University Press.

Palmer, J. (1998). *Environmental education in the 21st century: Theory, practice, progress and promise.* Routledge.

Rickinson, M., Dillon, J., Teamey, K., Morris, M., Choi, M.-Y., & Sanders, D. (2004). *A review of research on outdoor learning.* Field Studies Council.

Salazar, C., Jaime, M., Leiva, M., & González, N. (2024). Environmental education and children's pro-environmental behavior on plastic waste. Evidence from the green school certification program in Chile. *International Journal of Educational Development, 109*, 103106. https://doi.org/10.1016/j.ijedudev.2024.103106

Simmons, D., Durbin, L., & Keen, D. (2008). Learning for a sustainable future: Education, responsibility and a changing world. *The Canadian Journal of Environmental Education, 13*(2), 143–154.

Smyth, J., Dow, S., & McInerney, P. (2004). Education for sustainability: The role of teachers in promoting the green agenda. *Australian Journal of Environmental Education, 20*(1), 23–35.

Sterling, S. (2001). *Sustainable education: Re-visioning learning and change.* Green Books.

Tiemann, A., & Ring, I. (2025). Quantifying and mapping forest-related environmental education and training in Saxony, Germany. *Ecosystem Services, 73*, 101719. https://doi.org/10.1016/j.ecoser.2025.101719

Tilbury, D. (1995). Environmental education for sustainability: Defining the new focus of environmental education in the 1990s. *Australian Journal of Environmental Education, 11*(1), 4–12.

UNESCO. (2005). *United Nations decade of education for sustainable development (2005–2014): Framework for the international implementation scheme.* UNESCO.

Wals, A. E. J. (Ed.). (2007). *Social learning towards a sustainable world* (2nd ed.). Wageningen Academic Publishers.

Ziaka, Y., Robichon, P., & Souchon, C. (Eds.). (2002). *Éducation à l'environnement. Six propositions pour agir en citoyens* (pp. 29–36). Éditions Charles Léopold Mayer.

Zouine, M., El Adnani, M. J., Salhi, S. E., & El Anouar, E. M. (2024). Does higher education lead to lower environmental pollution? New evidence from MENA countries using econometric panel data. *Transnational Corporations Review, 16*(4), 200077. https://doi.org/10.1016/j.tncr.2024.200077

CHAPTER 20

The Role of Educational Institutions in Promoting Awareness of Sustainable Development Issues: A Systematic Review

Mohamed EL Hadri, *Ahmed Ait Aghzzaf*, *and Bouamama Cherai*

1 Introduction

Humanity currently faces a range of global challenges threatening its sustainability and the well-being of future generations, such as climate change, depletion of natural resources, poverty and inequality, pollution, lack of awareness, and insufficient education, alongside issues of poverty and political conflicts (A. Damiano, 2024). As these challenges increase, the need for innovative and sustainable solutions becomes more pressing to ensure comprehensive progress toward a sustainable future. In this

M. EL Hadri (✉) • A. Ait Aghzzaf
Nanomaterials, Technology and Innovation Group, Higher Normal School, Abdelmalek Essaadi University, Tetouan, Morocco
e-mail: mohamed.elhadri11@etu.uae.ac.ma; a.aitaghzzaf@uae.ac.ma

B. Cherai
Regional Center for Education and Training Professions of Tangier–Tetouan–Al Hoceima, Tetouan, Morocco
e-mail: bouamama.cherai@taalim.ma

K. Hirech (ed.), *Environmental Education and Sustainable Development*, https://doi.org/10.1007/978-3-032-05760-0_20

context, the concept of sustainable development has emerged as a holistic framework aiming to balance the needs of the present with those of future generations (Riabtsova, 2022). It is based on three dimensions focused on environmental conservation, social justice, and economic growth (Bucharest University of Economic Studies, Bucur, 2023).

Given the escalating challenges the world faces, sustainable development has become an urgent necessity. It seeks to find solutions that ensure environmental preservation, provide social justice, and achieve economic balance. In 2015, the United Nations established the Sustainable Development Goals (SDGs), comprising 17 goals covering environmental, social, and economic areas, aiming to secure long-term human well-being (De Felice et al., 2023). These goals have emphasized the importance of global partnerships and the collaboration of all parties to achieve a sustainable future for all (Apin & Hassan, 2023). Sustainable development, with its ambitious objectives and multidimensional approach, has thus become a priority for the international community to ensure the continuity of life and human well-being without compromising the ability of future generations to meet their needs (Mishra et al., 2024).

In this context, Goal 4 of the SDGs, which focuses on "ensuring inclusive and equitable quality education and promoting lifelong learning opportunities for all," stands out as a fundamental goal that enhances sustainability within communities (Karpowicz & Débora, 2023). This goal aims to achieve inclusive and fair education for all groups, enabling individuals to think critically, solve problems, and make sustainable decisions that contribute to addressing challenges. This is vividly reflected through redefining the prevailing educational discourse to focus more on social, economic, and environmental values, making them an integral part of the educational system (Akanbi & Adesina, 2024).

Education in institutions such as schools and universities is considered one of the main tools to enable individuals to acquire the knowledge and skills needed to build societies capable of tackling sustainability challenges (Griffin et al., 2022). It guides students toward a deeper understanding of concepts such as climate change, biodiversity loss, poverty, inequality, and the need for social justice (Leiva-Brondo et al., 2022). This is achieved through environmental activities, community projects, and volunteer work organized within educational settings (Z. Fadeeva & Y. Mochizuki, 2010).

Numerous studies have demonstrated the effectiveness of these activities in motivating learners to participate actively in achieving positive change within their communities, while also enhancing their awareness of environmental and social issues. For example, a study by Baron et al. (2024) highlighted the critical role higher education institutions play in sustainable development by integrating sustainability into curricula and partnering with industries to foster a culture of sustainability within communities and future generations. Another study by Faustino and Kaur (2023) pointed out the role of educational institutions in promoting critical thinking and fostering social transformation by equipping students with the skills and knowledge needed to effectively address environmental, social, and economic challenges.

Although some educational institutions have succeeded in raising awareness of sustainability concepts, many still face significant challenges that hinder their progress in this area (Mochizuki & Fadeeva, 2010). These challenges vary from a lack of financial resources and funding to insufficient training and awareness for teachers, and administrative or structural constraints in some educational systems, which may impede the integration of sustainable development concepts into curricula. In some cases, educational systems lack clear pedagogical visions that support the application of these concepts at all levels of education (Sterling, 2011).

Since educational institutions are among the key entities capable of guiding individuals toward a deeper understanding of the challenges to achieving sustainable development through curricula and extracurricular activities, these curricula help learners develop critical and analytical thinking skills about sustainable solutions that align with the needs of society and future generations, enhancing their ability to make responsible decisions that consider the three dimensions of sustainable development: environmental, social, and economic (Lasino et al., 2023; Thoriq & Mahmudah, 2023).

Extracurricular activities are also an effective means of motivating students to participate in local and global initiatives that promote sustainability, whether through community work, environmental projects, or volunteerism (Isac et al., 2023). This approach significantly contributes to developing a new generation of individuals who are aware of their responsibilities, ready to engage in sustainable development efforts at all levels, and contribute to building a cohesive society capable of facing future challenges with innovative and sustainable solutions (Décarpentrie et al., 2024).

This study aims to analyze the scientific articles published in 2024 on the role of educational institutions in promoting sustainable development concepts. It highlights various practices to identify strengths and weaknesses, providing recommendations to improve existing strategies. The study focuses on integrating these concepts into curricula and extracurricular activities while exploring the challenges that hinder their implementation. Additionally, it reviews best global practices, emphasizing educational programs that enhance environmental, social, and economic awareness among learners, enabling them to confront challenges and contribute to sustainable development within their communities. Through this study, we hope to provide a practical framework that enhances the effectiveness of education in spreading the culture of sustainability and empowering individuals to play an active role in building sustainable communities in the long term.

2 Methodology

This study aims to provide a comprehensive analysis focusing on the role of educational institutions in raising awareness and educating individuals about sustainable development issues in various aspects of life, such as the environment, economy, and society. It also explores how educational institutions use curricula and teaching methods that help disseminate sustainability principles and promote behaviors and practices that contribute to achieving the Sustainable Development Goals.

The analysis also includes how sustainability concepts are integrated into curricula, the training of teachers on the best methods for teaching these issues, as well as the role of universities and schools in implementing projects and initiatives that encourage attention to environmental, social, and economic issues. The focus of this research chapter revolves around addressing the following issue:

2.1 How Can Educational Institutions Contribute to Achieving the Sustainable Development Goals?

To ensure a comprehensive and accurate understanding of the main question of this research paper, a multi-method approach has been adopted, combining both qualitative and quantitative data in an integrated and coherent manner (Onwuegbuzie & Leech, 2005). This methodology contributes to enhancing the credibility of the results and providing comprehensive insights, as each method contributes a different type of data and information that complement each other. This combination helps provide a balanced analysis that covers all aspects of the research topic (J. W. Creswell, 2009).

The process of identifying and analyzing documents related to recent literature reviews addressing the role of educational institutions in raising awareness of sustainable development issues globally was carried out using a systematic literary review methodology based on the following steps:

Step 1: Document collection
Step 2: Eligibility assessment
Step 3: Quality evaluation
Step 4: Data coding

2.2 *Document Collection*

A timeframe for the research was established, encompassing all studies on the topic published in the year 2024, without geographical restrictions, focusing on studies relevant to the subject. The research was conducted by a team comprising two university professors specialized in educational sciences, along with a high school physics and chemistry teacher. The scope of the research was limited to previous reviews containing the following terms: "*sustainable development – awareness – institutions – schools – universities – education.*"

The research was carried out across a variety of academic databases to collect documents related to the topic. Among the most prominent of these databases, based on the quality of available documents, were "*SCIENCE DIRECT, SCOPUS, HAL.SCIENCE*" and others. During the initial research process, 392 documents were extracted as of November 1, 2024. The systematic review of the extracted documents then began, aimed at evaluating the literature on the topic comprehensively.

2.3 *Eligibility Assessment and Quality Evaluation*

The documents were examined using the decision tree technique, an analytical tool used to illustrate and analyze the decision-making process in various contexts. The decision tree graphically represents a structure that branches into several options, with each branch representing a possible choice or decision. This technique is used to compare different options based on a set of criteria or factors, helping to make the most effective decision based on available data and systematic analysis (Mittal et al.,

2017). It is also an effective tool for strategic decision-making and prioritization in the context of scientific research (Xu, 2023) (Fig. 20.1).

The research team relied on the decision tree to read and analyze all 392 documents. Through this technique, studies that aligned with the established study criteria were retained, helping focus on the most relevant and high-quality research works. After a thorough evaluation of the quality of the extracted studies, the team decided to exclude those that did not meet the established criteria, as they did not provide clear or accurate results or data directly related to the research topic.

2.4 *Data Coding*

The data was coded and classified using an Excel spreadsheet to facilitate the categorization of publications according to the criteria agreed upon by the research team. After conducting a thorough review, it was found that only 10 documents met the decision tree criteria.

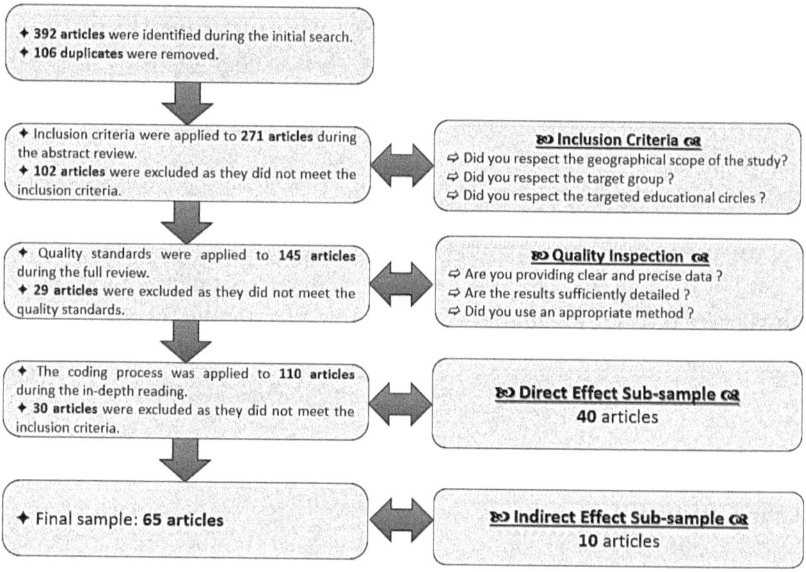

Fig. 20.1 Decision tree for the research process using decision tree (Mittal et al., 2017)

Through these steps, it was ensured that the research was based on reliable and extensive sources, which enhances the credibility of the results and contributes to providing accurate and objective scientific insights into the topic.

3 Results

The obtained documents were classified based on the following criteria: (*title, publication date, countries, study objective, methodology used, results, and recommendations*). They were organized into specific tables and formulated into explanatory texts to ensure better flow and ease of reading:

In a study titled "Research on the Level of Awareness of Pupils Regarding Sustainable Development Issues" (Taktakishvili et al., 2024), conducted in Georgia and published on March 31, 2024, the level of students' awareness of sustainable development issues in the schools of the Georgian capital was analyzed. The study aimed to assess students' awareness and understand the factors influencing their perceptions of these issues, along with providing recommendations to enhance this awareness. The study utilized a quantitative methodology, conducting interviews with 205 students using random sampling. The results revealed that the majority of students recognized the importance of sustainability and expressed a willingness to take responsibility; however, their actions did not fully reflect this awareness. The study also noted that traditional media is not an effective source of awareness for students, suggesting the use of alternative platforms such as social media. Furthermore, it was found that a significant proportion of students had not participated in any sustainability-related projects or activities. Based on these results, the study recommended implementing awareness programs and projects, either jointly or independently, to enhance students' awareness of sustainable development issues and encourage active participation in this field.

In a study titled "Status of Education for Sustainable Development (ESD) in Higher Secondary Schools of the Province of Sindh in Pakistan" (Jiskani et al., 2024), published in March 2024, the status of Education for Sustainable Development (ESD) in government secondary schools in Pakistan was evaluated. The study aimed to analyze the extent to which sustainable development concepts are integrated into educational curricula and identify the challenges faced by schools in this area. A qualitative methodology was employed, with interviews conducted with 20 school principals from government schools in the study area using random sampling. The results showed that key topics related to education for sustainable development were not included in the curricula. The study also revealed that schools do not provide sufficient knowledge and information about the fundamental concepts of sustainable development. Additionally, the study found a lack of awareness among teachers about the concept of education for sustainable development. Based on these results, the study recommended integrating ESD concepts into curricula and suggested organizing training programs for teachers to enhance their knowledge and skills, which would contribute to improving the quality of education and raising students' awareness of sustainability issues.

In a study titled "Exploring Teachers' Perceptions and Practices in Education for Sustainable Development (ESD) in Public Schools of Karachi" (Shakir et al., 2024), published on March 29, 2024, the focus was on the key role of teachers in integrating Sustainable Development Goals (SDGs) into curricula, the challenges they face, and how to teach students to address global challenges. The study used a quantitative methodology, collecting data from a sample of 100 teachers through a survey. The results showed that teachers expressed a willingness to integrate sustainable development principles into their teaching. However, the study pointed out that the rigidity of curricula is a major barrier to implementing ESD. It also observed that community and parental support for teachers plays a significant role in facilitating the integration of sustainability concepts in schools. Based on these results, the study recommended several measures, including integrating ESD principles across all curricula and educational levels, providing necessary funding and resources such as educational materials, technology, and appropriate infrastructure, and organizing comprehensive teacher training programs to enhance their competence in teaching sustainability concepts.

In a study titled "Evaluation of school administrators' initiatives for sustainable development goals" (Altinay et al., 2024), published in March 2024 in Turkey, the views of vocational high school principals regarding the SDGs were assessed to identify gaps and contribute to taking the right steps to achieve these goals. The study used a qualitative methodology with content analysis, conducting interviews with 15 principals from vocational high schools in Northern Cyprus. The participants emphasized the need to provide free and accessible healthcare services, ensure equal opportunities, and expand health insurance coverage. They also highlighted the importance of providing and protecting clean water, ensuring access to it for everyone, and the need for clean energy accessible to all.

Based on these results, the study presented several recommendations, including providing practical training on sustainable development to increase awareness and skills among teachers and school administrations, enhancing cooperation between teachers, school management, and the community, and developing policies and projects related to each SDG at the national level.

In a study titled "Assessing Students' Awareness, Attitudes, and Actions Towards Sustainable Development Goals (SDGs)" (Veeran et al., 2024), published on March 27, 2024, in Bahrain, students' awareness, attitudes, and behaviors toward sustainability were explored, with an evaluation of the factors influencing sustainability-related awareness and behaviors among students at the University of Applied Sciences in Bahrain. The study used a mixed methodology (qualitative and quantitative), conducting a survey among engineering students using random sampling. The results showed that many students engage in sustainable behaviors such as recycling and resource conservation, reflecting responsible environmental behavior. The research also revealed that students have a good understanding of sustainability concepts, are interested in energy and water conservation, and are aware of the SDGs. Based on these results, the study recommended increasing educational efforts to encourage student participation and motivate them to play an active role in promoting sustainability.

In a study titled "Evaluating Students' Awareness, Knowledge, and Access to Sustainable Development Goals" (Hdeib et al., 2024), published in January 2024 in India, the level of awareness and knowledge of higher education students in India regarding the 17 SDGs was evaluated, with a special focus on Goal 4 related to quality education. The study employed a mixed methodology (qualitative and quantitative) using a descriptive approach, gathering primary data through a survey. The results showed that higher education students had a higher level of awareness, knowledge, and access to the SDGs compared to students at lower educational levels. The study also indicated that the SDGs represent a relatively new educational approach, and that the education system plays an important role in teaching students to become change agents within society. Based on these results, the study made several recommendations, including increasing awareness, knowledge, and access to the SDGs through joint efforts between the government, higher education institutions, and NGOs, offering mandatory courses on SDGs in all higher education programs, and raising awareness among students by organizing programs and activities such as environmental clubs, Earth Day, Water Day, field trips, community service, and practical projects related to sustainability.

In a study titled "Development and Implementation of a Sustainable Development Goals (SDGs) Unit: Exploration of Middle School Students' SDG Knowledge"(Koçulu & Topçu, 2024), published on January 5, 2024, in Turkey, an educational unit on SDGs was designed and implemented to enhance middle school students' knowledge of these goals and prepare them as globally conscious citizens who contribute to building a sustainable future. The study used a qualitative methodology with content analysis, conducting interviews with 20 students in a public school in Turkey. The results showed a lack of practical educational materials related to the SDGs. The study also emphasized the important role of universities in enhancing university students' knowledge and attitudes toward these goals. Additionally, the study noted that teaching SDGs from an early age is essential, especially with the depletion of global resources. Based on these results, the study presented several recommendations, including developing education to prepare the next generation to address complex challenges such as poverty, hunger, inequality, and climate change, and developing integrated, practical educational materials on SDGs for use at all educational levels.

The study titled "Integrating Education for Sustainable Development into English for Computer Science Students and Lectures' Perspectives" (Wahyuni et al., 2024), published on January 1, 2024, in Indonesia, analyzed how to make education for sustainable development more relevant to students' needs and contribute to more sustainable teaching and learning. The study used a mixed methodology (quantitative and qualitative), conducting a survey with 54 students and interviews with 3 English lecturers specializing in computer science at a higher education institution in Indonesia. The results showed that integrating education for sustainable development helps prepare students to build a sustainable future by equipping them with the necessary knowledge and skills. However, the study pointed out that integrating sustainability education into technical fields, especially computer science and e-learning, is a complex challenge. The results also highlighted the importance of involving students to enhance their sense of responsibility toward sustainability, with most students and lecturers agreeing on the need to integrate sustainability education into all curricula. Based on these results, the study made several recommendations, including further research on the best ways to integrate sustainability education into different educational environments to enhance sustainability continuously, simplifying the integration of sustainability principles in technical fields such as computer science by using practical examples and involving students in sustainability-related projects and activities to enhance their sense of responsibility and deepen their understanding of sustainability concepts.

The study titled "Assessment of sustainability awareness and practice in a campus community" (Urbaniak et al., 2024), published on January 3, 2024, in the United States, assessed the knowledge and practices of students, staff, and faculty regarding sustainability at Central Michigan University. The study used a quantitative methodology, conducting separate surveys for students, staff, and faculty using a dedicated platform within the educational institution. The results showed that students, staff, and faculty are concerned with sustainability issues related to waste, water, energy, and food, and wish to promote these issues at the institutional level. However, despite students' interest in sustainability, they lack sufficient knowledge about it, while staff and faculty exhibit supportive sustainability knowledge and behaviors. The study also noted several barriers to achieving sustainable development, despite societal calls for investment and participation in sustainable projects. Based on these results, the study presented several recommendations, including using multiple mechanisms and activities to effectively engage all stakeholders in sustainability issues, and enhancing commitment to institutional projects, especially those related to waste and energy, such as improving recycling systems and composting.

The study titled "Fostering students' participation in the implementation of the sustainable development goals at higher education institutions" (Filho et al., 2024), published on March 4, 2024, in Germany, analyzed the level of knowledge and participation of university students in initiatives related to implementing the UN's SDGs. The study used a mixed methodology (quantitative and qualitative), through case studies and a survey that included a sample of 602 students from 53 countries. The results showed that student participation in sustainability-related activities reflects higher education institutions' commitment to implementing the SDGs. The study also noted that graduate students have greater awareness and knowledge of the SDGs compared to students in lower educational stages, emphasizing the role of advanced education in enhancing awareness. Furthermore, the study highlighted the crucial role of universities and media in spreading knowledge about sustainability and fostering awareness and dialogue on sustainability issues among future generations.

Based on these results, the study presented several recommendations, including strengthening higher education institutions' commitment to implementing the SDGs to encourage greater student participation in sustainable activities, providing intensive training programs for university students to enhance their knowledge of the SDGs and raise their awareness, and encouraging collaboration between universities and media to effectively disseminate sustainability messages, contributing to increasing societal awareness of sustainable development issues.

The reviewed studies emphasize the assessment of sustainability awareness among students, teachers, and administrators in various educational settings across multiple countries. These studies aim to explore the understanding and application of sustainability concepts in schools, universities, and vocational institutions.

In Georgia, the focus was on evaluating the awareness of students in the capital regarding sustainability issues, with a particular interest in identifying the factors influencing this awareness. The study in Pakistan examined the state of Education for Sustainable Development (ESD) in public higher secondary schools, highlighting the importance of teachers' roles in integrating sustainability into the curriculum, while also identifying challenges faced by educators. In Turkey, the study explored the perspectives of vocational school administrators regarding their understanding and implementation of sustainable development goals (SDGs), along with efforts to develop educational units to raise middle school students' awareness.

The study in Bahrain assessed the awareness, attitudes, and behaviors of engineering students at the University of Applied Sciences toward sustainability. In India, the research focused on evaluating the awareness of higher education students regarding the 17 SDGs, especially Goal 4, which focuses on quality education. Meanwhile, in Indonesia, a study investigated how Education for Sustainable Development (ESD) can be better aligned with students' needs, while a study in the United States assessed sustainability knowledge and practices among students, faculty, and staff at Central Michigan University.

Finally, in Germany, the study looked at the level of knowledge and participation of university students in SDG-related initiatives, involving a diverse global sample of students. The results indicate that while there is awareness and willingness to embrace sustainability issues among students, this awareness often fails to translate into meaningful action, with many students lacking participation in sustainability projects. In addition, although there is interest in sustainability, curricula remain rigid and unable to fully integrate sustainable development principles. Teachers, despite their interest, lack sufficient awareness and training to effectively teach sustainability concepts.

The study highlights the critical role of educational institutions, especially universities, in fostering a sustainable future by raising awareness and enhancing students' understanding of sustainability issues. It notes that higher education students tend to exhibit a greater understanding of sustainability compared to younger students, and that universities, along with media, play a significant role in shaping this awareness. Additionally, simplifying the integration of sustainability in technical fields, intensifying research, and creating practical, flexible learning materials are essential steps to prepare students for the challenges of sustainability and ensure they are equipped to address global issues such as climate change, poverty, and inequality.

4 Discussion

To implement Education for Sustainable Development in Morocco, it is essential to first raise awareness among students and teachers through specialized training programs. It is necessary to integrate sustainable development concepts into curricula across all subjects, taking into account local challenges such as water scarcity and desertification. Additionally, practical and field projects can be organized to engage students in real sustainable activities, such as awareness campaigns and tree planting initiatives. Universities and schools should also collaborate to provide flexible educational opportunities that enhance critical thinking and encourage students to participate in sustainable solutions. Finally, the media must play a prominent role in raising awareness about the Sustainable Development Goals through continuous media campaigns.

Based on the above, the following conclusions can be drawn:

4.1 Awareness Levels and Knowledge Disparities

Awareness and knowledge about sustainable development goals differ across educational groups (between school students and university students) and countries. For example, higher education students in India demonstrated high knowledge and awareness about sustainable development goals compared to younger students.

4.2 *The Importance of Integrating Sustainability into Curricula*

Many studies emphasize the essential role of curricula in enhancing students' understanding of sustainable development goals and preparing them as change agents in society. Early education on these goals increases the awareness of future generations and contributes to achieving sustainable development.

4.3 *Challenges in Achieving Sustainable Education*

There is a lack of practical educational materials about sustainable development goals, especially at the school level. For example, a study in Turkey showed a shortage of educational resources aimed at enhancing this knowledge among students.

4.4 *Variation in Research Methodologies and Its Impact on Study Results*

Research methodologies varied between quantitative, qualitative, or mixed methods. Studies that gathered both quantitative and qualitative data, such as those conducted in Indonesia and Germany, provided more comprehensive insights, contributing to a deeper understanding of how students and teachers engage with sustainability issues.

4.5 *Towards Sustainability as an Educational Imperative for the Future*

Teaching sustainable development goals from early education stages has become a necessity, especially given global challenges such as resource depletion.

5 Conclusion

In conclusion, this systematic review emphasizes the crucial role that educational institutions play in raising awareness about sustainable development, particularly in light of pressing global challenges such as climate change, resource depletion, and social inequalities. The review highlights that education goes beyond merely imparting academic knowledge; it is a powerful tool for empowering individuals to embrace sustainability values, engage in responsible behavior, and actively contribute to a more sustainable future. Despite the growing recognition of this role, there is still a significant gap between theoretical frameworks and practical application. To achieve meaningful progress, stronger coordination is needed between educational curricula and extracurricular activities, including field projects, community engagement, and environmental initiatives. Incorporating hands-on experiences into educational programs can greatly enhance students' understanding of sustainability concepts and their ability to apply them in real-world scenarios. Furthermore, educational institutions must adopt interdisciplinary approaches that connect subjects such as science, economics, and social studies, offering a comprehensive understanding of sustainable development. Teacher training programs should focus on innovative teaching strategies that encourage critical thinking, problem-solving, and collaboration skills. This review provides valuable insights for policymakers, curriculum developers, and educators by identifying effective strategies for integrating sustainable development principles into educational frameworks. By aligning educational policies with sustainability goals and promoting active learning methods, educational institutions can play a key role in preparing future generations to tackle global challenges and contribute to a more equitable and sustainable world.

References

Akanbi, G., & Adesina, A. E. (2024). Fostering sustainable development goal-4 through culturo-techno-contextual-approach in innovative steam education: A policy assessment. https://doi.org/10.32388/oc6ba0

Altinay, Z., Arica, S., Dagli, G., & Altinay, M. (2024). Evaluation of school administrators' initiatives for sustainable development goals. *Conhecimento & Diversidade, 16*(41), 92–120. https://doi.org/10.18316/rcd.v16i41.11237

Apin, L., & Hassan, M. S. (2023). Operationalising sustainable development practice for area-based sustainability implementing UN-SDGS (2015-2030) at the local district area. *Russian Law Journal, 11*(1S), 18–38. https://doi.org/10.52783/rlj.v11i1S.360

Baron, J., Lawrence, S. A., & Nevins-Bennett, C. (2024). Sustainable development goals. A case of the community college student initiative project. *European Journal of Education and Pedagogy, 5*(3), 24–35. https://doi.org/10.24018/ejedu.2024.5.3.833

Bucur, L.-M. (2023). Sustainable development: A viable solution to reduce poverty. *Management of Sustainable Development, 15*(2). https://doi.org/10.54989/msd-2023-0012

J. W. Creswell (2009). *Research design-qualitative, quantitative, and mixed methods approaches.* Sage.

De Felice, F., Petrillo, A., & Autorino, C. (2023). Development of a framework for sustainable outsourcing: Analytic balanced scorecard method (A-BSC) (vol 7, pg 8399, 2015). *Sustainability, 15*(13). https://doi.org/10.3390/su151310174

Décarpentrie, L., Bélanger, C., Rakotondramanana, B., Rakotondrasoa, S., Rakotomalala, O., Guirao, J. L., & Gadais, T. (2024). Long-term perspectives of participating in extracurricular activities in an extreme context in Madagascar. *Current Psychology*, 1–17. https://doi.org/10.31234/osf.io/msuwz

A. Damiano (2024). How its pursuit relates to environmental violence and why we should replace it with the concept of sustainable life. *Exploring Environmental Violence: Perspectives, Experience, Expression, and Engagement, 135.* https://doi.org/10.1017/9781009417150.009

Faustino, A., & Kaur, I. (2023). Education and sustainable development. *Asian Journal of Advanced Research and Reports, 17*(7), 97–101. https://doi.org/10.9734/ajarr/2023/v17i7497

Filho, W. L., Trevisan, L. V., Dinis, M. A. P., Ulmer, N., Paço, A., Borsari, B., Sierra, J., & Salvia, A. (2024). Fostering students' participation in the implementation of the sustainable development goals at higher education institutions. *Discover Sustainability, 5*(1), 22. https://doi.org/10.1007/s43621-024-00204-7

Griffin, M., Barona, J., & Gutierrez, C. F. (2022). Strategies to increase sustainability awareness in higher education: Experiences from Abu Dhabi Women's College. *International Journal of Sustainable Development and Planning, 17*(6), 1831–1838. https://doi.org/10.18280/ijsdp.170617

Hdeib, R., Aouad, M., Awawdeh, S., & Mahfoodh, S. A. (2024, January). Assessing students' awareness, attitudes, and actions towards sustainable development goals (SDGs). In *2024 ASU International Conference in Emerging Technologies for Sustainability and Intelligent Systems (ICETSIS)* (pp. 188–194). IEEE. https://doi.org/10.1109/ICETSIS61505.2024.10459476

Isac, C., Iordache, A. M. M., Baltador, L., Coculescu, C., & Niță, D. (2023). Enhancing students' entrepreneurial competencies through extracurricular activities a pragmatic approach to sustainability-oriented higher education. *Sustainability, 15*(11), 8708. https://doi.org/10.3390/su15118708

Jiskani, A., Guriro, S., & Fayyaz, S. (2024). Status of education for sustainable development (ESD) in higher secondary schools of the province of Sindh in Pakistan. *Global Educational Studies Review, 9*(1), 30–40. https://doi.org/10.31703/gesr.2024(IX-I).04

Karpowicz & Débora (2023). Impactos da pandemia na agenda 2030: um estudo sobre o ods 4 educação. *Revista Lusófona de Estudos Culturais e Comunicacionais (NAUS), 5*(1), 27–34. https://doi.org/10.29073/naus.v5i1.848

Koçulu, A., & Topçu, M. S. (2024). Development and implementation of a sustainable development goals (SDGs) unit: Exploration of middle school students' SDG knowledge. *Sustainability, 16*(2), 581. https://doi.org/10.3390/su16020581

Leiva-Brondo, M., Lajara-Camilleri, N., Vidal-Meló, A., Atarés, A., & Lull, C. (2022). Spanish university students' awareness and perception of sustainable development goals and sustainability literacy. *Sustainability, 14*(8), 4552. https://doi.org/10.3390/su14084552

Mishra, M., Desul, S., Santos, C. A. G., Mishra, S. K., Kamal, A. H. M., Goswami, S., Kalumba, A. M., Biswal, R., da Silva, R. M., dos Santos, C. A. C., & Baral, K. (2024). A bibliometric analysis of sustainable development goals (SDGs): A review of progress, challenges, and opportunities. *Environment, Development and Sustainability, 26*(5), 11101–11143. https://doi.org/10.1007/s10668-023-03225-w

Mittal, K., Khanduja, D., & Tewari, P. C. (2017). An insight into 'decision tree analysis'. *World Wide Journal of Multidisciplinary Research and Development, 3*(12), 111–115.

Z. Fadeeva & Y. Mochizuki (2010). Roles of regional centres of expertise on education for sustainable development. *Journal of Education for Sustainable Development, 4*, 51–59. https://doi.org/10.1177/097340820900400112

Onwuegbuzie, A. J., & Leech, N. L. (2005). On becoming a pragmatic researcher: The importance of combining quantitative and qualitative research methodologies. *International Journal of Social Research Methodology, 8*(5), 375–387. https://doi.org/10.1080/13645570500402447

Riabtsova, A. (2022). Sustainable development as a vector of economic and legal use of natural resources. *Entrepreneurship, Economy and Law, 1*, 17–22. https://doi.org/10.32849/2663-5313/2022.1.03

Lasino, Rukmana, A. Y., Mokodenseho, S., & Aziz, A. M. (2023). Environmental education for sustainable development: A bibliometric review of curriculum design and pedagogical approaches. *The Eastasouth Journal of Learning and Educations, 1*(2), 65–75. https://doi.org/10.58812/esle.v1i02.108

Shakir, F., Naz, A., & Ahmed, S. M. (2024). Exploring teachers' perceptions and practices in education for sustainable development (ESD) in public schools of Karachi. *Pakistan Journal of Humanities and Social Sciences, 12*(1), 902–914. https://doi.org/10.52131/pjhss.2024.v12i1.2090

Sterling, S. (2011). Transformative learning and sustainability: Sketching the conceptual ground. *Learning and Teaching in Higher Education, 5*(11), 17–33.

Taktakishvili, T., Charaia, V., Tvauri, G., Sachhaleli, N., & Lashkhi, M. (2024). Research on the level of awareness of pupils regarding sustainable development issues. *International Journal of Advanced Research, 12*(4), 1203–1214. https://doi.org/10.21474/IJAR01/18665

Thoriq, A., & Mahmudah, F. N. (2023). Education for Sustainable Development (ESD): A systematic literature review on curriculum development strategy design. *European Journal of Education Studies, 10*(5). https://doi.org/10.46827/ejes.v10i5.4803
Urbaniak, E., Uzarski, R., & Haidar, S. (2024). Assessment of sustainability awareness and practice in a campus community. *International Journal of Sustainability in Higher Education, 25*(9), 78–93. https://doi.org/10.1108/IJSHE-05-2023-0164
Veeran, S., Duraiswamy, A., Puwanenthiren, P., & Venkatesan, R. (2024). Assessment of students' awareness, knowledge, and accessibility of sustainable development goals (SDGs). *Multidisciplinary Reviews, 7*(7), 2024122–2024122. https://doi.org/10.31893/multirev.2024122
Wahyuni, D. S., Rozimela, Y., Ardi, H., & Mukhaiyar, M. (2024). Integrating education for sustainable development into English for Computer Science: Students and lectures' perspectives. *English Education Journal, 15*(1), 44–55. https://doi.org/10.24815/eej.v15i1.37809
Xu, J. (2023). Systematic analysis and application prospect of decision tree. *Highlights in Science, Engineering and Technology, 71*, 163–170. https://doi.org/10.54097/hset.v71i.12687

Index

K. Hirech (ed.), *Environmental Education and Sustainable Development*, https://doi.org/10.1007/978-3-032-05760-0

Kolibri 360 GmbH
Ferdinand-Jühlke-Straße 7
99095 Erfurt, Deutschland
produktsicherheit@kolibri360.de